GEOMORPHOLOGY IN
ENVIRONMENTAL MANAGEMENT

D0223573

Geomorphology in Environmental Management

AN INTRODUCTION

R. U. COOKE

LECTURER IN GEOGRAPHY
UNIVERSITY COLLEGE LONDON

AND

J. C. DOORNKAMP

SENIOR LECTURER IN GEOGRAPHY
UNIVERSITY OF NOTTINGHAM

CLARENDON PRESS · OXFORD
1974

Oxford University Press, Ely House, London W.1

GLASGOW NEW YORK TORONTO MELBOURNE WELLINGTON
CAPE TOWN IBADAN NAIROBI DAR ES SALAAM LUSAKA ADDIS ABABA
DELHI BOMBAY CALCUTTA MADRAS KARACHI LAHORE DACCA
KUALA LUMPUR SINGAPORE HONG KONG TOKYO

CASEBOUND ISBN 0 19 8740204
PAPERBACK ISBN 0 19 8740212

© R. U. Cooke and J. D. Doornkamp 1974

PRINTED IN GREAT BRITAIN
AT THE PITMAN PRESS, BATH

PREFACE

Geomorphology is the study of landforms, and in particular of their nature, origin, processes of development, and material composition. This book is concerned with those aspects of geomorphology that relate to man's use of the natural environment. Such 'applied' aspects of the subject have attracted some attention from each generation of geomorphologists, but attitudes towards them are changing, and for good reasons.

In the first place, there has been a general resurgence of interest in those aspects of the natural sciences relevant to man. This arises largely from a growing anxiety within many societies over human exploitation of the natural environment, and from practical steps taken to tackle environmental problems. The passing of the National Environmental Policy Act by the U.S. Congress in 1969, and the creation in Britain of the Department of the Environment in 1971 are two reflections of public concern.

A second, more specific reason for changed attitudes to studies relating geomorphology to human use of the natural environment arises from changing fashions within geomorphology as a whole. In recent years, a long-established emphasis on landform evolution has been challenged by a growing interest in the analysis of relations between contemporary forms and processes and their spatial variations. Perhaps not entirely coincidentally, this trend has placed research emphasis on those areas of the subject of greatest relevance to man. At the same time, it is becoming clear that some aspects of 'pure' and 'traditional' research in geomorphology are potentially of real practical value.

Traditionally, geomorphology has found its university place within the departmental palisades of geology (and especially engineering geology) and geography, and it is from these disciplines that the geomorphological response to public and professional concern with man–environment relations has been most pronounced. Of the many recent examples of discussions in environmental management where geomorphological considerations have figured prominently, we might cite several conference proceedings published in the United States: *The Importance of the Earth Sciences to the Public Works and Building Official* (Association of Engineering Geologists, 1966), *Geologic Hazards and Public Problems* (Office of Emergency Preparedness, Santa Rosa, California, 1969), *Environmental Planning and Geology* (U.S. Department of Housing and Urban Development, 1969), and *Environmental Geomorphology* (State University of New York, Binghamton, Publications in Geomorphology, 1971). While the sphere of man–environment relations is the legitimate concern of many, it has been a focus of activity in geography for over

100 years, and it is a focus that is increasingly attracting attention. Recent contributions have included studies as diverse as Tuan's philosophical exploration of *Man and Nature* (Association of American Geographers, 1971), and Hare and Jackson's pragmatic *Environment: A Geographical Perspective* (Department of the Environment, Canada, 1972).

Despite their traditional links with geology and geography, geomorphologists have always sought and encouraged links with those natural sciences, such as pedology, climatology, hydrology, and ecology, that can provide evidence essential to the solution of geomorphological problems. Indeed, significant geomorphological research has been done by practitioners in these fields. But for geomorphological work to be used effectively in environmental management, more is required of the geomorphologist—he must also be aware of the social, economic, and technical contexts in which his information is relevant.

In this book we have tried to bring together geomorphological material that has been, or could be of value in environmental management. The studies we quote come from a variety of disciplines and research organizations. Our aim is to provide a useful introductory survey for those interested in exploring the practical applications of geomorphology and for those who require a geomorphological component in their study or practice of environmental management. To the geomorphologist we offer a fresh look at parts of the subject, often using material that has not previously appeared in systematic geomorphology texts. We hope that the display of techniques, problems, and environmental relationships may be of value to geologists, physical geographers, and other natural scientists with an interest in geomorphology. To them, our occasional forays into social science may serve as an introduction to the practical realities of environmental management. To social scientists, planners, and others who increasingly have to consider the natural environment, this book may serve to introduce those aspects of geomorphology central to some of their problems. We fully realize that these are ambitious aims, and that any introductory volume of this kind must have many shortcomings. Experience over several years of teaching geomorphology to geologists, geographers, planners, and economists leaves us in no doubt that our task is difficult but rewarding, and that there is still an enormous amount of work to be done.

In the Introduction we discuss briefly some of the broader aspects of geomorphology in environmental management, and the role that geomorphologists have played or could play in it. Subsequent chapters are of two major types. The majority are concerned with selected systematic aspects of geomorphology relevant to environmental management. The approach in these chapters (1–9) has been to analyse the appropriate geomorphological systems, and especially the processes at work within them, to indicate how the systems can be modified or manipulated by man, and to explore some of the social and economic implications related to the actions of man. In the remaining chapters (10–14), methods of survey and classification to which a knowledge of geomorphology is actually or potentially important are examined. Many of the techniques discussed have been developed by geomorphologists, often working as part of research teams in international or government-sponsored organizations, in consultancy, or in universities. A catholic selection of case studies is an important ingredient of the book as a whole.

Our text has inevitably been limited to provide a volume of convenient dimensions. We have not attempted to be comprehensive. Rather, we have selected important themes and introduced them as fully as possible. Some of these themes are reviewed under several chapter headings. An example is the geomorphological contribution to solving problems in urban areas (Chaps. 2, 5, 6, 11, and 14). Our references are also selected. One of our discoveries during the preparation of this book was the extensive, valuable, but unpublished 'underground' literature which is not generally available; we have included some of this so as to give it wider attention, but we have also provided a substantial number of references to material available to students in most university or college libraries so that points raised in the text may easily be followed up.

We have greatly benefited from comments by acknowledged experts on particular topics. They have helped to eliminate errors of fact and to identify errors of emphasis and omission; we are, of course, alone responsible for any remaining inadequacies. We thank especially Dr. R. J. E. Brown (National Research Council of Canada), Dr. R. J. Chandler (Imperial College, London, England), Dr. K. J. Gregory (University of Exeter, England), G. E. Hollis (University College London, England), D. B. Honeyborne (Building Research Establishment, England), Dr. J. N. Hutchinson (Imperial College, London, England), B. E. Lofgren (U.S. Geological Survey, Sacramento, United States), Professor D. Lowenthal (University College London, England), Dr. M. Newson (Institute of Hydrology, England), Professor N. Woodruff (University of Kansas, Manhattan, United States), and Professor H. Th. Verstappen (I.T.C., Holland). Professor C. A. M. King has not only kindly written the chapter on coasts, but has also commented on much of the text. We would also like to thank Valerie Cawley (Cartographic Unit, Department of Geography, University College London) for her expert and sympathetic reconstitution of our primitive sketches into publishable diagrams, and Barbara Cooke and Rosemary Hudson, both of whom provided generous typing and secretarial assistance.

RONALD U. COOKE   1 *July* 1973   JOHN C. DOORNKAMP
University College London   *University of Nottingham*

CONTENTS

ACKNOWLEDGEMENTS

The authors and publisher are grateful for permission to reproduce tables and figures from other copyright works; details are given in the underlines and bibliography.

CONVERSION TO METRIC

Throughout this book metric units have been used whenever possible. This has meant the conversion of some data originally obtained in non-metric units. In a few cases, however, conversion has not been possible and the data, results, or equations are given in the units in which they were originally derived. When conversion was possible the following equivalents were used:

DISTANCES

    1 inch = 2·54 cm
    1 foot = 30·48 cm or 0·30 m
    1 yard = 91·44 cm or 0·91 m
    1 rod = 5 m
    1 mile = 1·61 km

AREAS

    1 sq inch = 6·45 sq cm (cm^2)
    1 sq foot = 0·09 sq m (m^2)
    1 sq yard = 0·84 sq m (m^2)
    1 sq mile = 2·59 sq km (km^2) or 259 hectares (ha)
    1 acre = 0·40 ha

VOLUME

    1 cu foot = 0·30 cu m (m^3)

WEIGHT

    1 ton (U.S.) = 907·18 kg
    1 lb = 453·59 g, or 0·45 kg

OTHERS

    1 ton per acre = 2·26795 × 10^3 kg per ha
    1 cu ft per sec = 101·94 m^3 per hour

INTRODUCTION:
GEOMORPHOLOGY AND ENVIRONMENTAL PROBLEMS

WITHIN the contemporary concern for environmental management many problems relate to the interaction between man, land, and water. Geomorphology normally involves a study of the two latter, and frequently recognizes that today man is the most important geomorphological agent in some parts of the world. Wherever man uses land he has to accommodate its relief, materials, and water resources to his purposes. The study of these things falls within the domain of the hydrologist, engineering geologist, pedologist, and agriculturalist, and yet they maintain their identity under the title of geomorphology as well. It is with the form, materials, and processes of the earth's surface that geomorphology is concerned. It has its practitioners in many fields.

An awareness of the role of geomorphology in environmental management is growing rapidly after a very slow start. During the first half of this century the main stream of academic geomorphologists tended to overlook the work of Gilbert (1880) and others whose concern for the explanation of landform evolution through observable processes within the framework of geological materials laid down a potentially fruitful basis for geomorphological studies in environmental management. The geomorphological literature of the first half of this century does contain, however, a few applied studies. Glenn's (1911) paper illustrating accelerated erosion induced by man, Sherlock's substantial *Man as a Geological Agent* (1922), Bryan's (1925) extended analysis of erosion and sedimentation in the south-western United States, and the study of man-induced soil erosion by Jacks and Whyte (1939) are examples of these. In addition to these published works there are also a number of studies which were undertaken for private consultation. These include examinations of construction sites, evidence provided in legal disputes, and reports on the search for mineral resources. Most of them are not generally available.

Nevertheless, the study of contemporary processes by geomorphologists was relatively unimportant during these decades, with the significant exceptions of work on glaciers (e.g. Johnson, 1904; Gilbert, 1906; Thorarinsson, 1939; see Embleton, 1972), on shoreline processes (e.g. Beach Erosion Board, 1933; Lewis, 1931; Gaillard, 1904; Johnson, 1919), and on aeolian processes (Bagnold, 1941). By and large the analysis of present-day processes, especially fluvial processes, became the responsibility of hydrology and hydraulic engineering. Within these fields there now exist textbooks of immense relevance to geomorphologists (e.g. Linsley, Kohler, and Paulhus, 1949; Chow, 1964).

Despite sporadic attempts during the 1950s to develop further the field of process studies, the major advance came with the publication of *Fluvial Processes in Geomorphology* by Leopold, Wolman, and Miller (1964). This work is replete with information about present-day fluvial processes and by implication has much relevance to a number of environmental problems. More direct relevance was implied by the later publication of *Water, Earth and Man* (Chorley, 1969), and of *Environmental Geomorphology* (Coates, 1971). In fact much work in geomorphology since 1950 is of direct relevance to environmental problems; but it has not often been applied to them.

In a presidential address to the South African Geographical Society, Dixey (1962) stressed the need for more geomorphological studies to be orientated towards the needs of man. Yet, in a broad review of the subject seven years later, Dury (1969) still had to admit that geomorphology does not record numerous applied contributions, though he prophesied that many such contributions would come, and recognized that great scope for applied work exists in the investigation and prediction of geomorphological processes and their effects. It may seem surprising that Dury should still have to say this in 1969, when fifteen years earlier Thornbury (1954) devoted a convincing chapter to applied geomorphology in a widely-read textbook. Using specific examples to illustrate his view, Thornbury reasoned the case for applying geomorphology to hydrological studies, mineral exploration, engineering projects including dam-site and airfield-site selection, as well as to oil exploration and to military needs.

The necessity to understand geomorphological processes has now been amply demonstrated in situations that involve such things as flooding, landsliding, soil erosion by wind or water, coastal erosion and deposition, and the weathering of building stones. Geomorphologists are increasingly realizing the value of their work in the solution of applied problems; and they are making some singularly important contributions.

By 1950 there were also new developments in landform evaluation taking place, most notably those made by members of the Commonwealth Scientific and Industrial Research Organisation (CSIRO) in Australia. These developments initially amounted to little more than the making of an inventory in which a landform classification provided the framework for storing knowledge about the physical landscape. This became known as the land-systems approach. At the same time, particularly in Poland, attention was being directed towards a more comprehensive scheme for mapping land-surface form, materials, and processes as a basis for planning. The technique of geomorphological mapping which arose out of this work demanded the identification and classification of all slopes, the recognition of processes both past and present, and the relationship of form and process to bedrock and regolith materials. Its clear purpose was to provide a comprehensive analysis of the physical properties of an area, usually with a view to assessing its natural resources or its physical limitations for development.

During the 1960s, therefore, research activity became redirected towards geomorphological studies with an emphasis on contemporary forms and processes (e.g. Ward, 1971); and at the same time, resource surveys in the developing countries and in otherwise uncharted terrain were based on mapping procedures in which geomorphology played a leading (and some-times the dominant) role. There is little evidence at present of cross-fertiliza-tion taking place between these two trends, despite their common relevance to many practical problems. The way out of this situation may lie in recog-nizing that both land-systems mapping and geomorphological mapping are valuable in effectively describing present conditions and in revealing much of the nature and context of contemporary processes. For some management problems little more may be required than a catalogue of data upon which decisions can be based. In many problems, however, the dynamic element of temporal and spatial variations in processes may be critical for sensible management.

Douglas (1971) reinforces this view by two case studies. The first concerns the River Aare, Switzerland, where engineering works and agricultural activity resulted in river erosion and deposition problems. In order to alleviate flooding engineering works were carried out in the mid-nineteenth century which thereby reclaimed land for agriculture. However, without recharge from flood waters there was a drop in the water table causing the peat of the reclaimed land to dry out and its surface to subside. This brought about renewed flooding, and the need for a second phase of engineering work, which took much more account of the dynamic relationships of fluvial processes. The second case study is of a coastal resort area in Belgium where harbour construction and the erection of buildings on sand dunes led to severe beach erosion. The stability of the dune coastline had already been partly threatened by the fall in level, through artificial drainage, of fields on the landward side. Construction on the dunes broke down their dynamic equilibrium with the beach by restricting the supply of sand from the dunes to the beach. This increased the steepness of the beach and erosion accelerated. Douglas con-cluded that man's challenges to nature occur quite frequently, with an altera-tion of one aspect of the physical environment leading to a succession of readjustments. Both of these case studies show that management of the environment requires a thorough understanding of physical process–response systems, not least if these systems are to be considered in cost-benefit analyses.

The concept of land management taking place with an informed awareness of the nature of geomorphological systems is not entirely new. The exponents of geomorphological mapping, even in the 1950s, often implied the importance of linking the mapping of form to an understanding of process. Indeed, by 1957, Dylik, who was involved in geomorphological mapping, was able to talk about the importance of dynamic geomorphology to the economy of a country:

The earth surface whose diversification constitutes the object of geomorphological studies, is the meeting point of lithosphere, hydrosphere and atmosphere. Furthermore it is closely related to the biosphere. . . . Land-form surfaces and their elements are . . . also surfaces of human activities, surfaces of actual or future economic exploitation. Economic activities, although sometimes adapting themselves to the existing land forms of the area, are anyhow always compelled to take them into account (Dylik, 1957, p. 6).

Dylik added that since the land surface is continually undergoing change it is evident that for economic ends a knowledge of processes and the possibility of anticipating the direction of future relief development are much more important than a mere description of the present-day relief pattern. Similar positions were adopted by Klimaszewski, Tricart, and others (see Chap. 14).

The complete understanding of a process can never come, however, from studying that process in isolation. A geomorphological process functions within a physical system. This will embrace not only landform and earth materials but also processes involving the exchange of energy and the movement of materials, and it will include the operations of man (Chorley and Kennedy, 1971). Man, as the two case studies referred to above have shown, has to be seen not only as someone who is affected by geomorphological events, but also as someone who can change, control, and generally manipulate the environment so as to influence the effect of geomorphological processes. The fact that man does not always understand the consequences of his manipulations is amply demonstrated by recurring man-induced catastrophies (Brown, 1970). To avoid such events a knowledge is required of the whole system within which the processes are operating. In Clayton's words: 'The environment is extremely complex, and anyone tempted to interfere with it, even in an attempt to reverse man-induced deteriorations, should wait until he has enough knowledge to predict the full results of his activity' (Clayton, 1971, 84). This is not an easy task. Much progress towards this end may be made, however, by continually examining the context within which the process is operating. A knowledge of physical principles and dynamic interrelationships makes possible an understanding of the processes themselves. Land-systems mapping helps to define the situation of a site problem. Geomorphological mapping isolates the relevant components of the geomorphological system.

A second fundamental consideration is that the physical systems of an area can never be fully understood in isolation from the social, cultural, and economic attitudes and conditions of the people of that area: 'The interaction between man and environment involves aspects of both the physical and social sciences, and an approach based on one alone is unsatisfactory' (Clayton, 1971, 84). Taken in the context of an exploding world population and an ever-increasing pressure on land, geomorphology is obliged to include more and more events which occur on a human time scale. At the same time pioneer work in man's perception of the physical environment (Craik, 1970) indicates that the relevance of geomorphology has to be seen in the context of man's response to what he *thinks* the environment is like rather than what it is *actually* like.

Many geomorphological systems embrace influences that extend beyond those traditionally claimed as part of geomorphology. By the same token geomorphologists may be seen by some to be laying claim to matters that they have long ignored; but subject boundaries may provide artificial barriers to progress. It is surely better to devote attention and energy to serious problems and not to be over-concerned with the name of the discipline, or with the antecedents of its practitioners. Thus the geomorphologist is likely to develop contacts with a variety of environmental managers and technicians, especially engineers, farmers, foresters, planners, and politicians, and he will probably find himself adding to his staple reading material studies in economics, engineering, geography, planning, law, and even psychology. In short, disciplinary boundaries have little meaning to the applied geomorphologist, except in so far as he must recognize them in order that he may cross them with caution.

The terms 'environmental management', 'resource management', and 'conservation' are fashionable today. The effective control in such activity rests with an enormous number of individuals, many of whom are not trained environmental scientists. The problem is to bring the scientist and the environmental decision-maker together. Frye (1967) recognized, for instance, that too many geologists (and by implication he included geomorphologists) are talking to each other rather than to the public or professional managers. In the management of the environment the public is extensively involved. In this context the public includes publicly appointed policy-making personnel. It also includes the farmer, the tourist, the building contractor, the forester, and indeed everyone who makes significant use of the land.

Many private firms and government agencies exist which deal with some part or another of the the earth's geomorphological systems. At the international level UNESCO, in its concern for the welfare of mankind, has acknowledged the important role of applied geomorphology (see, for example, the UNESCO publication *Nature and Resources*). It has also given much practical support to the International Hydrological Decade (IHD), in which geomorphologists have made a number of significant contributions; and, to name but one other example, to the postgraduate course in Applied Geomorphology at the University of Sheffield, England. On the national level there exist bodies such as the U.S. Soil Conservation Service, the Land Research Division of the CSIRO in Australia, and departments of hydrology and agricultural ministries in almost every country in the world. The list is a long one, and references to such organizations are made in the following chapters.

Clearly the stage is set for a very rapid development in applied geomorphology. This in its turn may bring many positive advantages to the theoretical side of the subject. As Verstappen (1968) indicated, applied geomorphology involves the practical testing of theories, and it produces hard facts that can be fed back into pure research. Vague and doubtful theories will not survive any attempts to apply them. In addition geomorphologists will be brought more and more into contact with scientists in other, often overlapping and cognate disciplines, to the benefit of both.

The relevance of geomorphology to many environmental problems can be demonstrated in almost every part of the world, but the degree of relevance varies with the environment and with the people who occupy or who are developing the area. Two ends of the spectrum are indicated by comparing the needs in an established urban area in a highly industrial, economically advanced society and those of an area of virgin territory in comparatively undeveloped and unpopulated lands. In the former case, as in Los Angeles, geomorphology may contribute much to environmental management especially in the context of engineering geology. In the latter case the geomorphologist has a vital part to play in land planning so long as his work can be understood by those who are to make the decisions. In between these two extremes are situations in which geomorphology may conceivably have a contribution to make in environmental management but much depends on the inhabitant of the land in question. A well-educated farmer in Western Europe or North America who is aware of environmental dangers, particularly those relating to mismanagement, but who is without the specific knowledge to interpret landscape for himself, may wish to clear an old woodland and want to know of the potential dangers of soil erosion. In order to do this he will call for expert advice. By way of contrast the subsistence crop farmer with a large family in one of the developing countries of Africa, and of necessity putting heavy pressures on his land for the crops and pastorage which he needs, may be suspicious of advisory intervention. To take another set of contrasting examples, one city planner who is concerned with the allocation of new areas for residential, industrial and recreational uses may appreciate that regolith and bedrock vary within the area of proposed expansion. He knows that he must fully take into account also the conditions of drainage and slope stability within the area. For this he will seek professional advice. Another city planner in a similar situation, unaware of these questions, may make planning decisions based upon his own superficial examination. He may also pass on the responsibility of the actual site problems to the site engineer who has to cope with the consequences of ill-advised site allocations. An advance knowledge of the materials, forms, and processes present at the site could ultimately save such a developer a great deal of time and money.

Summary

Geomorphology, which is concerned with landform, materials, and their related processes, is pertinent in all aspects of environmental management involving these physical phenomena. Although there is a general impression to the contrary, several important studies of geomorphological processes were carried out in the first half of this century, and it is frequently upon these that work in environmental geomorphology is now based. Significant advances in the mapping of a land's geomorphological characteristics have added another invaluable tool in assessing the potential of land for development, and in providing a valid basis for land-use planning. Geomorphological processes do not operate in isolation. They are generally a part of a whole system of interacting phenomena. In addition, their significance to environmental problems has to be assessed in the light of the social, economic, and cultural conditions of the people in that area. Environmental geomorphology knows no traditional subject boundaries and embraces parts of such subjects as hydrology, engineering geology, pedology, and geography. The potential for applying geomorphological knowledge depends not only on the problem but also on the willingness of the environmental manager to appreciate the value of this knowledge.

1

THE DRAINAGE BASIN IN ENVIRONMENTAL MANAGEMENT

1.1. Introduction

Man tips his refuse into a stream and expects it to be removed without any further consequence to himself, and without difficulty. Houses are built on conveniently flat land, and at times are flooded. Dams are built in the mountains and supply man with water or generate electric power for his use. A road is constructed, and the hillside on which it is built gives way to landsliding. All of these situations arise in the context of the drainage basins in which they take place, for each is related to water in the landscape. This book includes many such instances where man's use of the environment must take account of its setting within particular fluvial systems.

River networks occur within drainage basins (catchments), which frequently have finite boundaries (divides or watersheds). There is a natural input into any drainage basin from precipitation, and a natural output by way of water and sediment discharged along the river channel. Within any drainage basin there may be a variety of energy exchanges, but in general the dominant processes are those induced by gravity (e.g. soil creep and landslides) and by water either on the hillsides or in the rivers. There are many other natural events taking place within drainage basins, such as biological activity, which are related to the fluvial system or the hydrological cycle. In particular, although man has the power to act independently of the constraints of a drainage basin, much of what he does is related to, or is influenced by, the nature of the basin in which he operates.

The concept of the drainage basin as an appropriate areal unit for the organization of human activity is not a new one. Early forms of political organization were centred upon drainage basins, and stream channels in particular, so as to obtain water supplies, fish, and game, and to use the rivers for transport, while the direct use of water as a source of power tied industrial societies to valley locations. Certain of the early irrigation works led to an economic, social, and political structure which developed in the context of a drainage-basin unit. This included some early Chinese dynasties whose power lay in their control of the water resources in the drainage basin (Smith, 1969).

Industrial development in many areas saw man 'breaking free' from some of the earlier physical controls of the environment, and drainage basins have not continued to exert a restrictive influence. Nevertheless, it can still be argued that drainage basins form the natural unit for physical, economic, industrial, and social planning. However, this is 'an ideal to which planners have aspired rather than a principle which has been firmly established in water management policies anywhere in the world' (Sewell, 1965, p. x). A study of the Fraser River Basin (Sewell, 1965) confirmed an increasingly well-established principle of man's reaction to the environment, namely that man's response (in this case to the proposed adoption of an integrated approach to water management) hinges upon the existence of a powerful incentive, such as a natural crisis. Environmental planning should not wait for crises to arise.

It is pertinent at this point to examine the structure of the drainage basin as a system. In this way its physical make-up can be appreciated before considering any further its role as a planning unit in environmental management.

1.2. The Structure of the System

Fundamental to the concept of a drainage basin are the valley and channel networks that it contains, and the divide that surrounds it and serves to separate it from adjacent basins. The organization of the network is important because it reflects the efficiency of the main lines of energy and material flow through the system (e.g. Chorley, 1969). The plan characteristics of the channels in a drainage basin are also important in a different sense. Many such networks have been shown to have a pattern which is not significantly different from random (Shreve, 1967), and it is difficult to associate them with any particular origin or development. However, when a distinctive organized pattern is discernible then strong environmental control (e.g. bedrock structure) is implied.

The real importance to applied geomorphology of a study of the structure of a drainage basin and its channel network lies in the integrated nature of that structure. Many of the morphometric characteristics of a basin (e.g. its size, the length of its streams, or the drainage density) can be directly related to fluvial characteristics such as the discharge of water from the basin (Chap. 4). Easily measured basin properties tend to have definable predictive qualities that can be used to infer the likely effects of any interference by man with the system. Similarly these properties enable estimates to be made of the dynamic characteristics of a river basin in a remote area, a requirement in a natural resource survey, or in an unmonitored part of an already developed area. This is not to imply, however, that it is enough to apply regression equations (e.g. of the type given in Chap. 4) developed in one area to any other area, or that the correlations between morphometric and dynamic properties are always predictable. Expertise is required to assess the relevance of particular applications, not least in terms of the history and geomorphological variability of the basin concerned.

Although predictive equations may be directly applicable in some parts of a system it is seldom the case that they also apply to the whole system. Careful geomorphological analysis is always required to assess the relevance of a predictive statement to a particular situation. Equally important is the need to recognize those portions of the basin (hillsides or river channels) where as much water and materials are being exported as are being imported. Those portions are acting solely as units across which transport is taking place, without net loss or gain. Such sections are in a *steady state* (Schumm and Lichty, 1965), and any interference with them by man must lead to an excess of either erosion or deposition, for the temporary balance of forces will have been destroyed.

The achievement of a steady state, in which landforms and the rate of supply and removal of water and sediment are mutually adjusted, may depend on scale. For example, a study in the upland areas of mid-Wales suggested that a steady state applied to meso-scale (e.g. slope and stream reaches) and micro-scale features (e.g. sites) in fluvial landforms, but macro-scale features (e.g. whole river basins) cannot be considered as independent of the time factor in their development since they may still carry signs of earlier periods of geomorphological activity (Slaymaker, 1972).

The analysis of drainage basins has usually been based on an ordering system applied to the channel network whereby it receives a comparative rating according to the number and order values of its tributary branches. For example, Strahler (1952) proposed an ordering system (Fig. 1.1) in which

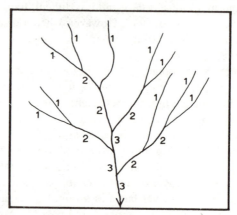

Fig. 1.1. A stream-ordering system (based on Strahler, 1952)

all streams having no tributaries are designated first-order streams. When two first-order streams join, the resulting stream has second-order status, and two second-order streams on joining give rise to a third-order stream. Thus a unit increase in order takes place downstream of the junction of two streams having the same order. No change in order takes place when streams of differing orders meet. In this system small events, such as the addition of just one first-order stream can have a marked effect on the final order given to the stream leaving the basin, and thereby on the order of the basin (Doornkamp and King, 1971).

For any catchment a variety of properties can be measured. These help to define the character of the network and of its basin. Some of those commonly measured are listed in Table 1.1., together with their length (**L**), area

TABLE 1.1

Morphometric and related variables, their dimensions, and symbols used

Variable	*Symbol*	*Dimensions*
Drainage network		
Stream order (used as subscript)	u	0
Number of streams of order u	N_u	0
Total number of streams within basin order u	$(\Sigma\,N)_u$	0
Bifurcation ratio	$R_b = N_u/N_{u+1}$	0
Total length of streams of order u	L_u	L
Mean length[1] of streams of order u	$\bar{L}_u = L_u/N_u$	L
Total stream length within basin of order u	$(\Sigma\,L)_u = L_1 + L_2 \ldots + L_u$	L
Stream length ratio	$R_l = \bar{L}_u/\bar{L}_{u-1}$	0
Basin geometry		
Area of basin	A_u	L^2
Length of basin	L_b	L
Width of basin	Br	L
Basin perimeter	P	L
Basin circularity	$R_c = A_u$/area of circle having same P	0
Basin elongation	R_e = diameter of circle having same P/L_b	0
Measures of intensity of dissection		
Drainage density	$D_u = (\Sigma\,L)_u/A_u$	L^{-1}
Constant of channel maintenance	$C = 1/D_u$	L
Stream frequency	$F_u = N_u/A_u$	L^{-2}
Texture ratio	$T_u = N_u/P_u$	L^{-1}
Measures involving heights		
Stream channel slope	θ_c	0
Valley-side slope	θ_g	0
Maximum valley-side slope	θ_{max}	0
Height of basin mouth	z	L
Height of highest point on watershed	Z	L
Total basin relief	$H = Z - z$	L
Local relative relief of valley side	h	L
Relief ratio	$R_h = H/L_b$	0
Ruggedness number	$R_n = D \times H/5280$	0

[1] The superscript bar indicates (here and throughout) a mean value.
Source: Modified from Doornkamp and King (1971).

(L^2), and volume (L^3) dimensions. From these, dimensionless numbers (e.g. bifurcation ratio) can be obtained. Statistical analysis has shown that most of the variation in the morphometric parameters of drainage basins can be accounted for by variations in basin area, the total number of streams, total basin relief, and stream frequency (or alternatively drainage density) (Doornkamp and King, 1971). Significantly high correlations have been found to exist (Doornkamp and King, 1971) between (i) basin area and total stream length, the lengths of streams at each order, and the mean lengths of streams at each order; (ii) total number of streams and the number of streams at each order; (iii) stream frequency (or drainage density) and drainage density (or stream frequency), relief ratio, and total basin relief; (iv) total relief and local relative relief of valley-sides. Melton (1958) also included angle of valley-side slope, and a ruggedness number in his analysis, and these can be incorporated in a summary of correlation linkages as shown in Figure 1.2.

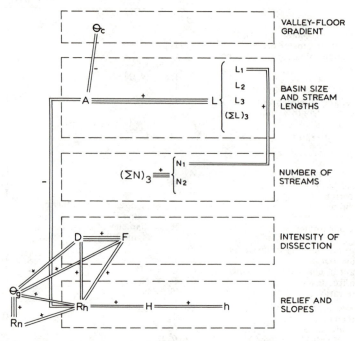

Fig. 1.2. Summary of correlation linkages between the morphometric parameters of drainage basins (for explanation of symbols see Table 1.1)

The practical significance of such a structure only comes when it can be related to process variables, such as the hydrological characteristics (e.g. river discharge) of a basin. This theme is expanded, with examples, in Chapter 4.

An important component of any drainage basin system is provided by its valley-side slopes. These direct the surface water towards the river channels, and are a major source for river sediments. A simple but effective functional subdivision of valley-side slopes (Hack and Goodlett, 1960) recognizes five types (Fig. 1.3):

(i) *Nose*—the driest area, whose contours are convex outwards from the valley (as on a spur). It receives least moisture from up-slope and provides shedding sites. Runoff is proportional to a function of the radius of curvature of the contours.

(ii) *Side slope*—this has straight contours, receives water from the nose, and has wetter sites than those on the nose. Runoff is proportional to a linear function of slope length.

(iii) *Hollow*—the contours are concave outwards and demarcate the central part of the valley. All slopes converge on the stream, and it comprises very moist slopes, increasing in water content towards the stream. Runoff is proportional to a power function of slope length.

(iv) *Foot slope*—the gentle lower part of a side slope, and may be composed of regolith rather than based on solid rock.

(v) *Channelway or valley floor*—the river course plus any adjacent flood-plain.

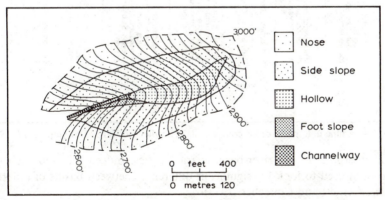

FIG. 1.3. A functional subdivision of slopes within a drainage basin (after Hack and Goodlett, 1960)

Dalrymple *et al.* (1968) proposed a functional nine-unit subdivision of the valley-side slopes which combines their form with a statement of the contemporary geomorphological processes (Fig. 1.4). This model can be adapted to the character of most areas, and local details of bedrock and soils may be added to make it relate to a specific drainage basin. The integrated view of form, materials, and processes implied by this approach is essential to the

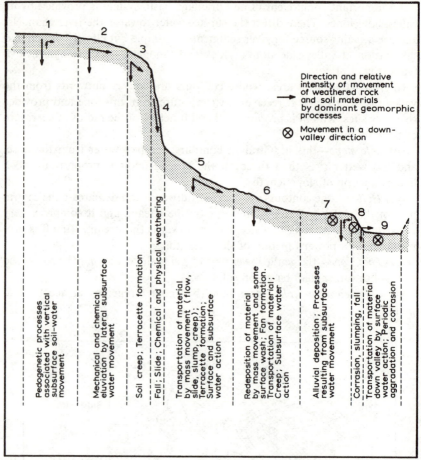

Direction and relative intensity of movement of weathered rock and soil materials by dominant geomorphic processes

Movement in a down-valley direction

1

2

3

4

5

6

7

8

9

Pedogenetic processes associated with vertical subsurface soil-water movement

Mechanical and chemical eluviation by lateral subsurface water movement

Soil creep; Terracette formation

Fall; Slide; Chemical and physical weathering

Transportation of material by mass movement (flow, slide, slump, creep); Terracette formation; Surface and subsurface water action

Redeposition of material by mass movement and some surface wash; Fan formation. Transportation of material; Creep; Subsurface water action

Alluvial deposition; Processes resulting from subsurface water movement

Corrasion, slumping, fall

Transportation of material down valley by surface water action; Periodic aggradation and corrasion

Fig. 1.4. A nine-unit model of the components of a valley side (after Dalrymple *et al.*, 1968)

valid interpretation of river-basin characteristics. Indeed this approach has also been used to look for significant differences between basins of different orders (Arnett and Conacher, 1973)

1.3. Man's Interference with Fluvial Systems

Although some apparently simple relations exist within the drainage-basin system as a whole, the system may be a complex one composed of many different attributes of form, materials, and process. Any interference by man with one of these, whether it be deliberate or accidental, can set in motion a sequence of events that may extend throughout the whole system. Ruhe (1971) described a case where the straightening of a channel had repercussions throughout the basin. In Willow River (Crawford County, Iowa, U.S.)

straightening led to channel deepening and widening. In addition, new deeply-entrenched gullies extended for many miles up tributary streams, and developed on hillsides, disrupting agricultural land and public roads.

When urbanization takes place large-scale changes are induced within the catchment both in terms of its hydrological characteristics and its sediment yield (see Chaps. 4 and 5). When new ground is broken for urban development, vegetation is removed and the ground is disturbed by construction equipment. Runoff and storm flow increase, land erosion is accelerated and the sediment load of the streams is often dramatically increased. Wolman and Schick (1967) recorded up to 50,000 tons/km²/yr on one site, as compared with 80–200 tons/km²/yr under natural conditions. Thus urban development may cause large sedimentation problems further down-valley. For example, excessive amounts of sediment induce the building up of channel bars, erosion of channel banks as a result of deposition within the channel, obstruction of flow and increased flooding, shifting configuration of the channel bottom, blanketing of bottom-dwelling flora and fauna, and alterations in these due to changes in light transmission and abrasive effects of sediment, as well as alterations of species of fish due to changes produced in the flora and fauna upon which the fish depend.

When the construction of houses and roads has taken place there is a decrease in the natural infiltration, groundwater level may be lowered, and artificial drainage lines are introduced. Local interference with minor streams may cause these to flood or to undermine their banks. Streams may be diverted for public water needs, decreasing the runoff down the old stream course. Untreated waste, including sewage, may be discharged into streams, causing pollution which in turn may be lethal for aquatic life and detrimental to the use of the water downstream either as a source of supply or for recreational purposes. Streets and gutters act as storm drains, often creating higher flood peaks in the local streams. As towns continue to grow, the demands of industry upon the water supply increase, and pollutants may be discharged into the natural drainage. Water demands may necessitate the building of dams upstream, or cause water to be brought in from adjacent basins. This leads to further water-flow control. The channels become increasingly restricted or confined, and flood damage may increase since peak flows will require a high level of water in these artificial channels. This in itself may then lead to an improvement in the storm drainage system (Savini and Kammerer, 1961). All of these changes may be studied and their control effected in the context of the physical characteristics of the catchment in which they occur. Steps are being taken in some areas to monitor basins threatened by urbanization so that comparative data exist which can be used when measuring the effects of construction work (Miller, Troxell, and Leopold, 1971; Gregory and Walling, 1973).

In practice many of man's influences may extend no further than, or only a limited distance from, the site of his activities. Some changes, however, such as dam construction, may be much more drastic, influencing not only the hydrological conditions above the dam site but also, by controlling the amount of outflow from the dam, influencing the conditions downstream. When this is done on a major river it may even be necessary to introduce international agreements by which one nation's vital water supplies are not held back by the presence of a controlling dam in another country (e.g. Owen Falls dam across the Nile, Uganda).

1.4. The Drainage Basin in Land Evaluation

(a) Irrigation

A drainage-basin context for land-evaluation procedures has much to commend it either where the area and problem of interest are contained within one catchment, or where the problem is intimately linked to fluvial processes. For example, land irrigation is normally dependent on adequate water supplies within the catchment, and ideally the use of this water in one area should have no damaging physical or economic consequence on areas downstream. The same is true of an industrial complex which will consume substantial amounts of the water available within a drainage basin. An evaluation of the water-supply budget within a basin will soon show if additions are required from other basins. In each case, evaluation must take place within the context of the water-supply regions, namely drainage basins.

The physical planning of an irrigation project should include a full assessment of its economic, social, and political consequences. Small-scale irrigation works may be undertaken by a single farmer entirely on his own land. Projects at the regional or national level frequently need to be considered as part of an integrated, multi-purpose, water resources project (e.g. Fig. 1.6). At this scale there are many major factors to be considered at the planning stage including the geomorphology and geology of the area, its hydrological characteristics, design problems for flood control, drainage, and water-quality control, and the ultimate use of the water for agriculture (irrigation), hydro-electric power, navigation, domestic and industrial water supply, recreation, fish, and wildlife (Dixon, 1964). Major programmes of this type have included the Tennessee Valley scheme and the Missouri River project in the United States, and the Orange River scheme in South Africa. The physical context of each water resource project, whether it be solely for irrigation or a multi-purpose project, has to be the drainage basin. On occasions the potential economic return may be sufficient to justify engineering works that carry water across or through drainage divides (e.g. the Snowy Mountain scheme in Australia); but it is always done with a consideration for the physical characteristics of both the supplying basin and the receiving basin (see Douglas, 1973).

There was an early though indirect recognition in the United States of the drainage basin as a management unit through the influence of irrigation programmes. As long ago as 1902 Congress enacted the Reclamation Act, giving the U.S. Bureau of Reclamation responsibility for undertaking irrigation development. This was done with an eye towards settlement in the arid West. The concept involved was that irrigation projects, presumably in the context of specific catchments, would lead to the growth of associated industries and thus form the basis for regional economic development (Ciriacy-Wantrup, 1964).

A recent example of an integrated approach to basin planning relying heavily on the needs for irrigation is that of the Indus Basin in Pakistan,

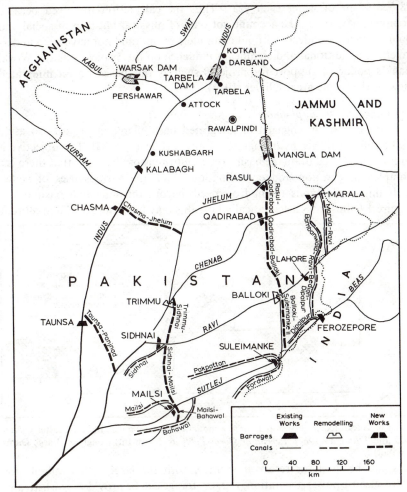

FIG. 1.5. Irrigation planning, the Indus and Tarbela projects (after Olivier, 1972)

the aim of the project being to bring the national average nutrient level of the population up from a low level to one of 2,200 calories per head per day by 1975.

By means of storage dams upstream and canals between rivers downstream (Fig. 1.5), vital water supplies are made available to the southern parts of Pakistan, where agriculture depends almost entirely on irrigation. The available groundwater supplies, which at 370×10^8 m³ amounted to twice the mean annual runoff from the basin, were integrated into the project. They are used for crops at the peak periods of water requirement.

Planning had to take account of the local population's sensibilities about the existing network of irrigation canals, covering 162,000 km² and using 98.6×10^9 m³ of water each year. Thus the calculations involved became complex, having to take account not only of physical but also of social and political factors. Indeed, without computers these could not have been undertaken. An additional and significant influence on this project was the World Bank's policy of support for projects such as this based on an integrated view of the whole area (Olivier, 1972).

(b) Integrated basin planning

Spain is actively engaged in integrated basin-planning projects so as to make full use of her hydrological resources (Olivier, 1972). The objective is to use the selected drainage basins on a multi-purpose basis, so that there is a complete integration and co-ordination of all the various uses of water, including the demands arising from needs beyond the immediate basin limits. Figure 1.6 illustrates schematically how this may be organized. High-altitude

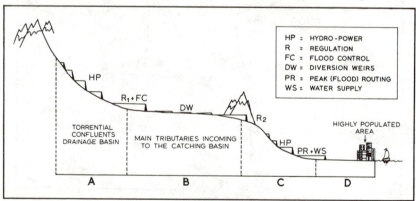

FIG. 1.6. Integrated water-use planning, a model for developments in Spain (after Olivier, 1972 and Toran, J. (1967), *Revista de Obras Publicas*, Ninth Int. Congr. on Large Dams, Istanbul)

sites can be used for reservoir storage, with the large falls available for hydro-electric power generation. Larger storage can be provided downstream, in flatter terrain, but above the wide, more populated valleys. More hydro-electric power can be generated downstream if necessary, physical conditions allowing. At any stage in the system controls can be included to deal with flood conditions or the maintenance of adequate water supplies for the highly populated lowlands.

There are, however, many planning projects that are largely unrelated to watershed boundaries. Economic planning regions are seldom conceived as tidy physical units. Often policies are laid down and plans conceived that may extend across many drainage basins. In such cases the context of a drainage basin as a planning unit may become somewhat irrelevant, except at the execution level of site evaluation and site controls.

Hamilton *et al.* (1969) described their attempts to define a simulation model for the regional economic planning of the Susquehanna River (United States). The river basin has an area of 70,000 km², and receives a mean annual precipitation of 1,000 mm. The major factors in selecting the study area were the physical characteristics of the basin, but it was necessary that the boundaries of existing economic regions were not unduly distorted. That is to say areas which are tied to the economy of the basin, but which lie physically outside it, were included for the purposes of the model. Data could only be collected on a County basis, so County boundaries had to be used to mark the limits of the area. In other words, desirable as it may have been to limit the analysis to the area within a specified drainage divide, economic and administrative (data source) constraints prevented this from being entirely possible.

The authors of this study stated their belief that, with regard to regional model-building and river-basin planning, both regional economic growth and river-works construction are inextricable parts of the same general system. They stress that the regional economic analysis of river basins should take into account the effects of the economy upon the water as well as the possible feedbacks from the water sector to the economy. The Susquehanna model thus includes: (i) the demographic, economic, and water sectors of the planning scheme; (ii) a computerized simulation model in which it is possible to insert a wide variety of assumptions and parameters; (iii) a dynamic system whereby explicit feedbacks and lagged variables are included both within and between the various sectors. One of the feedbacks, for example, concerns the effect of the construction of various types of river works on the employment sector of the economy. By means of this approach neither the physical nor the economic systems are studied in isolation, and the dynamic element of the drainage-basin concept is explicitly involved.

(c) Ecosystem studies

Ecosystems in the natural environment are extremely difficult to monitor as so many energy exchanges are taking place at the same time. These include climatic inputs to the system, such as precipitation and chemical elements, biological exchanges between plants (or animals) and both the air and the soil, pedological changes as weathering continues, and changes due to water within the system. In addition there are the many outputs from the system which can be very difficult to record. Some of these are by way of water and sediments (whether dissolved or solid) along the rivers, while other water losses may occur by infiltration beneath the ground.

In an assessment of the inputs and outputs of an ecosystem Bormann and Likens (1969) considered the possibility of using the drainage basin as a suitable evaluation unit. They based their argument on the concept that nutrient cycles are strongly related to the hydrological cycle. By careful instrumentation of some catchments in the Hubbard Brook Experimental Forest in the White Mountains of central New Hampshire they were able, on this basis, to arrive at absolute values concerning input–output relationships within the ecosystem of that area.

1.5. Conclusion

Drainage basins form natural geomorphological units frequently displaying organized relations among their constituent parameters of land forms and drainage. Many of man's activities must be seen in the context of the drainage basins in which they occur. In fact man is not only influenced by, but can himself severely affect behaviour of the processes operating within a drainage basin. River catchments form the natural unit for planning where the physical system itself is directly involved (e.g. irrigation projects, integrated water-development schemes), but they may be of lesser significance in other forms of planning (e.g. regional economic planning schemes).

Many geomorphological events cannot be understood out of the context of the basins in which they occur. Certainly the drainage basin forms the natural unit for the environmental management of events tied to the whole fluvial system. Some of these are described in the following chapters.

2
SOIL EROSION BY WATER

2.1. Introduction

The spectre of declining agricultural productivity arising from erosion of soil by wind and water has haunted conservationists for much of the twentieth century. Intimations of impending catastrophe and appeals for action have been broadcast for many years through such books as *The Rape of the Earth* by Jacks and Whyte (1939), Bennett's *Soil Conservation* (1939), and the F.A.O.'s *Soil Erosion by Water* (1965). Disasters in the Dust Bowl, the Tennessee Valley, and elsewhere disclosed the seriousness of the problems and symbolized the purpose of the crusade. The research response was published in a plethora of periodicals which include the *Journal of Soil and Water Conservation, Agricultural Engineering*, and the *Proceedings of the Soil Science Society of America*. Administrative and political responses were made manifest in such organizations as the Civilian Conservation Corps, the Soil Conservation Service and the Bureau of Land Management in the United States, and many comparable institutions elsewhere.

Soil erosion figures less prominently in contemporary environmental debates than it did in similar debates before the Second World War. Today, 'pollution' has tended to replace 'erosion'. But this change in no way reflects the declining importance of soil erosion. Perhaps the false notion is widespread that because conservation practices have been developed to combat soil loss, the problem has been solved. This notion is unjustified for two reasons: in many areas of soil erosion, conservation practices are not used; and conservation practices have not always been successful where they have been employed.

2.2. Rates of Soil Erosion

Earth scientists are concerned with many types of soil erosion, and they normally draw an important distinction between 'geological' and 'accelerated' erosion. Geological erosion is the rate at which the land would normally be eroded without disturbance by human activity. Accelerated erosion is the increased rate of erosion that often arises when man alters the natural system by various land-use practices, and as such it falls squarely into the subject of environmental management. The 'problem of soil erosion' is really the 'problem of accelerated erosion'.

Present erosion rates vary greatly from place to place. In order to study the pattern of world erosion, standard measures are required and many are available. Sediment yield, or one of its many variants, is the most widely used (e.g. Holeman, 1968). It expresses in tons/km²/per annum (or similar units) the removal of sediment from specified drainage basins. It may be based on measurements of river load being discharged past a measuring point, or upon the amount of sediment which has accumulated in a reservoir. Because bedload and dissolved load in rivers are rarely measured, many estimates of sediment yield are based on analysis of the sediment suspended in flowing water, which in any case may account for up to 90 per cent of the total mechanical load of the river (Chap. 4). Detailed studies, based on field measurements, have been made of the sediment yield of individual rivers; at the other extreme, estimates of sediment yield on the continental scale arise out of theoretical analysis, as discussed below. Unfortunately, the data relating to erosion rates are often not comparable because so many different techniques have been used. Another problem is that, in order to obtain reasonable averages, data are required for long periods, and this requirement is often not satisfied.

Any single measurement of erosion rates is affected by numerous variables, of which rock type, climate, vegetation, and drainage-basin characteristics (such as area, steepness of slope, drainage density, relief, and length of slope) have been found to be the most important (Chow, 1964; Stoddart, 1969; Young, 1969). There have been several attempts to determine the pattern of world erosion rates using sediment-yield data and taking into account some or all of these variables. Fournier (1960), for example, used the following empirical equation to predict sediment yield from a knowledge of relief and climate:

$$\log E = 2 \cdot 65 \log (p^2/P) + 0 \cdot 46 \log \bar{H}. \tan \phi - 1 \cdot 56 \qquad (2.1)$$

where

E = suspended sediment yield (tons/km²/p.a.);
p = rainfall in the month with greatest precipitation (mm);
P = mean annual precipitation (mm);
$\bar{H}$ = mean height of basin (m);
ϕ = mean slope in a basin;

and p^2/P is used as a measure of precipitation seasonality, or the incidence of rainfall concentration.

The Fournier equation (2.1) requires the calculation of both the mean height and the mean slope of a drainage basin, it is tedious and time-consuming to use, and the results are at times unreliable. In constructing the map showing the world distribution of erosion rates based on suspended sediment yield data (Fig. 2.1), therefore, Fournier first classified drainage basins according to relief and climate into four categories:

Ia   Low relief, temperate climate;

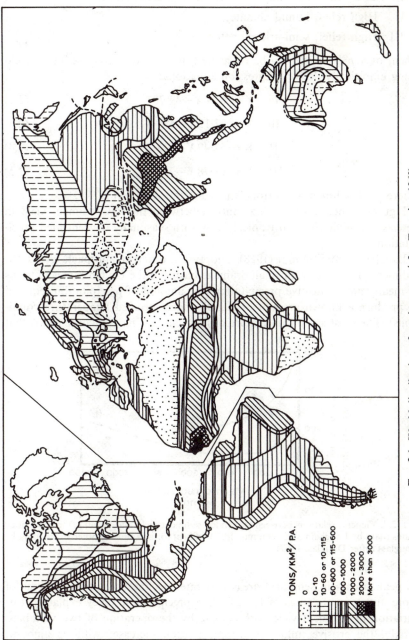

TONS/KM²/ P.A.

0
0-10
10-60 or 10-115
60-600 or 115-600
600-1000
1000-2000
2000-3000
More than 3000

Fig. 2.1. World distribution of erosion rates (after Fournier, 1960)

Ib Low relief, tropical, subtropical, semi-arid climate;

II High relief, humid climate;

III High relief, semi-arid climate.

Then, only p^2/P needed to be calculated in order to predict sediment yield using empirically derived regression equations:

$$Ia \quad Y = 6 \cdot 14X - 49 \cdot 78 \tag{2.2}$$

$$Ib \quad Y = 27 \cdot 12X - 475 \cdot 4 \tag{2.3}$$

$$II \quad Y = 52 \cdot 49X - 513 \cdot 21 \tag{2.4}$$

$$III \quad Y = 91 \cdot 78X - 737 \cdot 62 \tag{2.5}$$

where Y = sediment yield (tons/km²/p.a.) and $X = p^2/P$.

Figure 2.1 demonstrates the contrast between the low sediment yield in deserts and high yields in tropical areas with seasonally concentrated precipitation.

Langbein and Schumm (1958), adopting a different approach, evaluated sediment yield data from small drainage basins in the United States in terms of mean annual effective precipitation (the amount of precipitation required to produce a known amount of runoff under specified temperature conditions). The most significant feature of their curve (Fig. 2.2, curve (a)) is that

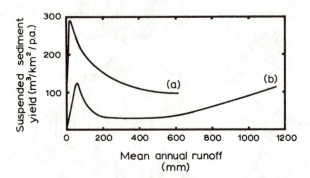

Fig. 2.2. General relations between suspended sediment yield and runoff (a) in the United States (data by Langbein and Schumm, 1958) and (b) for a sample of other rivers (data by Douglas) (after Douglas, 1967)

its peak occurs in semi-arid areas and rates decline both towards areas of more and towards areas of less effective precipitation. They explained this variation of sediment yield with climate by the operation of two variables. Firstly, the erosive influence of precipitation increases with its amount. Secondly, and opposing this influence, is the protective effect of vegetation which also increases with precipitation. These factors can be summarized in an equation which, when solved empirically and converted to metric units, states (Douglas, 1967):

$$E = \frac{1 \cdot 631 \; (0 \cdot 03937 \; P)^{2 \cdot 3}}{1 + 0 \cdot 0007 \; (0 \cdot 03937 \; P)^{3 \cdot 3}} \tag{2.6}$$

where E = suspended sediment yield in $m^3/km^2/p.a.$, P = effective precipitation (mm), and where the numerator represents the erosive influence and the denominator, the vegetation-protection factor. Douglas (1967) compiled another graph (Fig. 2.2, curve (b)) based on data from a selection of rivers, including major Asian rivers, which may be compared with Langbein and Schumm's graph. Apart from the generally lower sediment yields, the second curve shows a rise in suspended sediment yield as mean annual runoff increases over 600 mm. Douglas concluded from this and related information that highest sediment yields are to be expected in areas of marked wet and dry seasons—not only the semi-arid regions, but also areas of monsoon climates—where intense seasonal precipitation effectively erodes the soil and drought prevents the growth of a dense protective vegetation cover.

It is most important to emphasize that Fournier's map and Langbein and Schumm's graph are probably based on data from areas where man has significantly increased erosion rates. The patterns in part, therefore, are the patterns of accelerated erosion. How would the patterns of geological erosion rates differ? Douglas (1967) indicated that geological erosion rates are probably less, and in some cases very much less than present rates. In eastern Australia, for instance, Fournier's prediction of erosion rates is an order of magnitude greater than that measured by Douglas, and the difference probably arises in part from the fact that Fournier's original data are affected by greater human activity than are the Australian data. Similarly, the peak of Langbein and Schumm's curve may be exaggerated because in the semi-arid lands of the United States man has greatly altered the vulnerable environment and erosion rates are very much higher than in areas with similar climate but little human modification. Unfortunately a detailed map of world accelerated erosion is not available, but one has been produced for the United States (Fig. 2.3).

The mainly empirical methods of estimating sediment yield, such as that used by Fournier, thus provide a useful yardstick for the examination of erosion rates on a world-wide scale. They do no more, however, than define those major areas that are exceptionally prone to erosion and where man has to exercise the greatest care in his interference with the geomorphological system. In practice, and in order to be effective, the control of soil erosion requires close examination of soil movement and sediment transport within manageable areas, and preferably within well-defined physical systems, such as the drainage basin (see Chap. 1). The resulting predictive equations all too often apply only to the areas for which they were derived and are not

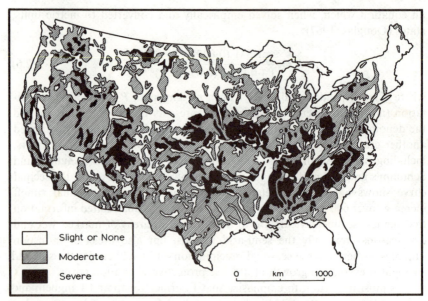

FIG. 2.3. Distribution of accelerated soil erosion in the United States, 1948 (after U.S. Department of Agriculture, 1957)
Moderate erosion = 25–75 per cent of topsoil lost, with some gullies
Severe erosion = over 75 per cent of topsoil lost, with numerous gullies

universally applicable. However, they help to identify the important predictive parameters in the soil erosion problem.

Detailed studies of sediment yield in small areas can be exemplified by a preliminary statistical examination of the controls on sediment yield from eleven drainage basins in Queensland, Australia (Douglas, 1968). Sediment yield was shown to be significantly correlated (at the 0·02 level) with mean annual precipitation, mean annual runoff, the relief/length ratio of the drainage basins and the p^2/P ratio (used by Fournier, see above). Some of these parameters are highly correlated amongst themselves, and if the effects of drainage density, ruggedness of the relief, the proportion of the area under rain forest, and relative relief are held constant (by partial correlation) then the association of the p^2/P ratio and the mean annual precipitation (P) with sediment yield is increased. Taken alone the p^2/P ratio tends to provide the single best predictor of sediment yield, thus supporting Fournier's work. Multiple regression, based on a group of three parameters drawn from the discrete sets of intercorrelated parameters and representing the individual groupings under the components in a principal components analysis, showed that:

$$\log_{10} ss = -8\cdot41 + 2\cdot704 \log \frac{p^2}{P} + 5\cdot603 R_b + 2\cdot967 \log D \qquad (2.7)$$

where

    ss = suspended sediment yield;

    p^2/P = Fournier's rainfall concentration ratio (see above);

    R_b = the bifurcation ratio (geometric mean R_b in this case, see Table 1.1);

    D = drainage density.

This multiple regression equation accounts for 82·34 per cent of the variance in ss, with the multiple correlation coefficient of 0·91 being significant at the 0·01 level.

2.3. Raindrop Erosion

Soil erosion by water involves two important sequential events: the detachment of particles and their subsequent transportation. The two principal agents for this work are raindrops and flowing water. *Raindrop erosion* involves the detachment of particles from soil clods by impact and their movement by splashing. *Runoff erosion* (Sect. 2.4) largely concerns the transportation of loose material (much of it often prepared by raindrop erosion) by turbulent water flowing in sheets, rills, or gullies, although some detachment of particles can occur in runoff erosion.

Each of these two types of erosion comprises its own set of forces and resistances: some forces tend to promote particle detachment, for instance, and others tend to resist it. Put slightly differently, the nature of soil erosion depends on the relationship between *erosivity* of raindrops and running water on the one hand, and the *erodibility* (i.e. detachability and transportability) of soil material on the other. Much of soil erosion research is concerned with measuring and comparing the variables that determine these forces in order to be able to predict the likelihood of erosion and reduce soil loss. In the following discussion, emphasis is placed on soil erosion of agricultural land. But it is important to remember that soil erosion may also be a problem, for example, on forest lands after a fire or extensive tree felling or where vegetation-covered land is laid bare, as by the practice of grassland burning in many tropical lands, or by site preparation in urban development.

As research progressed in the 1930s, it was realized that the amount of soil in runoff increased rapidly with raindrop energy, and it was noted that erosion could be very greatly reduced by preventing raindrop impact. Since that time, the detachment and movement of soil particles as a result of raindrop impact has come to be recognized as a fundamentally important and often initial phase of soil erosion. It is no exaggeration to say that in certain circumstances as much as 90 per cent of erosion on agricultural land may result from this process. Results of research on this subject have been reviewed by Ellison (1947), Fournier (1972), Hudson (1971), and by Smith and Wischmeier (1962).

The nature of particle detachment and movement is, of course, a reflection of the relations between the characteristics of the rainfall and the characteristics of the soil and ground surface. The major properties of rainfall of importance to its erosivity are drop mass, size, size distribution, direction, rainfall intensity, and raindrop terminal velocity. From these variables, the kinetic energy [0·5 mass(velocity)²] and momentum (mass × velocity) can be determined.

The median size of raindrops increases with rainfall intensity for low- and medium–intensity falls in the form:

$$D_{50} = aI^b \tag{2.8}$$

where

D_{50} = median size of raindrops (mm);
  I = intensity (e.g. mm per hour);
a,b = constants.

For high-intensity rainfalls, drop size declines slightly (Hudson, 1971), and maximum drop sizes appear to be in the order of 5–6 mm diameter.

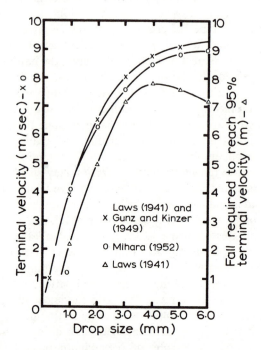

FIG. 2.4. Relations between drop size, terminal velocity, and fall required to attain 95 per cent of terminal velocity (after Smith and Wischmeier, 1962, from which the data sources may be obtained)

The impact velocity of raindrops is also related to drop size. A free-falling raindrop accelerates under the influence of gravity until the gravitational force is equal to the frictional resistance of the air, when the drop will be falling at its *terminal velocity*. The relations between drop size, terminal velocity, and the distance of fall required for 95 per cent of terminal velocity to be attained under laboratory conditions are shown in Figure 2.4. It is clear from this graph that, because fall distances to maximum fall velocity are so short, most drops will strike the surface at their terminal velocity. Air turbulence and winds will affect the terminal velocity of raindrops in natural rain.

Once size, terminal velocity, and intensity of drops are known, momentum and the kinetic energy of rainfall can be calculated by summing the values for individual drops. Rainfall intensity can easily be derived by recording rain-gauge data, and this intensity may be compared with calculated kinetic energy (Fig. 2.5). The kinetic energy of rain varies, of course, from storm to storm, and therefore in order to obtain a reasonable average annual figure (or similar general statement) attention must be given to the frequency and duration of storms as well as their intensity.

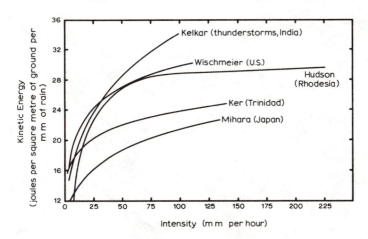

FIG. 2.5. The relation between kinetic energy and intensity of rainfall, from studies by various workers in different localities (after Hudson, 1971)

Having established the nature of raindrop impact, the next step is to determine the relations between rainfall and soil detachment and transportation. Several empirical studies describe these relations. Free (1960) found that splash erosion of sand was proportional to (kinetic energy)$^{0.9}$ and for soil, it was proportional to (kinetic energy)$^{1.46}$. Bisal (1960) concluded that:

$$G = KDV^{1.4} \tag{2.9}$$

where

    G = weight of soil splashed (gm);
    K = a constant for soil type;
    D = drop diameter (mm);
    V = impact velocity (m/sec).

Multivariate analysis of field-plot data from the United States led to the adoption of a compound parameter to explain most satisfactorily soil loss in terms of rainfall (Smith and Wischmeier, 1962). This parameter, known as the EI_{30} index, is the product of kinetic energy, E (in foot tons per acre inch) and intensity, I (inches per hour), where I_{30} is specifically the 30-minute intensity (i.e. the greatest average intensity in any 30-minute period during a storm). The spatial pattern of this index in the eastern United States is shown in Figure 2.6. It reveals that the potential danger of accelerated erosion, using this measure as a guide, is greater in the Gulf Coast areas than in the western and northern parts of the country.

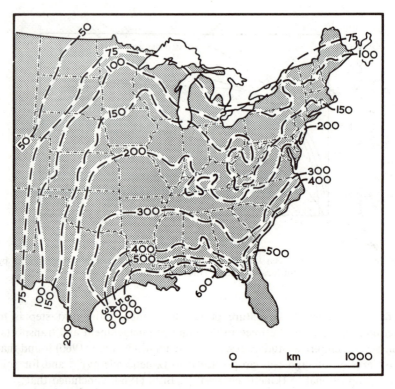

FIG. 2.6. Average values of the rainfall-erosion index for the eastern United States (after Smith and Wischmeier, 1962)

Soil splash also varies with the detachability and transportability of the soil and the vulnerability to splash of the surface involved. For example, soil loss is greatly reduced if the soil is protected by vegetation, and the degree of protection varies according to the type of cover. Similarly, the ease with which soil clods can be destroyed by raindrops will depend on factors affecting soil consolidation, such as clay, stone, base, and organic-matter content. Again, the slope of the surface is important. If the surface is horizontal then splash erosion may break up the soil and move particles through the air, but there will be no overall loss of soil from the field, despite the fact that individual particles may be splashed up to 60 cm into the air, and may be moved up to 1·52 m horizontally. Ellison (1947) showed, however, that there can be net erosion simply by rainsplash erosion on a sloping surface. For example, on a 10 per cent (6 degrees) slope he found that 75 per cent of soil splash was *downslope*. Ellison also showed that maximum splash occurs shortly after the surface is wetted and thereafter decreases, probably because the surface film of water increases in thickness (thus reducing impact on soil particles) and because the most easily eroded particles are soon removed.

The process of soil splash, which attacks the whole exposed surface uniformly, produces several types of damage. Firstly it detaches particles from clods, destroys soil structure, and prepares debris for transport by runoff. Secondly, it transports debris in splash and results in soil loss from sloping surfaces. Thirdly this erosion elutriates the soil—it removes clays, humus, and other soil nutrients and leaves an impoverished soil behind. A fourth problem is that raindrop impact may disperse surface clay particles and lead to the formation on drying of a crust which reduces infiltration capacity, thus promoting surface runoff and the second type of water erosion, runoff erosion.

2.4. Explanatory Models of Runoff Erosion

(a) Horton's model—overland flow

In an argument of impeccable logic and impressive elegance, Horton (1945) formulated a theory of surface runoff and erosion which provided a sound basis for an understanding of the causes and control of runoff erosion and for much subsequent research (Strahler, 1956; Leopold, Wolman, and Miller, 1964; Kirkby, 1969; Emmett, 1970).

According to Horton, runoff does not occur immediately rain falls on a bare soil surface. If the soil is unsaturated, water will infiltrate into the ground at a rate determined by soil structure, soil texture, vegetational cover, biological structures in the soil, soil-moisture content, and condition of the surface. *Infiltration capacity* (f_p) (the maximum sustained rate at which a particular soil can transmit water) varies during a rainstorm—initially it may be quite rapid (f_o), but with time it declines to a constant value (f_c). This decline is due to 'rain packing'—in-washing of fine material, swelling of colloids, and a breakdown of the surface soil structure. It is this minimum infiltration capacity which is predominant during most long rainstorms. If the rate of precipitation exceeds this value of infiltration capacity, water will begin to accumulate in and on the soil and runoff may result. If the rain continues for a period t_R, then:

$$f_p = f_c + (f_o - f_c)e^{-kt_R} \qquad (2.10)$$

where k is a constant and $e = 2 \cdot 71828$ (the base of Naperian logarithms). As f_p decreases, with time during rain, so runoff must increase. In practice it is difficult to measure either the infiltration capacity or the runoff properties of a drainage basin at any one moment in time. This is because in most basins there are considerable variations in the infiltration characteristics of the soils that it contains. At the same time there are continual changes taking place in the amounts of precipitation being stored by the ground.

Infiltration is also controlled by the conditions that exist prior to the onset of rain. An earlier rainfall may have left the soil partially saturated. In addition, infiltration varies with the seasons as these govern the state of the vegetation, the condition of the land if agriculture is being practiced, and the temperatures which in turn control evaporation rates. Areas with vegetation have a higher infiltration capacity than barren areas, for the vegetation retards surface flow, the roots make the soil more pervious, and the ground is shielded from some of the effects of raindrop impact, so that the amount of soil packing is reduced.

There are several methods for arriving at infiltration values. For example, field measurements can be made on experimental plots of the amount of runoff resulting from a known quantity of artificially induced rainfall; then from the relationship:

Precipitation = Infiltration + Runoff

an approximate value for infiltration can be obtained. (The amounts intercepted by vegetation and held in storage on the ground surface during the experiment are assumed to be small.)

The difference between rainfall intensity and infiltration capacity is called the supply rate (σ). Once there is a positive supply rate, water will begin to collect in the storage area provided by ubiquitous surface depressions. The next stage is for water to overflow between depressions and for a thin layer of water to develop. Ultimately, after a critical thickness of this layer is exceeded the water will begin to flow.

Initially the flowing water will have insufficient energy to pick up and transport soil, but as it proceeds down-slope its force will increase to the point where it exceeds the resistance of the surface material and erosion can begin. Up-slope of this critical distance (x_c) there is a zone of no runoff

erosion where only raindrop erosion can occur. It is clearly desirable to calculate where x_c will occur on a slope, as it is here that gully erosion may
begin.

The force per unit area exerted by flow parallel with the soil surface, F, can
be defined by the DuBoys formula (in Imperial units) for shear stress:

$$F = w \frac{d}{12} \sin \theta \qquad (2.11)$$

where

$F =$ eroding force in pounds per square foot;
$w =$ weight of a cubic foot of water;
$d =$ depth of overland flow (in);
$\theta =$ angle of slope.

Of the variables in this equation, the most difficult to determine is d, depth
of flow. Horton showed that the depth of flow at a point any distance from
the top of the slope can be described (in Imperial units) as follows:

$$d_x = \left(\frac{\sigma n x}{1020} \right)^{3/5} \left(\frac{1}{s^{0 \cdot 3}} \right) \qquad (2.12)$$

$d_x =$ depth of flow at a given distance from the slope divide (in);
$\sigma =$ supply rate (in/hr);
$n =$ a factor describing surface roughness;
$x =$ distance (ft) down-slope;
$s =$ tangent of slope.

A value of d_x can therefore be substituted into Equation 2.11 to give an
expression of the total eroding force at x:

$$F = \frac{w}{12} \left(\frac{q_s n x}{1020} \right)^{3/5} \left(\frac{\sin \theta}{\tan^{0 \cdot 3} \theta} \right) \qquad (2.13)$$

where $F =$ eroding force (lb per sq ft), and $q_s =$ runoff intensity (in/hr),
being equal to σ for steady overland flow.

It remains to determine x_c, the critical downslope distance where runoff
erosion begins. This can be done using Equation 2.13, substituting R_i (the
threshold value of resistance of the surface) for F, making $x = x_c$, assuming
$w = 62 \cdot 4$ lb per cu ft, and solving for x_c, thus:

$$x_c = \frac{65}{q_s n} \left(R_i \frac{\tan^{0 \cdot 3} \theta}{\sin \theta} \right)^{5/3} . \qquad (2.14)$$

That is to say, the width of the zone of no runoff erosion is inversely proportional to the intensity of runoff and surface roughness, and directly proportional to the 5/3 power of the resistance. For slopes of less than 20 degrees,

the width of the belt of no runoff erosion decreases with increase in slope. Naturally, raindrop erosion can be, and usually is, effective in this watershed zone.

One of the most important features of this demonstration is that it clearly identifies those variables that determine runoff erosion and which therefore must be modified in order to reduce erosion. The list of variables is worth repeating: rain intensity and infiltration capacity (i) as expressed in supply rate (σ) or runoff intensity (q_s); length of overland flow; slope (θ); and surface roughness (n).

Once the critical threshold has been crossed, flowing water will begin to incorporate sediment and, all other things being equal and assuming uniform material, the erodibility of sediment can be expressed in terms of the relations between particle size and flow velocity (Fig. 2.7). A point to notice on this figure is that the most easily eroded particles are between 0·1 and 0·5 mm (medium and fine sand), and that higher velocities are required to transport smaller as well as larger particles. The curve for settling velocity, below which deposition will occur, is also shown on Figure 2.7.

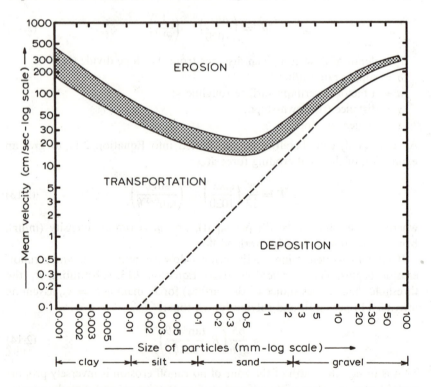

FIG. 2.7. Relations between particle size, erosion velocity, and settling velocity for uniform sediments (after Hjülstrom, 1939)

Normally runoff will be in rills (defined as channels that can easily be removed by ploughing) or in gullies. As water is concentrated into rills its depth and velocity increase, and detachment and transport of particles both increase as a result. Erosion proceeds largely by the water-and-sediment mixture scouring bottoms and sides of the rills, by waterfall erosion at the head of channels, and by mass movement into the channels (through, for example, slumping and the weathering of soil). The initial rill system usually develops rapidly, with some rills enlarging faster than others, until a simpler, more stable pattern is produced. Such rill and gully erosion is common only on vegetation-free surfaces, such as those in semi-arid and arable lands or where the vegetation cover has been seriously disrupted (Sect. 2.7d). The initial pattern may be guided by already established plough lines and cattle paths, especially where they have a down-slope orientation.

(b) The saturated throughflow model

As an alternative to the overland flow model the saturated throughflow model has been developed in which emphasis is placed on the movement of water down-slope through the upper soil horizons (Kirkby and Chorley, 1967; Kirkby, 1969).

The velocity of groundwater moving through soils (throughflow) is slow (about 20 cm/hr) compared with that of overland flow (about 27,000 cm/hr) (Kirkby and Chorley, 1967). Velocity of throughflow, and its discharge, vary with the permeability of the material through which it passes. Darcy's Law (Fig. 2.8) can be applied to this situation whereby:

$$Q = \frac{k(h_t - h_b)A}{L} \qquad (2.15)$$

where

Q = discharge;
h_t = hydraulic head at the top;
h_b = hydraulic head at bottom (both h_t and h_b are measured with piezo-
        meters);
A = cross-sectional area of column of soil;
L = length of column of soil;

and

$$k = \frac{k'\rho g}{\eta}$$

k' = permeability;
ρ = density of water;
g = acceleration of gravity;
η = viscosity of water;

and also:

$$\text{velocity of flow} = \frac{Q}{A} = ki$$

where

$$i = \frac{(h_t - h_b)}{L}, \text{ the hydraulic gradient.}$$

Since k can be calculated for any one soil, and since h_t, h_b, A, and L can be measured, it is thus possible to calculate Q. An important assumption has to be made, however, which is that the soil is uniform in its permeability characteristics. No drainage basin will ever contain soils which are all of the same permeability. In order to calculate Q, for any basin, it is necessary to discover the variations in permeability that exist. In practice a first estimate of variations in permeability would come from an analysis of soil textures. Permeability is very low in unweathered clays, low in fine sands, silts, and stratified clays, higher for mixed sands and gravels, and high for sorted gravels.

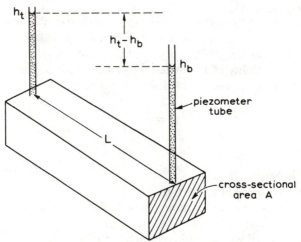

FIG. 2.8. To illustrate the parameters used in Darcy's Law

Kirkby (1969) evaluated the mathematics of the throughflow model in the following way. Throughflow, travelling through soil-pore spaces, never attains a steady state, and its discharge per unit contour length (q_T) is given by:

$$q_T = (P - f_*) \, vt \qquad (2.16)$$

where

$P = $ rate of surface percolation;

f_* = rate of infiltration at base of the permeable soil;
v = velocity of throughflow;
t = time elapsed.

In order to calculate v, Kirkby combined Darcy's Law, restated as:

$$Q = z \cos \theta \left\{ K.m.\sin \theta - D.\frac{\partial m}{\partial x} \right\} \qquad (2.17)$$

and the continuity equation, which states that differences between inflow and outflow must be accompanied by changes in moisture content:

$$\frac{\partial m}{\partial t} + \frac{\partial Q}{\partial x} = i \qquad (2.18)$$

where

Q = down-slope discharge measured in a horizontal direction;
x = distance down-slope;
K = soil permeability;
D = soil diffusivity;
θ = surface angle of slope;
m = soil moisture content;
z = thickness of soil layer;
i = rainfall intensity;
t = the time elapsed.

Under conditions of high moisture content of the soil (i.e. with throughflow) the soil moisture is constant (i.e. does not vary with time and $\partial m = 0$), and thus equation 2.18 becomes:

$$\frac{\partial Q}{\partial x} = i. \qquad (2.19)$$

In other words near-saturated soils respond very rapidly to changes in rainfall intensity, even before overland flow beings.

From the above it can be seen that the important controls on throughflow are soil characteristics, distance down-slope, angle of slope, and the intensity of the rainfall. The attainment of a fully saturated state occurs most readily in soils adjacent to streams, on concave slopes, in hollows or where the soils themselves are either thin or impermeable. This process generally involves an up-slope extension of the existing channel system (Kirkby, 1969) and is independent of distance from the hill crest, which is an important control in the case of overland flow.

As shown in the discussion of the Horton model for overland flow, the erosive force of runoff increases both with distance down-slope and the angle of slope. This means that on a hillside with a convexo–concave profile the erosive force approaches its maximum on the steepest, and often central, part of the profile. Most erosion then takes place just below this region of steepest gradient, and it is here that gullies are initiated and channel flow begins.

Overland flow and throughflow are likely to be two extremes of a continuous sequence of possible conditions for gully development, with overland flow more common in semi-arid areas and throughflow more common in humid areas. However, if any group of gullies adopts either of the preferred sets of locations, this will indicate which of the two causes of gully formation is likely to be dominant.

2.5. Major Variables in the Soil Erosion System

From the preceding discussion it will be apparent that soil erosion is largely controlled by variables which relate to climate, topography, soil characteristics, vegetation, and land-use practices. The climatic variables, it has been shown, can be summarized by the rainfall index (EI_{30}). The role of soil characteristics, the effect of vegetation, and topographic variables must be examined in a little more detail before comprehensive relations between soil loss and explanatory variables can be presented.

(a) Soil erodibility

The erodibility of soils by water depends on several properties, notably those that affect detachment and transportability by rainsplash and runoff, and those that influence infiltration capacity.

One approach to measuring soil erodibility is to monitor the loss of material from plots of standard dimensions on different soil types. In this empirical research, plots commonly have an arbitrary slope of 9 per cent (5 degrees) and a length of 72·6 feet (22·1 m), so that every 6 feet (1·82 m) of width corresponds to an increase of 0·01 acres (40·03 sq m) in the area of the plot (see Table 2.1). A second approach is to use experimentally validated soil erodibility indices. These have the advantages that they can be measured relatively quickly and require little special equipment. The indices generally fall into one of two categories (Bryan, 1968): those that measure soil-dispersion properties, and those combining soil-dispersion properties with a measure of a soil's water-transmission properties. An early example of the former is Middleton's dispersion ratio: the amount of silt-plus-clay in an undispersed sample compared with that in a dispersed sample, expressed as a percentage. This ratio is based on the assumption that only dispersed material can be eroded, and Middleton found it to be normally above 15 per cent for erodible soils. However, it makes no allowance for the dispersion of previously undispersed material by raindrop impact, and it does not accurately reflect the erodibility of soils with a high sand content.

Chorley (1959) developed an index which exemplifies those based on combined soil-dispersion and water-transmission properties. He combined a measure of mean shearing resistance (based on soil penetrometer readings) and a figure for permeability as follows:

Index of erodibility = (mean shearing resistance × permeability)$^{-1}$. Shearing resistance was found to be significantly related to soil density, range of grain size, and soil-moisture content. This index has been criticized on the grounds that shearing resistance is more important in gravity-controlled mass movements of material than in erosion dominated by raindrop impact and surface runoff.

Experimental evaluation of several indices by Bryan (1968) led to the conclusion that none of them was ideally suited for universal application. This is largely because they do not include efficient and precise measures of the effect of raindrop impact. The most satisfactory index tested was that based on the percentage weight of water-stable aggregates >3 mm in diameter.

(b) The role of vegetation

Vegetation cover is perhaps the greatest deterrent to soil erosion for it tends to protect the surface from raindrop impact, reduce the amount of water available for runoff by consuming it and by improving infiltration capacity, and (by increasing surface roughness) decrease the velocity of runoff. In addition, plant roots help to keep the soil in place. A grass cover is often the most efficient defence against soil erosion by running water. Thus, in general, vegetation tends to reduce both runoff and erosion. Two examples will suffice to illustrate these fundamental points. Results typical of experiments designed to measure runoff from experimental plots with different land uses are shown in Figure 2.9. This figure suggests that runoff is inversely related to the density of vegetation and the frequency of cultivation. Relations between soil loss

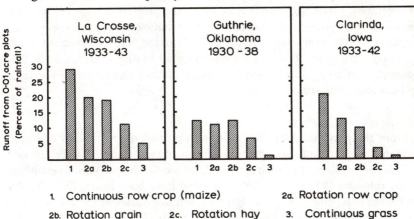

FIG. 2.9. Effect of different farm practices on runoff from 0·01 acre (40·03 m²) test plots (after Glymph and Holtan, 1969)

and crops are illustrated in Table 2.1. The high soil loss from row crops arises partly because they require the preparation of a seedbed which tends to break up the soil, and partly because planting in rows leaves a relatively high

TABLE 2.1

Soil loss from silt-loam soil 1933–1942 (United States)

(Plots: 9 per cent slope, 72·6 feet long)

	Soil loss	
Cropping system	*tons per acre*	*kg × 10² per hectare*
Continuous maize (unfertilized rows up and down slope)	38	861·82
Continuous maize (as above, but on subsoil)	52	1179·33
Rotation		
a. Maize	18	408·23
b. Oats	10	226·79
c. Clover	5	113·40
Continuous lucerne	0·1	2·27
Continuous blue grass	0·03	0·68

Source: F.A.O. (1965)

proportion of the soil bare. The beneficial effects of vegetation cover are not confined to living plants; mulches of crop residue, such as chopped maize stalks, provide protection from raindrops and as they decompose the residues temporarily improve soil aggregation.

The relations between plants and soil erosion are considerably more subtle than these two general examples suggest. For example, the protective effect of vegetation varies with the time of year—the protection afforded by seedlings is naturally less than that provided by mature crops, and the greatest vegetative protection may not coincide with the most destructive rains. Similarly, certain plants may encourage rainfall to drip from leaves or flow down stems, thus perhaps locally increasing raindrop impact on the one hand and concentrating runoff on the other.

(c) *The effect of topography*

From the previous discussion of raindrop erosion and models of runoff erosion it is clear that topographic variables of importance in water erosion include surface slope, surface length, and surface roughness. Various estimates of the relations between soil loss and slope have been made. Zingg (1940) concluded that, other things being equal, soil loss varies as the 1·4 power of the per cent slope; and Smith and Wischmeier (1962) described a similar relationship. Zingg (1940) also observed that soil loss varied as the 1·6 power of slope length.

(d) *Universal equations for predicting soil loss*

Once the major variables affecting soil loss have been isolated and evaluated, it should be possible to derive a general equation that describes predicted total soil loss in terms of these variables. Early empirical research by Musgrave (1947), for instance, showed that soil loss varied in the following way:

$$E = I R S^{1\cdot35} L^{0\cdot35} P_{30}{}^{1\cdot75} \tag{2.20}$$

where (in Imperial units)

E = soil loss (acre/in);
I = inherent erodibility of the soil (in);
R = cover factor;
S = slope (per cent);
L = length of slope (feet);
P_{30} = maximum 30-minute amount of rainfall, 2-year frequency (in).

Since 1947 further progress has led to the development of a *universal equation* for predicting soil loss, the main value of which is to define the nature of the most important controlling factors:

$$A = R K L S C P \tag{2.21}$$

where

A = average annual soil loss (tons/acre);
R = rainfall factor;
K = soil-erodibility factor;
LS = slope length–steepness factor;
C = cropping and management factor;
P = conservation practice.

Briefly, the value of these variables may be calculated as follows: $R = \Sigma EI_{30}/100$ (EI_{30} as defined above); K is expressed in terms of soil loss in tons per acre for each unit of R for the locality and for continuous fallow on a 9 per cent slope, 72·6 feet long; LS is obtained by dividing actual length and steepness by the standard 9 per cent slope 72·6 feet in length; the cropping factor, C, is the expected ratio of soil loss from land cropped under specific conditions to soil loss from tilled fallow under identical conditions; P, a factor that takes the benefits of contour ploughing etc. into account, equals 1 where there are no contour practices and cultivation has been straight up- and down-slope—conservation practices reduce P below 1.

Of these factors, R relates to the trigger mechanism supplied by rainfall, and the others are natural and human environmental factors that determine the effect of rainfall. (For further details of this equation see Chow, 1964; F.A.O., 1965; Hudson, 1971; or Smith and Wischmeier, 1962.)

2.6. Control of Soil Erosion by Water

The aim of soil-erosion control is to reduce soil loss so that soil productivity is economically maintained. In systems terms, the aim is often to restore equilibrium in the soil system so that rate of loss is similar to rate of regeneration. At present, tolerable rates of loss have to be estimated, and clearly they will vary from place to place because both rates of regeneration and the economics of control will vary. For practical purposes tolerable rates of loss range between one and five tons per acre per annum (22.6×10^2 and 113.4×10^2 kg/ha). In terms of the universal soil-loss equation, the tolerance level can be substituted for A (Equation 2.21) and the equation can be solved in terms of C and P.

Soil-conservation practices fall into three main groups—crop-management practices, supporting erosion-prevention practices, and practices designed to restore eroded land. Each practice relates to the manipulation of one or more of the variables in the water-erosion system. In the following brief summary, the relations between practice and principle are emphasized. For greater detail on this subject the reader is referred to the works by F.A.O. (1965), Fournier (1972), Frevert *et al.* (1955), Hudson (1971), and U.S. Department of Agriculture (1957).

(a) Crop-management practices

As much erosion control must be carried out on land in agricultural use, it is clearly desirable to exploit the beneficial effects of vegetation and healthy soil in the form of intelligent cropping practices appropriate to the particular environment. The use of a legume or grass crop in rotation at least one year in five, for instance, often gives a high degree of protection from raindrop erosion. At the same time it bestows additional advantages by providing for a period of soil recuperation, improving soil structure and nitrogen content, and increasing soil organic content (thus tending to reduce soil detachability). Similarly, the judicious application of fertilizers and manures may not only help to improve crop yields but may also encourage soil conditions that decrease detachability and increase infiltration capacity. A third technique involves the planting of 'cover crops'. Cover crops may be grown when the main crop has been harvested in areas where the growing season is long enough to sustain them. Such crops serve to protect the soil at times when they would be exposed to rainfall erosion; in addition they may be ploughed in to provide a beneficial 'green manure'. An alternative approach is to 'interplant' protective crops of grass or a legume between the main crop if it tends to be associated with soil loss. 'Mulch tillage' involves the covering of the surface with crop residues (such as grain straw or stubble). This practice provides protection from raindrop impact when the field would normally be bare, and at the same time the residues hold water, retard surface flow and promote the formation of erosion-resistant aggregates. These are only a few examples of crop-management practices which are designed to protect the soil by maintaining or improving soil structure and using the protective effect of vegetation to advantage.

(b) *Supporting erosion-control practices*

The most important supporting measures of erosion control are 'contour farming', 'strip cropping', and 'terracing'. Contour farming involves planting rows of crops and using farm machinery along the contours of the land (normal to the direction of slope). This simple device—which is not as widely used as might be expected, especially in view of evidence that it is often more economical of effort—disrupts surface flow and reduces its velocity by increasing surface roughness, restricts the loci for rill development and, in some circumstances, conserves water. Contour farming is most effective on medium slopes and on deep, permeable soils. It also has its dangers, for if soil ridges are breached, water may be concentrated in the breaches, giving rise to large gullies.

'Strip cropping' consists of creating alternating strips of crops and grass or legume parallel to the contours. The dense and complete cover in the sod strips serves to trap sediment carried from crop strips, filter runoff from up-slope and reduce its velocity, increase infiltration rate, and protect the soil from raindrop impact. Normally the strips form part of a crop-rotation system. Strip cropping is especially appropriate on slopes too steep for terracing (e.g. 6–15 per cent, or 3·5–8·5 degrees), and on farms where forage is an important part of the economy. The width of strips varies with ground slope. For example, recommended strip widths on soil with fairly high water intake are as follows (F.A.O., 1965, 92):

Percent slope	Slope in degrees	Strip width (m)
2–5	1–3	30–33
6–9	$3\frac{1}{2}$–5	24
10–14	$5\frac{1}{2}$–8	21
15–20	$8\frac{1}{2}$–$11\frac{1}{2}$	15

Terracing normally requires the creation by earth-moving equipment of an embankment (with or without a channel up-slope of it) parallel to the contours. Most terraces reduce slope gradient, break the original slope up into shorter units, conserve soil moisture, and remove runoff in a controlled fashion. They also allow a field to be given over to a single crop. The spacing of terraces is related to ground slope, in the same way as strip-crop widths, and will be constant for given soil and climatic conditions. Of critical importance in the design of efficient terraces is the geometry of the drainage channel which removes runoff from the terrace base: this must not be susceptible to erosion and should be able to accommodate the largest runoff likely to occur in a reasonable time period (such as a decade). The problem of drainage channel design is complex, but its solution rests on well-established and fundamental hydraulic principles (e.g. Chow, 1964; Frevert *et al.*, 1955; Henderson, 1966).

(c) *Restoration of gullied land*

The removal of runoff from terraces usually requires the construction of waterways, and if these are inappropriately designed they can easily become gullies. Natural waterways can also become gullied by increased runoff from areas of poor land-use practice. In both of these cases, measures are required to remove the gullies and to restore the drainage channels. The most commonly used method is to cover the waterway with grass or to encourage natural vegetation. A second approach is to convert the gully into a stable artificial channel with dimensions appropriate for the discharge of water. A third method is to reduce water supply by conservation practices in the tributary lands. An alternative is to eliminate flow from the gully by diverting it into an artificial channel. A fifth method is to reduce erosive flow velocities in gullies by building structures such as spillways and weirs which dissipate the flow energy, and by creating stable channel sections between the structures.

Many of the points made above are supported by Table 2.2. Rainfall, runoff, and erosion were measured for three experimental plots. On the first no soil conservation was practised. On the second contour ploughing was accompanied by a control of drainage outflow; while on the third contour ploughing was accompanied by terracing. The rows show that increased conservation practice decreased the amount of erosion, and usually runoff as well. The columns show the effectiveness of grassland in providing a protective land cover.

2.7. Case Study: Economic Implications in an Australian Context

(a) *Introduction*

The previous discussion of the soil-erosion system has been concerned mainly with the loss of soil from agricultural land and with measures for preventing it. The consequences of such soil loss vary greatly from area to area: here, there may be a problem of silting and loss of feed, fertilizers, and seed; there, sprinkler irrigation may be prevented because it exacerbates raindrop impact and raindrop erosion. These consequences and those of the appropriate control measures are reflected in the economics of farm management, but the precise economic implications of soil erosion are extremely difficult to disentangle from the complex variety of conditions which affect a

TABLE 2.2

Rainfall, runoff, and erosion on three cultivated experimental areas

Column 1—without soil conservation
Column 2—with contour ploughing and drainage control
Column 3—with contour ploughing and terraces

Year	Rain (inches)	Rain (mm)	Crop	Runoff (%) 1	2	3	Erosion (tons/acre) 1	2	3	Erosion ($kg \times 10^2/ha$) 1	2	3
1934	21·26	540·00	Oats	19·8	27·9	25·3	40·01	24·04		907·41	545·22	
1935	36·07	916·17	Corn	32·0	28·6	22·1	47·38	16·96		1074·55	384·64	
1936	24·77	629·15	Grassland	8·2	7·8	5·6	0·88	0·11		19·96	2·49	
1937	21·63	549·40	Maize	16·1	18·8	13·8	19·71	1·82		447·01	41·28	
1938	26·13	663·70	Oats	4·7	2·7	2·7	2·62	0·16	0·05	59·42	3·63	1·13
1939	26·68	677·67	Corn	20·0	13·7	13·2	25·84	0·77	0·63	586·04	17·46	14·29
1940	27·62	701·54	Grassland	7·7	7·0	6·0	3·27	0·07	0·04	74·16	1·59	0·91
1941	35·04	890·01	Maize	16·3	16·3	13·4	63·93	0·37	1·12	1449·90	8·39	25·40
1942	32·89	835·40	Oats	17·7	17·7	10·0	38·44	0·78	0·58	871·80	17·69	13·15

Source: Fournier (1972) with metric equivalents added.

farmer's budget sheet. The case study examines some of these difficulties in an Australian context.

Decades of research into water erosion of soil have led to a partial understanding of the system and to the design of various measures to prevent excessive soil loss. There remain large areas of ignorance among scientists: for example, too little is known about the 'tolerable soil loss' in different circumstances, and basic data on soil erodibility are scarce in most countries. Nevertheless it is possible to reduce soil loss significantly, if somewhat imprecisely, given the present state of technical knowledge. But it is much more difficult to disseminate the knowledge and to have soil conservation practices adopted by farmers. Perhaps as important, land managers must be able to identify symptoms of erosion before the most valuable, surface humic horizons are lost; and ideally conservation practices must be implemented before erosion begins.

(b) Economic consequences in Australia

In a series of studies, Molnar (e.g. 1964) has attempted to measure the loss of production and the decline of land values due to soil erosion by means of economic models, experiments, and farm surveys in Australia.

It can be argued that in order to maintain output from a soil that has been eroded an increase in input will be required. Such change is illustrated graphically in Figure 2.10, where X_2 would need to be increased to X_3 to

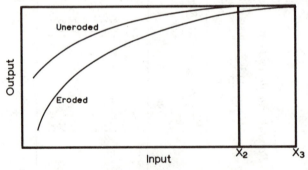

FIG. 2.10. Model of input and output on uneroded and eroded soils for soils with similar surface and subsurface conditions (after Shroder, from Molnar, 1964)

maintain output on an eroded soil. This model was tested experimentally in New South Wales and Victoria by comparing production of wheat in a wheat-fallow rotation on uneroded plots, and on plots from which 7·6 cm or 15·42 cm of topsoil had been removed (to simulate sheet erosion). All plots received 20·41 kg of super-phosphate with the wheat. The results are shown in Table 2.3. Clearly these experiments indicate that since inputs are the same regardless of yield obtained, there is less return on capital and labour on land where the yield is lower.

TABLE 2.3

The effect of loss of topsoil on wheat yield and protein content at Wellington and Gunnedah, New South Wales, and on soil nitrogen at Gunnedah

Depth of soil removed (cm)	Wellington (1955–63 av.)		Gunnedah (1955)		
	wheat yield (kg/ha)	wheat protein %	wheat yield (kg/ha)	wheat protein %	soil nitrogen %
0	1325	11·3	1450	10·8	0·16
7·6	965	10·6	990	8·4	0·12
15·42	800	10·2	620	9·1	0·10

Source: Molnar (1964), with metric conversion.

Differences in the value of annual production due to differences in erosion and soil type in two areas of Victoria can also be determined approximately by farm surveys, as Table 2.4 shows.

TABLE 2.4

Differences in the value of annual production due to differences in erosion and soils at Dookie and Coleraine, Victoria

Locality	Years of data	Difference in value (in Australian £) of production per acre (0·4 ha.) between:	
		(a) Uneroded and worst eroded land	(b) Best and worst soils
Dookie	1951–2	£1 9s. 6d.	—
Coleraine	1954–5	£2 16s. 0d.	£5 0s. 0d.

Note: These data must be used with caution because of several problems of putting a cash value on soil quality and erosion. Values for different years need to be adjusted to a common base, and are in any case only of relative significance; the figures are understated because the quality of lambs could not be valued (e.g. lambs may not have been kept as long on eroded land); the analysis is not based on complete information (e.g. cereal production at Dookie was not considered); other effects of erosion, such as increased working hours involved in cultivating eroded land, were also not considered.

Source: Molnar (1964).

It is important to realize that economic implications of soil conservation must be viewed in the light of the *economic aims* of the conservation measures. Molnar pointed out that both in Australia and in the United States, conservation measures are in general supposed to halt the decline in production and it is not normally assumed that an *increase* in production will follow.

Economic models designed to describe production trends need to recognize the possibility that income may be maintained, increased, or decreased. Some of the relations between expected income and time, using different assumptions, are shown on Figure 2.11. In *a*, conservation practices restore

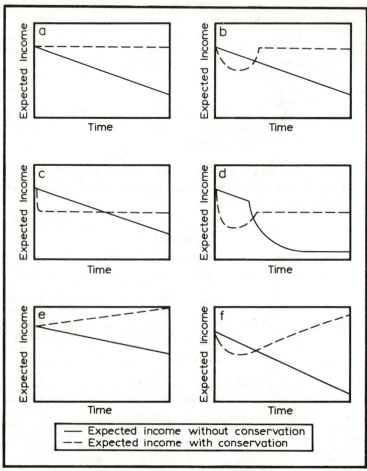

FIG. 2.11. Relations between expected incomes and time on land with and without soil-conservation practices, using various assumptions (after Barlowe, from Molnar, 1964)

income to the level before erosion began; *b* shows a temporary initial drop in income; in *c* income is eventually stabilized at a lower level than before the onset of erosion; and *d* corresponds to *b*, except that the 'stable' income is lower than initial income. In *e* and *f*, the alternative assumption is made that income will eventually rise. It is difficult to validate these models because the interesting variables cannot easily be isolated from other aspects of the farm economy. One approach is through the comparison of 'before and after' incomes from adequate farm records. For example, Figure 2.12 implicitly suggests the effect of soil conservation works, begun in 1951, on a farm at Armstrong, Victoria. The analysis indicates a clear trend, but unfortunately it is based on incomplete records and shows only gross production.

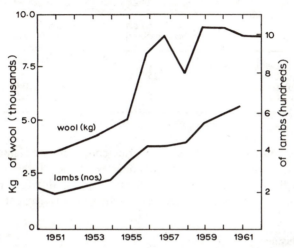

Fig. 2.12. The effect of soil conservation on production for a farm at Armstrong, Victoria (after Garside, from Molnar, 1964)
Conservation works began in 1951. Between 1951 and 1963 the improved pasture increased from 88 ha to 264 ha, the number of sheep increased from 700 to 2,000, and wool cut per hectare increased from 1·78 kg to 4·99 kg

Molnar also attempted to determine decline in land values due to erosion in the Dookie and Coleraine areas of Victoria. He found that degree of erosion and quality of soil affected both prices paid by buyers of land and values of land assessed from farm production data (Table 2.5). At Dookie, soil erosion does not affect prices paid for land, although it seriously affects productivity. This is probably because it is mainly in the form of sheet erosion in this area and is relatively unspectacular. At Coleraine, however, where more visible gully erosion predominates, soil erosion does affect prices paid for land. Nevertheless, computed differences in land values due to erosion are much greater than the differences between prices paid—farmers apparently do not realize the consequences of erosion on productivity.

Accelerated soil erosion arises from mismanagement. It is therefore important to know why mismanagement occurs. In Australia, sheet erosion occurs most frequently on land that has been overstocked in successive years. Molnar pointed out that overstocking can arise from ignorance, or because farmers have been unable to earn an adequate income with fewer stock. Reasons for the latter situation may have complex origins in lack of capital,

TABLE 2.5

Effects of soil and erosion (a) on prices paid by buyers of land and (b) on the values of land assessed from farm production data

Locality	Differences between best and worst soils [£(*Austr.*) per acre (0·4 ha)]	Differences between areas of no erosion and worst erosion [£(*Austr.*) per acre (0·4 ha)]
(a) Dookie	£21	—
Coleraine	£25	£6
(b) Dookie	—	£12 6s. 0d.
Coleraine	£41	£23 6s. 0d.

Source: Molnar (1964).

high interest rates, drought, adverse price conditions, too small farms, or inappropriate farm location and boundaries. None of these problems is necessarily the responsibility of the farmer. Responsibility for the solution of soil-erosion problems must surely rest in part with government agencies.

2.8. Conclusion

The case study of the previous section provides a necessary counter-balance to the examination of soil-erosion systems in the earlier part of this chapter. It is clear that the geomorphologist and other earth scientists have learnt much about soil erosion and the means to control it. In terms of environmental management it is important for the present state of erosion in an area to be assessed. In addition information is required about the relative vulnerability of soil to erosion. In this both process studies and theoretical work can provide information valuable to the solution of soil-erosion problems; but, as the case study has shown, soil erosion is as much a cultural problem as it is a physical one (see also Blase and Timmons, 1961). No equation will solve the farmer's problem if he sees no problem. Ultimately, professional knowledge of soil erosion must be transmitted to farmers and other resource users in an intelligible and acceptable way; and such trans-mission is as much a task for the geomorphologist as the study of the physical problem itself. For an extended and fascinating analysis of this and related soil-erosion problems, the reader is referred to Held and Clawson (1965).

3
SOIL EROSION BY WIND

3.1. Introduction

The removal of soil by the wind is a problem that has attracted much attention, especially since the disastrous years between the two World Wars of the 'Dust Bowl' in the Great Plains of the United States. The hazard has several faces. During dust storms drifting soil may block roads, fill ditches and canals, and bury fences, illness resulting from the inhaling of dust is increased, and weeds and insects may be widely dispersed. Wind erosion may cause damage to crops by, for example, removing seeds, exposing plant roots, and blasting leaves. In addition, the soil is depleted of the more easily removed particles which include organic matter and smaller grains; such depletion, especially of the surface humic horizons, may seriously and progressively reduce the soils' water-retention properties and productivity.

Despite much research, the problem is still widespread. It is particularly prevalent in arid and semi-arid lands where the delicately balanced equilibrium of the relatively dry natural environment can easily be upset by inappropriate agricultural and grazing practices. Although the *threat* of wind erosion hangs over most dry lands (Fig. 3.1), the problem is in fact greatest in those dry areas where inappropriate development has accompanied recent 'pioneer' settlement or some degree of over-population. These areas include parts of the Great Plains of the United States and Canada, the steppes of western Siberia and Kazakhstan (Zhirkov, 1964), the fringes of arid Africa, and parts of Australia and India. At a smaller scale, wind erosion can be a local difficulty in more humid areas where dry, windy conditions may cause erosion of ominously deteriorating soils, as in East Yorkshire, England, and parts of Poland.

As with water erosion of soils, scientists have adopted several approaches to the study of the problem. Some have sought to *simulate* natural conditions in the laboratory by carrying out wind-tunnel experiments. Others have devised special equipment for *monitoring* soil loss by wind erosion in the field, although this task is more difficult than that of monitoring water erosion. A third group, often drawing empirical evidence from field and laboratory studies, has attempted to *predict* wind erosion through the use of several easily measured controlling variables. Undoubtedly much of the credit for our present knowledge of the wind-erosion system must go to Chepil and his co-workers who carried out pioneer studies in the United States and Canada for over thirty years. This work continues under Woodruff at Manhattan, Kansas. In this account we inevitably draw heavily upon the seminal work done in Kansas. One of the most important papers on wind

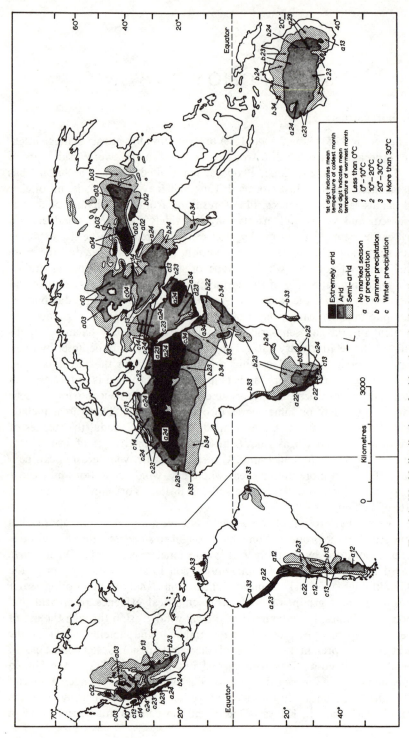

Fig. 3.1. World distribution of arid and semi-arid lands (after Meigs, 1953)

erosion is that by Chepil and Woodruff (1963), which includes references to most of the significant papers by Chepil and his associates, such as those of Chepil (1945, 1951, 1955) and of Chepil and Milne (1941).

Interestingly separate from this research, but equally fundamental, is Bagnold's outstanding analysis of *The Physics of Blown Sand and Desert Dunes* (1941). For detailed consideration of Bagnold's contribution and other studies of the wind-erosion system in deserts the reader is referred to Cooke and Warren (1973).

The Food and Agriculture Organization (F.A.O.) of the United Nations Organization has indicated the importance of wind erosion in land resource management by various publications (e.g. F.A.O. 1960). Similarly the subject of wind erosion has been considered as important by engineers (Task Committee, 1965).

3.2. The Wind Erosion System

(a) Variables in the system

Fundamentally, the wind-erosion system comprises variables related to surface winds and climate, surface materials, and surface conditions. Wind erosion begins when air pressure on surface soil particles overcomes the force of gravity on the particles. Initially, the particles are moved through the air with a bouncing motion, known as *saltation*; their impact on other surface particles may promote further movement, by saltation, *surface creep*, or *suspension*. Chepil estimated that, for most of the soils he examined, 50–75 per cent of the weight of the eroded soil was carried in saltation, 3–40 per cent in suspension, and 5–25 per cent in surface creep. Clearly the process depends on the availability of particles which can be moved in saltation, and on the attainment of wind velocities at or above which particle movement can be initiated by saltation (*fluid threshold*) or by impact (*impact threshold*).

(*i*) *Surface winds*, capable of initiating particle movement, are turbulent. The velocity of surface winds increases with height above the surface and the following equation is one description of such increases:

$$V_z = 5 \cdot 75 V_* \log \frac{z}{k} \tag{3.1}$$

where V is mean velocity at any height z, V_* is drag velocity, and k is a roughness constant. As saltating material becomes incorporated into the wind, the momentum, and hence surface wind velocity, are reduced. Chepil and Milne (1941) demonstrated that the more erodible the soil, the greater the concentration of moving grains, and the greater the reduction of surface wind velocity.

Mean wind drag per unit horizontal area of ground surface, $\bar{\tau}$, (expressed, for example, in dynes/cm^2), is related to drag velocity and fluid density. In general the relations can be expressed by the equation:

$$\bar{\tau} = \rho V_*{}^2 \tag{3.2}$$

where ρ is fluid density.

At the threshold of grain movement, three types of pressure are exerted on the surface soil particle: *impact* or *velocity pressure* (positive) on the windward area of the particle; *viscosity pressure* (negative) on the leeward area of the particle; and *static pressure*, a negative pressure on the top of the particle, caused by the so-called Bernoulli effect (which arises from pressure reduction where fluid [air] velocity is increased, as at the top of the particle). *Drag* on the top soil particle is due to the pressure difference against its windward and leeward sides, and *lift* is caused by a decrease of static pressure at the top of a particle compared with that at the bottom. The values of drag and lift required to initiate movement are, of course, affected by the character of the grains.

Once the particle has been entrained, drag and lift change rapidly. The trajectories of saltating particles (Fig. 3.2) reflect the relations between the

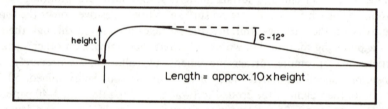

height

6 - 12°

Length = approx. 10 x height

FIG. 3.2. The trajectory of a saltating grain

forward force of the wind, the mass of the grains, and the pull of gravity. Initially, grains usually rise almost vertically; as they enter more rapidly moving air their paths are flattened; and as the initial force of upward movement is dissipated, grains begin to fall under the influence of gravity. On striking the surface, grains may bounce back into the air and/or strike other particles, causing them to be pushed along or saltated. The height of the saltating grains' trajectory may be up to two metres, depending mainly on particle size and bed roughness. In general saltation height is inversely related to particle size and directly related to bed roughness. The length of the trajectory is normally approximately ten times the height (Fig. 3.2).

The critical characteristic of the wind is its velocity. Additional characteristics of importance are its turbulence, direction, frequency, and duration. Other important climatic considerations concern the availability of water, which is in turn related to atmospheric, evaporation and evapotranspiration conditions. In general, the drier the soil the more vulnerable it is to wind erosion. Thus wind erosion is usually most serious where wind velocities and evaporation rates are high, precipitation is low, and drought is common.

(ii) *Surface materials and surface conditions*. The principal characteristics of soil particles with respect to wind erosion are size and density. Several general statements can be made about these characteristics. Many soil particles are composed of quartz, which has a specific gravity of 2·65. The most erodible particles of 2·65 density are about 0·1–0·15 mm in diameter. Threshold wind velocities required to move grains larger than 0·1 mm are defined by the square-root law:

$$V_{*t} = A\left(\frac{\sigma - \rho}{\rho}\right)^{1/2} - gD \qquad (3.3)$$

where V_{*t} is the threshold drag velocity, σ is the specific gravity of the grain, ρ the specific gravity of the air, g the gravity constant, D the diameter of the grain in centimetres, and A is a coefficient the value of which in air for particles above 0·1 mm in diameter was found to be 0·1 for the fluid threshold, and 0·084 for the impact threshold (Chepil, 1945). For smaller particles, threshold velocities do not conform to this law: the velocities increase with decrease in grain size, owing probably to cohesion between finer particles and to the fact that particles may be too small to protrude into the turbulent flow of air. The relations between particle size, specific gravity, and threshold drag velocity for a clay-loam soil are summarized in Figure 3.3. Experimental studies suggest that particles less than 0·1 mm in diameter may be moved in suspension; those between 0·1 and 0·5 mm are commonly moved by saltation; and particles larger than 0·5 mm tend to be moved by creep.

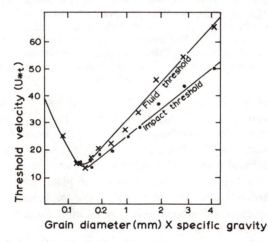

FIG. 3.3. Relations between particle size, specific gravity, and threshold drag velocity for a clay–loam soil (after Chepil, 1945, © 1945 The Williams & Wilkins Co.)

The *equivalent diameter* of a particle with a density of 2·65 gm/cc is defined as $peD/2·65$ where pe is the bulk density of erodible soil particles, and D is their diameter. Chepil found that very few particles with equivalent diameters exceeding 0·5 mm (actual diameter = 0·84 mm) are eroded. In general, the potential erosion of material increases as the percentage of soil fractions greater than 0·84 mm in diameter declines (Woodruff and Siddoway, 1965).

So far this discussion of surface materials has only been concerned with individual loose particles. But such particles are frequently combined in various ways to produce soil structures resistant to erosion. Chepil and Woodruff (1963) distinguished four major types: primary (water-stable) aggregates; secondary aggregates, or clods; fine material among clods; and surface crusts. Primary aggregates are held together by water-soluble cements of clay and colloids. Clods are held together in a dry state by cements comprising mainly water-dispersible particles smaller than 0·02 mm in diameter. Cohesion between clods is provided largely by water-dispersible silt-and-clay-sized particles. Surface crusts arise from (a) raindrop impact which tends to reorientate clay particles parallel to the surface (Chap. 2) and (b) washing of fine particles into near-surface pores (McIntyre, 1958a, b).

The binding agents for these dry structures include chiefly silt, clay, and decomposing organic matter, and it is these agents that determine the mechanical stability of the structures. As Chepil and Woodruff (1963, p. 262) reported:

the relative effectiveness of silt and clay as binding agents depends somewhat on their relative proportions to each other and to the sand fraction. The first five per cent of silt or clay mixed with sand is about equally effective in creating cloddiness, but the quality of the clods is different. Those formed with clay and sand are harder and less subject to abrasion by windborne sand than those formed from silt and sand. For proportions greater than five per cent and up to 100 per cent the silt fraction creates more clods, but clods are softer and more readily abraded than those formed from clay and sand. The greatest proportion of non-erodible clods exhibiting a high degree of mechanical stability and low abradability is obtained in soils having 20 to 30 per cent of clay, 40 to 50 per cent of silt, and 20 to 40 per cent of sand.

The decomposition of organic matter on and in soils is also associated with the creation of temporary cementing substances. These substances are derived from the decomposition products of plant residues, the decomposer micro-organisms, and their secretory products, and they serve to bind particles together and improve the soil structure. Although the effect of decomposition products is temporary, it may last in places for up to five years. In contrast, calcium carbonate tends to decrease mechanical stability, and the erodibility of some calcareous soils may be high. Such is the case in many arid and semi-arid lands.

The importance of structural units is that many of them act as non-erodible or less-erodible obstacles to wind erosion. For wind erosion to proceed beyond the removal of existing loose particles, the structural units must be broken down by weathering forces, raindrop impact, or wind abrasion. The structures vary in their resistance to these forces. Their susceptibility to wind abrasion (sometimes called 'abradability'), for example, varies inversely with their mechanical stability and thus with soil type (Fig. 3.4c). Mechanical stability is a function of inter-particle cohesion (Smalley, 1970). Generally, in a dry state, primary aggregates are most stable, and clods, crusts, and fine material among clods are less stable (in that order). Wind abrasion may cause the progressive breakdown of soil structure as erosion continues.

Non-erodible particles (including stones) seriously restrict the progress of wind erosion, for the amount of material removed is limited by the height, number, and distribution of the non-erodible particles. As erosion proceeds, the height and number per unit area of non-erodible particles increases until ultimately the non-erodible particles completely shelter erodible material from the wind, and a *windstable surface* is created. This final stage is defined by the *critical surface-barrier ratio* (the ratio of height of non-erodible surface projections to distance between projections which will barely prevent movement of erodible fractions by the wind).

Three further characteristics of the surface are important variables in the wind-erosion system: soil moisture, surface roughness, and surface length. Only dry soil particles are readily erodible by the wind: *soil moisture* promotes particle cohesion and restricts erodibility (Fig. 3.4a). As described by Woodruff and Siddoway (1965), the rate of soil movement varies approximately inversely as the square of effective surface soil moisture. The soil moisture at any particular time, of course, is determined by the properties of the soil and the particular weather conditions. *Surface roughness* is composed of the physical elements comprising the surface configuration—clods, ridges, pieces of plant residue, etc. In general, a rough surface is more effective in reducing wind velocity than a smooth one and is thus less susceptible to erosion, provided the material contains non-erodible particles. Finally, the greater the length of the surface across which uninterrupted airflow occurs (the so-called 'fetch' of the wind), the more likely wind erosion is to reach its optimum efficiency.

Vegetation cover influences the nature of wind erosion in several ways. Firstly, the quantity of vegetation, as represented by the proportion of covered ground, governs the extent to which the surface is exposed to erosion. Secondly, vegetation tends to increase surface roughness, and hence reduces wind erosion. In general, the taller the crop, the finer the vegetative material, and the greater its surface area, the more wind velocity is reduced (Fig. 3.4b). Thirdly, plant residue is important in protecting the surface and in adding organic material to it (Fig. 3.4d).

Of the numerous variables in the wind-erosion system, some are permanent,

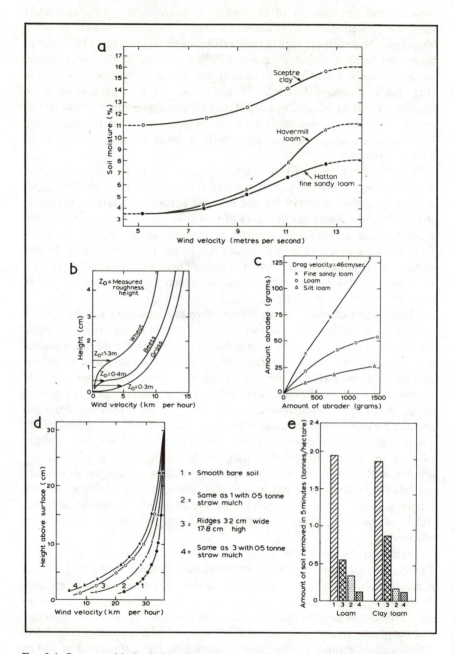

Fig. 3.4. Some empirical relationships between variables in the wind-erosion system
a. Influence of soil moisture on the threshold velocity of soil movement by wind (after
Bisal and Hsieh, 1966). Only the erosive fractions were used (<0·84 mm diameter) in

others change. The characteristics of wind, structural units, organic residues, soil moisture, and vegetation may all change over short periods, and especially seasonally. In contrast, textural properties of surface material tend to be fairly constant, unless they are progressively modified by weathering, erosion, and husbandry.

(b) Initiation and progress of wind erosion

Wind erosion may begin when the equilibrium of the system is disrupted by a change in one or more of the component variables to the extent that the saltation process starts. Changes that cause such disruption are numerous and include reduction of precipitation, increase of temperature, increase of wind velocity, destruction of soil aggregates, reduction of surface roughness, and reduction of vegetation cover.

For saltation to start, it is necessary that the fluid threshold velocity pertinent to the most easily erodible loose surface material be reached. If, as is often the case, there is a hard surface crust which has to be broken, then the initial fluid threshold velocity is higher than that required to move the most erodible particles. Impacts from saltating particles initiate movement of other particles if their impact threshold velocity is achieved. The fluid and impact threshold velocities are similar for most particles, but the impact threshold velocity becomes lower than the fluid threshold velocity for grains of increasingly greater size (Fig. 3.3.). As erosion progresses across a surface, the quantity of debris in motion increases until it is at the maximum sustainable by the wind; the increase is known as avalanching. Avalanching is accompanied by increased surface abrasion—which tends to destroy soil structure and increase the supply of erodible particles—and by particle sorting.

The rate of soil movement, q, (in g/cm/sec) may be described as follows:

$$q = a(D_e)^{1/2} \frac{\rho}{g}(V'_*)^3 \tag{3.4}$$

---

these wind-tunnel experiments. The percentages of sand, silt and clay in these three soils were:

	Sand	Silt	Clay
Fine sandy loam	75·8	15·0	9·2
Loam	33·4	48·2	18·4
Clay	25·5	32·3	42·2

b. Wind velocity distribution over different types of vegetation and a snow-covered surface (after Frevert, Schwab, Edminster, and Barnes, 1955)
c. Loss of weight of dry clods 25–50 mm in diameter of different soil types subjected to abrasion in a wind tunnel by drifting soil composed of particles less than 0·42 mm in diameter at a drag velocity of 46·0 cm/sec (after Chepil, 1951)
d. Influence of surface treatment on wind velocity and (e) on soil erosion. 1, 2, 3, and 4 are the same in both (d) and (e) (after Frevert, Schwab, Edminster, and Barnes, 1955)

where V'_* is drag velocity above an eroding surface, D_e is the average equivalent diameter of soil particles moved by the wind, ρ/g is the mass density of air. Coefficient a varies with different conditions, notably 'the size distribution of the erodible particles, the proportion of fine dust particles present in the mixture, the proportion and size of nonerodible fractions, position of the field, and the amount of moisture in the soil' (Chepil and Woodruff, 1963, p. 245).

The quantity of material, X (in tons/acre) removable from a given area may be defined as follows:

$$X = a(V'_*)^5. \tag{3.5}$$

Under suitable circumstances, erosion will continue once it is initiated until a wind-stable surface is produced.

(c) The wind erosion equation

The major factors involved in wind erosion have been expressed by Chepil and others in terms of a functional equation of predictive value, an equation similar to the universal equation for soil loss by water described in Chapter 2:

$$E = f(I, K, C, L, V) \tag{3.6}$$

where

E = erosion in tons per acre per annum;
I = soil and knoll [slope] erodibility index;
K = soil ridge roughness factor;
C = local wind erosion climatic factor;
L = field length (or equivalent) along prevailing wind-erosion direction;
V = equivalent quantity of vegetative cover.

The derivation of the elements in this equation, and methods of solving the equation graphically and by computer work are described in several publications (e.g. U.S. Dept. Agriculture, 1970b; Woodruff and Siddoway, 1965). Because the relations between variables in the equation are complex, an estimate of wind erosion cannot simply be made by multiplying together different values of the variables, and various charts and tables, or computer programs, are required to reach a solution.

A simple example of the value of this type of functional equation comes from Israel, where Yaalon and Ganor (1966) used the climatic factor, C, to delimit the relative wind-erosion conditions. C is defined by the equation:

$$C = \frac{v^3}{(P-E)^2} \tag{3.7}$$

where v is average annual wind velocity in miles per hour at a standard height of 10 m, and $(P-E)$ is Thornthwaite's measure of precipitation effectiveness

(Thornthwaite, 1931). The climatic index is based on the fact that rate of soil movement varies directly as the cube of wind velocity (Equation 3.4), and inversely as the square of effective moisture, which is taken to be proportional to the $P-E$ index. The base point for determining values of the climatic index is the annual average value (2·9) of it at Garden City, Kansas. Values at other stations are expressed as a percentage of this figure, so that for any given locality:

$$C' = \frac{100\, v^3}{2 \cdot 9 (P-E)^2} \tag{3.8}$$

Figure 3.5 shows the pattern of wind erodibility in Israel as revealed by this index in 1958 and 1960–1. Yaalon and Ganor's data show that the boundaries of the climatic wind-erosion index coincide well with the arid and semi-arid climatic zones of Israel.

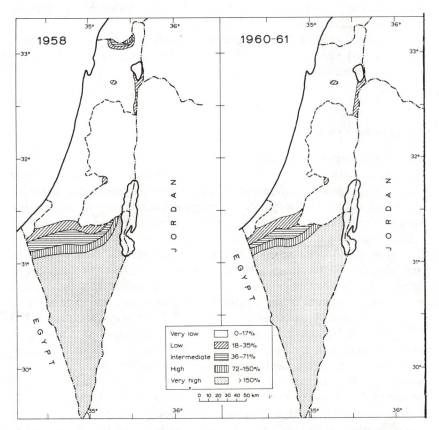

FIG. 3.5. The pattern of wind erosion in Israel, based on the Climatic Index, C (after Yaalon and Ganor, 1966)

3.3. Control and Prevention of Wind Erosion

It is clear from the preceding discussion that wind erosion can be controlled or prevented by the manipulation of key variables in the wind-erosion system. Fundamentally, the solution to the wind-erosion problem lies in the reduction of wind velocity and/or the improvement of ground surface conditions so that it is impossible for saltation to begin. The most important key variables and the alteration of them required to reduce wind erosion are summarized in Figure 3.6. Empirical relationships between some of these variables are shown in Figure 3.4.

Having identified the direction of the changes required in the variables, the question arises 'which is the *most efficient* way of altering the system in order to reduce wind erosion to an acceptable level?' Clearly, the desired result could be achieved in many different ways. The best answer will depend partly on the local environment, cost, and on the technical and financial resources of those making the adjustments.

The various control measures may be grouped conveniently under three headings: (a) vegetative methods, (b) ploughing practices, and (c) soil conditioning methods.

(a) Vegetative methods

(i) *Windbreaks.* Perhaps the most familiar solution to the wind-erosion problem is to place a barrier across the path of the wind and thus reduce wind velocity at the ground surface both in front and behind the feature, and at the same time reduce field length. Sometimes the barriers may be of materials such as netting, stakes or rows of palm fronds, but more commonly they are relatively permanent, growing vegetational structures.

A great deal of research has been carried out into the design, location, and effectiveness of windbreaks and the larger shelterbelts, and several important points arising from it should be emphasized. Firstly, because vegetational windbreaks are relatively permanent their location must be carefully planned. It is essential, for example, that they should be set as nearly as possible at right angles to the most damaging winds. Spacing is also important, and it should be related to the degree of shelter afforded by each obstacle. Secondly, the effectiveness of wind barriers depends mainly on wind characteristics, and the width, height, and porosity of the barriers. The protected distances either side of a barrier are usually measured in terms of barrier heights. For instance, percentage velocity reductions for average tree shelterbelts with winds approaching at right angles to the belts are as follows (F.A.O., 1960):

WIND VARIABLES

VELOCITY	−
FREQUENCY	−
DURATION	−
MAGNITUDE	−

DEBRIS VARIABLES

PARTICLE SIZE	±
SOIL CLODS AND COHESIVE PROPERTIES	+
ABRADABILITY	−
TRANSPORTABILITY	−
ORGANIC MATTER	+

SURFACE VARIABLES

VEGETATION −	residue	+
	height	+
	orientation	+
	density	+
	fineness	+
	cover	+
SOIL AND MOISTURE		+
SURFACE ROUGHNESS		+
SURFACE LENGTH − (distance from shelter)		−
(SURFACE SLOPE)		±

FIG. 3.6. Key variables in the wind-erosion system. Wind erosion will normally be reduced if the values of some variables are increased (+) and if other variables are reduced (−)

Percentage velocity reduction	Distance from barrier, downwind (in barrier heights)
60–80	0
20	20
0	30–40

On a percentage basis, these reductions are relatively constant regardless of open wind velocities. In general, wind velocity is affected to about 5–10 times windbreak height on the windward side of the barrier and about 10–30 times on the leeward side. (It should be noted, however, that wind velocities might be *increased* at the ends of a barrier as a result of wind funnelling, and hence long barriers are generally preferable to short ones.) In terms of soil erosion, barrier effectiveness depends on the relations between open wind velocities, sheltered area velocities, and fluid threshold velocities (e.g. U.S. Dept. Agriculture, 1972). For example, the fluid threshold velocity for many soils is between 19·3 and 24·1 kph. A 50 per cent reduction of a 32 kph wind would provide complete control, but erosion might occur in the same area if open wind velocity was 80 kph. Clearly, the fully protected zone of any barrier is reduced as open wind velocity increases.

A third point is that the most effective barrier is not completely impermeable because such barriers create diffusion and eddying effects on their lee sides. Semipermeable barriers are most effective, for although they provide smaller velocity reductions their influences extend further downwind. Permeability can be controlled through vegetation density and width of the barrier. The shape of shelterbelts is also important. Experience suggests that a triangular cross-section is preferable to a streamlined shape or an abrupt vertical front. Shape is controlled by carefully selecting the plants to be grown and planning their distribution.

Finally, windbreaks serve purposes other than wind-erosion control. Under suitable management they may provide such things as timber, refuge for wildlife, and protection for buildings.

The disadvantages of windbreaks are that they require land, they are expensive, they take time to mature, the shelter they provide is limited, and they compete with other crops for moisture and soil nutrients in environments where both may be scarce. Windbreaks also require a degree of forward planning.

(ii) *Field-cropping practices.* Vegetative methods related to field-cropping practices are often simpler, cheaper, and more effective than windbreaks, and some of them have the additional advantage that they can be used as emergency measures.

One aim of wind-erosion control is to trap moving particles, especially saltating particles, and another is to protect the surface from attack. Various vegetative methods have been employed to achieve these ends. One is to plant 'cover' crops which grow to protect the surface when it would normally be exposed to erosion, notably in the period prior to spring planting. In some erosion-prone areas of Australia oats or barley are sown for this purpose. A second technique is to mix erosion-vulnerable and erosion-resistant crops in alternating strips normal to the prevailing winds. The erosion-resistant crops trap particles, protect the surface beneath them and, in some cases, shelter the vulnerable crops. An example of this 'strip-cropping' technique (see also Chap. 2), is the alternation of wheat, sorghum, and fallow strips in fields in parts of the semi-arid United States. Other things being equal, width of strip increases as soil texture becomes finer (except for clays subject to granulation). Another variable is the angle of the wind to the strips (Table 3.1).

TABLE 3.1

Strip dimensions for the control of wind erosion

Soil class	Width of strips					
	Wind at right angles		Wind deviating 20° from a right angle		Wind deviating 45° from a right angle	
	feet	metres	feet	metres	feet	metres
Sand	20	6·09	18	5·48	14	4·26
Loamy sand	25	7·62	22	6·70	18	5·48
Granulated clay	80	24·39	75	22·86	54	16·46
Sand loam	100	30·49	92	28·04	70	21·34
Silty clay	150	45·73	140	42·68	110	33·53
Loam	250	76·21	235	71·64	170	51·82
Silt loam	280	85·36	260	79·26	190	57·92
Clay loam	350	106·70	325	99·08	250	76·21

Note: The table shows average width of strips required to control wind erosion equally on different soil classes and for different wind directions, for conditions of negligible surface roughness, average soil cloddiness, no crop residue, 1-foot high erosion-resistant stubble to windward, 40 m.p.h. (64·4 km/h) wind at 50 feet (15·24 m) height, and a tolerable maximum rate of soil flow of 0·2 tons per rod (203·2 kg/5 m) width per hour.

Source: After Chepil and Woodruff (1963) with metric conversion.

A general principle, of course, is to select as far as possible crops that provide the best surface protection: small-grain crops, legumes, and grasses are all fairly effective, once they are established. As wind erosion occurs on dry land, the choice of crop may be limited by the availability of water, although the choice may be extended if supplementary irrigation is available. The main advantage of strip cropping is that it restricts soil avalanching; disadvantages include possibilities of weed and insect infestation and the difficulties of grazing appropriate strips.

Closely allied to the cover-crop principle is that of stubble and crop-residue management. Stubble and other field-crop residues protect the surface from erosion (Fig. 3.4d, e), trap moving particles, and provide organic matter for the soil. In some cases stubble may be left between strips of ploughed land normal to the prevailing wind. The two variables of stubble height and strip width can be manipulated according to wind intensity and soil erodibility. In general, for the same degree of protection, strip width decreases with stubble height, and thus the area of tilled land can be maximized by leaving tall stubble. The principles are that the strip should be wide enough to prevent saltating particles from jumping it and receptive enough to stop saltating particles that strike it. Stubbles vary in their effectiveness. In general, those of small-grain crops are more effective than those of large-grain crops. And some residues last longer than others: for instance, wheat and rye straw is more durable than legume residue. It is important that the stubble be disturbed as little as possible; thus, in areas where the ground has to be tilled, stubble cover may be retained by subsurface ploughing.

(b) Ploughing practices

Because soil erosion is approximately inversely proportional to surface roughness, it is clearly desirable to create a rough, cloddy surface when ploughing or creating a seedbed. The effect of ridging cloddy soils by ploughing normal to the eroding wind is to reduce wind velocities and perhaps even reverse wind direction between ridges, thus promoting deposition of particles within hollows (Fig. 3.4d, e). If the spacing is correct with respect to saltation wavelength, then saltation may be stopped completely. If the ridging is done after a rain, large clods may also be produced in suitable soil. But ridging is temporary, and the surface roughness may be reduced by wind erosion and other processes (such as frost action); the hope is that ridging will last through the period when the soil is vulnerable to wind erosion. Another problem of ridging is that it may promote drying of the soil, thus increasing the possibility of wind erosion.

Considerable attention has been given to the design of agricultural equipment suitable for the cultivation of soils potentially vulnerable to wind erosion. One difficulty is that the conventional mould-board plough may break up the soil and bury plant residues, thus exacerbating the problem of wind erosion. Alternative tools have been employed, each with a particular purpose and a general concern about wind erosion in mind. For example, the lister or ridger creates a roughened surface and erosion-resistant ridges, and it is suitable for the emergency control of wind erosion in some areas. The disc harrow and similar equipment that tends to break down soil structure and leave a smooth surface are suitable for the cultivation of light stubble but not for the cultivation of stubble-free soils. Various field and chisel cultivators—usually comprising points fitted to curved or straight shanks and pulled through the soil at depth—are often effective because they control weeds, crop residues are not greatly reduced at the surface, and a rough, ridged surface is created. These tools, and others, are described by the F.A.O. (1960).

(c) Soil conditioning methods

As wind erosion predominantly affects dry soil, it is usually desirable to conserve and maintain soil moisture. These aims can be achieved by reducing evaporation, reducing unnecessary plant growth (by weeding, etc.) and by reducing runoff. Surface mulching, for instance, conserves soil moisture and promotes infiltration. Rolling of a soil that is damp below the surface might in certain circumstances serve to moisten the surface; and occasional irrigation might serve the same purpose. In addition, infiltration of available rainfall may be increased by such practices as strip cropping and terracing (Frevert et al., 1955; Zingg, 1954).

Other methods of soil conditioning are directed towards the creation or maintenance of erosion-resistant clods. This aim may be achieved, for example, by timing ploughing so that it occurs soon after a rain, or by increasing soil organic matter. Finally, a soil-conservation practice that holds some promise for the future is the application by spraying of certain artificial compounds, such as water-based emulsions of resin, which protect the ground surface (Armbrust and Dickerson, 1971; U.S. Dept. Agriculture, 1972).

(d) Conclusion

Any fully informed farmer is likely to use a combination of techniques to limit the loss of his soil. The combination he chooses will depend on many considerations, but he will probably try as far as possible to create and maintain soil aggregates, roughen the surface, provide barriers to wind and to particle movement, and maintain soil moisture, vegetative cover, and plant residues at the soil surface. One fundamental consideration is how he perceives the problem of wind erosion, and this is introduced below.

3.4. Perception of Wind Erosion in the Great Plains

In the last analysis, the control or prevention of wind erosion requires not only a full understanding of the wind-erosion system but also a recognition of the problem and a desire to solve it. Unfortunately, it often happens that these requirements are not satisfied. There are many reasons for this. In the first place, farmers frequently fail either to recognize the phenomenon of wind erosion or to recognize it as a problem. Or, even if the problem is recognized, there may be no desire or motivation to solve it. These responses may be especially true in areas where there are numerous other farming difficulties, such as drought, insufficient capital, and uncertain market conditions. Secondly, although the wind-erosion system is now almost fully understood by a small number of scientists, those responsible for wind-erosion control, such as farmers, may be unaware of the remedial measures or the physical principles underlying them. Thirdly, even if the desire to control wind erosion exists, the decision to act against the problem may be delayed by pressing financial circumstances or by the exigencies of the farming calendar. Often aid from insurance or governmental agencies is not available. And in any case the farmer *may* be unconcerned about the long-term future of his soil, especially if he is a pioneer, transient cultivator, or speculative land purchaser. Finally, remarkably little is known about the ways in which farmers perceive and respond to the hazard of wind erosion.

One relevant study of considerable interest is that by Saarinen (1966) in the Great Plains of the United States. He used a variant of the Thematic Apperception Test (Murray, 1943) in which individuals in a sample of ninety-six male farmers (Fig. 3.7) were asked to tell stories about 'ambiguous'

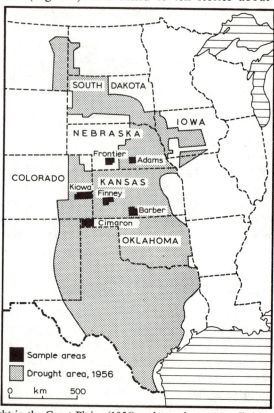

Fig. 3.7. Drought in the Great Plains (1956) and sample areas studied by Saarinen (after Borchert, 1971 and Saarinen, 1966)

photographs. In telling the stories the farmers attributed thoughts, feelings, attitudes, and actions to people in the photographs and thus betrayed something of their own personalities. The stories were analysed by the Schaw–Henry technique in which the dominant theme is abstracted and divided into three parts: the initial stage-setting phase, the manipulatory phase of action, and the resolution or outcome phase.

One of the photographs used showed a lone figure walking in a dust storm, a scene familiar to many Great Plains' farmers. Responses to this picture tell something of the farmers' personalities and of the ways in which they respond to wind erosion. Over 90 per cent of the stories began with a recognition of the facts that the scene results from a lack of moisture and that the lone figure faces serious problems. The farmers revealed a variety of coping mechanisms. Almost half adopted active approaches to the problem: about one-quarter gave restricted solutions, such as the need for action to salvage what remains of a crop; about 20 per cent stressed the need for a survey of the problem. But in most active approaches the farmers merely contemplated a continuation of what they had been doing in the past, a renewal of effort without change of practice. Some 37 per cent of the farmers adopted passive coping mechanisms. Twenty-five per cent stated the need for environmental aid: some, for example, prayed for rain, and at the same time revealed their stoicism. A small proportion offered no coping mechanism. From this study and others Saarinen drew some tentative conclusions about Great Plains' farmers. For example, they seem to respond rather passively to environmental threat, showing tenacity and stoicism rather than a desire to find solutions. For these reasons, it might be difficult to introduce new farming practices.

Saarinen's study provokes several observations. Firstly, results of the Thematic Apperception Test and other techniques for probing environmental perception are difficult to interpret, mainly because much depends on the perception of the interviewer as well as on the interviewee and because the human personality is in any case infinitely complex. Secondly, the study reveals sufficient diversity within a group of superficially similar farmers to suggest that gross generalizations concerning groups and comparisons between groups are extremely difficult to make. Who can say how grain farmers on drought-prone lands of the Russian Steppe or of central Africa will respond to the threat of wind erosion? It is not unreasonable to expect that different cultural groups will respond in different ways, and policy-makers need to take this into account. A third observation is that although scientists may understand the wind-erosion system and the means of controlling it, there is likely to be little progress until the resource users and managers perceive the problem in a way that promotes the adoption of remedial measures. Attention has to be given to bridging the gap between the scientist and the farmer, for example, by education or through particularly innovative farmers.

Finally, it is clear that even in this part of the United States, where more research has been undertaken and more aid supplied than anywhere else in the world, the problem has still to be completely solved. Despite the experience of the 1930s and the post-1945 years, the acreage of cultivated land is still increased when climatic, economic, and political circumstances are favourable in areas quite unsuitable for crops (Zingg, 1954). And the Soil Conservation Service estimated in 1955 that there were still 5·6 million hectares in the Great Plains under cultivation that should be returned to grass (Dasmann, 1968). For other studies on this subject see Knight and Rickard (1971) as well as Sims and Saarinen (1969).

3.5. Case Study: Wind Erosion in Lincolnshire, England

The lighter soils of eastern England have always been potentially vulnerable to wind erosion, and soil erosion by wind has often been recorded locally, especially in areas of reclaimed fen peat and in the sandy uplands (e.g. Pollard and Millar, 1968; Radley and Sims, 1967; Robinson, 1969; Agricultural Advisory Council, 1970). Until recently wind erosion has been regarded as a relatively minor inconvenience; but there are now signs that it is becoming a more serious problem.

In Lincolnshire, areas potentially susceptible to wind erosion include the sandy loams and alluvial soils of the Trent Valley, the regions of Cover Sands, the calcareous and sandy soils of the Lincolnshire Heights and Wolds, and the zones of fen peat and alluvium (Fig. 3.8). Much of the agricultural landscape of the county away from the fen peat and alluvium is recent, having been enclosed and 'improved' during the eighteenth and early nineteenth centuries when extensive 'wastes' were transformed with varying degrees of success from heath and rough grazing into neatly geometric landscapes dominated by farms with mixed economies. Rotations involving grass, root crops, and cereals have generally been used on the improved land, and such rotations included among their many advantages some degree of protection to the soil: the fields were bare for only part of the time and the rotation system helped to maintain soil structure. In addition, the hedgerows around fields afforded shelter.

In recent years, however, there have been significant changes in agricultural practices that may have tended to increase the erodibility of soil material. The widespread and continuous use of chemical manures and the burning of straw and stubble on arable land may have reduced the cohesive properties of the soils; the disc harrowing of light soils and the continuous cultivation of barley may have had similar effects. In addition, the percentage of grassland has declined and the proportion of arable crops, especially cereals, has increased so that, for example, more land is now worked to a fine vulnerable seedbed in spring, and the ability of the soil to maintain its

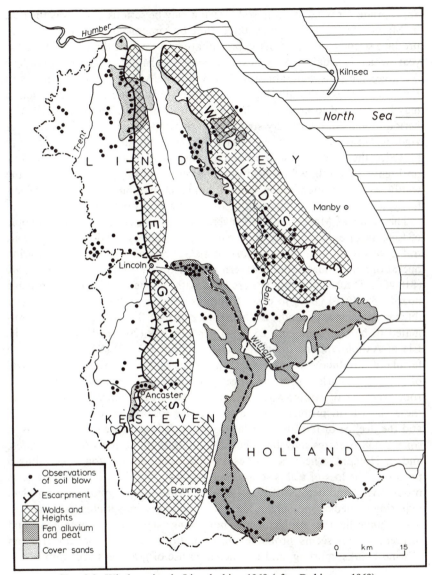

FIG. 3.8. Wind erosion in Lincolnshire, 1968 (after Robinson, 1969)

humus content and structure has been reduced by the reduction of grass.
Finally, the demands of mechanical cultivation and the high value of land
have led to the removal of trees and hedgerows, thus exposing more land to
the attack of winds and increasing the fetch of eroding winds.

Under given climatic conditions, it seems probable that these changes
lead to increased wind erosion, and when climatic conditions are most

appropriate for wind erosion, soil blowing may be serious. Such was the case in March 1968. At that time, three climatic circumstances commingled to produce serious erosion. Firstly, precipitation in the first three months of the year, which is usually low in any case, was only 54 per cent of the Standard Average: the soils were abnormally dry. Secondly, the number of frosts in February was significantly higher than average in some areas: frost action may have helped to produce a finer soil tilth than usual. Thirdly, there was a period in mid-March of very strong westerly winds associated with a series of vigorous low-pressure troughs which followed the dry, frosty winter and preceded the growth of soil-protecting crops. The observations of soil blow on Figure 3.8 are based on field observations and on reports made after the windy spell: the information is not comprehensive, but it probably reflects fairly accurately the distribution of wind erosion (Robinson, 1969).

The events of March 1968 had several serious consequences for the people of Lincolnshire. Many roads were partially blocked by wind-blown material: traffic was disrupted, and the cost of clearance was considerable. Clearing operations in Lindsey (the northern administrative unit) alone cost £4,000 ($10,400). Ditches and drains were filled with sediment: it cost one drainage board £5,000 ($13,000) to clear 10 drains. And many farmers had to clear ditches on their own lands, at their own expense: the average cost was estimated to be approximately £5 ($13) per 22 yards (20·1 metres), and in the Isle of Axholme alone the cost may have been £17,500 ($45,550). Also on farmland, productivity was reduced in places by the uncovering or removal of seeds (e.g. barley, peas, and beet), and by the 'scorching' of leaves and root-exposure of winter-wheat plants. Perhaps even more important than these short-term problems is the progressive loss of topsoil from the land and the decline of fertility associated with it. It is estimated that in Lindsey, some 2,438 hectares may be affected; in Kesteven, perhaps 6,477 hectares; and in Holland, where no figures are available, soil erosion is certainly a problem. The loss of soil is serious because it has implications for long-term productivity, and because it is not always identified by farmers as a problem requiring remedial measures. The recent changes in agricultural practices which underlie the problem are not being significantly modified, and further erosion seems probable until solutions such as the construction of wind breaks, marling, and the increased use of grass leys are adopted.

3.6. Conclusion

Wind erosion is a process affecting dry soils, especially in arid and semi-arid regions and in more temperate areas that suffer periodic droughts. It is not confined to agricultural lands, but it certainly has its most serious human consequences on such lands. Since the disastrous 'blows' in the infamous Dust Bowl area of the central United States in the 1930s, there has been much progress in the understanding of the wind-erosion system, thanks largely to the work of agricultural engineers and desert geomorphologists. As a result, general prediction of potential soil loss is possible, and numerous strategies for reducing soil loss are widely employed. Future research is likely to focus on refinement of predictive indices, wind-erosion forecasting, perception of wind erosion, and the formulation and implementation of farm-management plans which employ the results of wind-erosion research.

4
RIVERS AND RIVER CHANNELS

4.1. Introduction

Rivers have figured prominently in the affairs of men. Through their use in irrigation schemes they have provided the basis for powerful bureaucratic territorial government, such as that in the Wei Ho Valley with its capital at Chang'an (Smith, 1969). They have formed political boundaries, both between and within nations, and their vicissitudes have not infrequently led to territorial disputes. Rivers have provided man with drinking water, and they also provide the basis for freshwater fisheries from which he may also obtain a food supply that in some primitive economies is an important factor either in man's survival or in the provision of a balanced diet. At many economic levels rivers provide a means of waste disposal. In industrial societies this may amount to the generation of a significant level of water pollution, killing fish and other life, and militating against the downstream use of the same river by other men. Rivers normally carry seawards a sediment load of materials washed off the valley-sides or acquired from the channel floor and banks. In areas of accelerated erosion induced by man's interference with the valley-sides (Chap. 2) rivers carry an excessive amount of sediment which may accumulate downstream in reservoirs, along river channels, or in harbours, each of which is expensive to clear. Where channels are sufficiently large they may be used for inland navigation.

Rivers thus form an essential element in man's environment, and their management involves many professions. The sanitary engineer is concerned with the sediment borne by rivers as a pollutant, and with its ecological implications. The civil engineer handles river control schemes, reservoir designs, harbour development, and bridge construction. The administration of rivers varies very much from country to country, and at times appears to be somewhat confusing. As Ciriacy–Wantrup (1964) reported, public district water responsibilities can lead to difficulties of administration, as is the case for example in the Santa Clara Valley, California, where there exist the South Santa Clara Valley Water Conservation District, the Santa Clara County Parks and Recreation Commission, and the Santa Clara County Flood Control and Water Conservation District. Generally within the United States, however, the Constitution divides Federal responsibility in the water field between the legislative, executive, and judicial branches. There is a similar subdivision within the individual states. The Constitution also distinguishes between Federal and state-authority responsibilities. Most of these responsibilities lie with the states, but Federal powers are reserved for water policies involving international or interstate relations. This includes the power to regulate interstate commerce. The actual use of rivers may in practice involve many organizations. To give but one example, the use of rivers in the United States for navigation primarily involves planning, construction, and operation of the project by the U.S. Army Corps of Engineers. However, since water-use is usually planned for multiple purposes, any one project might also involve the U.S. Bureau of Reclamation, the U.S. Coast Guard, and the U.S. Geological Survey (Ciriacy–Wantrup, 1964).

The very wide interdisciplinary interest in river management has led to many conferences, and publication of their proceedings. To name but two, there was the Federal Inter-Agency Sedimentation Conference held at Jackson, Mississippi, in 1963 (U.S. Dept. of Agriculture, 1965), and a conference on River Management (Isaac, 1967), held in Newcastle upon Tyne, England, in 1966. There are many books relating to river management, and the pertinent hydrological characteristics of rivers. Of these the most comprehensive is the *Handbook of Applied Hydrology* (Chow, 1964). Fluvial geomorphology is treated by Leopold, Wolman, and Miller (1964), with channel hydraulics examined in Blench (1969), Henderson (1966), and Yalin (1972). Water resources development is discussed by Kuiper (1965). A number of selected papers, some of historical interest but many of practical importance, are reproduced in Schumm (1972). *Applied Hydrology* by Linsley, Kohler, and Paulhus (1949) remains a useful study.

The following discussion does not attempt to cover all of the many facets of river management mentioned above. Some basic aspects of fluvial geomorphology are introduced together with indications of their practical relevance. The case studies illustrate selected types of river management problems.

4.2. Processes in the River System

The two dominant processes within river channels are those of erosion and deposition. Erosion may take place on the walls or bed of the channel under conditions of turbulent flow. Deposition occurs when the calibre of the material to be moved by the river is too large for movement by a particular set of velocity and flow conditions. Upward eddies under turbulent flow conditions will tend to carry the material away, and cause erosion, while downward eddies will bring about deposition.

The power of the water to dislodge material on the stream bed or bank is related to the forces generated and their general efficiency in overcoming the forces resisting movement. The force tending to produce movement is known as the shear stress (τ).

(a) River-bed erosion

The shear stress that can be generated on the bed of a river (τ_o) is a function of the specific weight of the fluid (γ, which for clear water is 1,000 kg/m^3 or 62·4 lb/ft^3), and both the hydraulic mean radius (R) and the slope (S) of the channel. These latter are defined in Figure 4.1 (where R = cross-sectional area $\div$ wetted perimeter):

$$\tau_o = \gamma RS \tag{4.1}$$

τ_o is also known as the *unit tractive force*, in that it is a measure of the drag (shear) generated per unit wetted area by the flowing water.

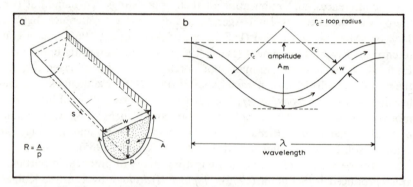

FIG. 4.1. The geometric properties of river channels (a) cross-section (b) plan Where w = width, d = depth, p = wetted perimeter, A = cross-sectional area, A_m = meander amplitude, λ = meander wavelength, and r_c = loop radius

Thus minimum bed erosion occurs when channel slope (S) is low, and the wetted perimeter is large compared with the cross-sectional area of the channel. These constraints are important in the design, for example, of an irrigation channel in which erosion has to be kept to a minimum.

Turbulence is generated by a number of factors which include the roughness of the river bed. This roughness is defined in empirical equations by such parameters as Manning's roughness coefficient, n. It appears in the Manning formula in which mean river velocity (V) is given as:

$$V = \frac{1\cdot486}{n} . R^{2/3} . S^{\frac{1}{2}} \tag{4.2}$$

where V is in ft/sec, R is in feet, and S is the rate of loss of head per foot of channel (channel slope).

In practice this means that for a channel of given dimensions and gradient, velocity of flow increases as roughness decreases. Values for n have been assessed for different bed conditions, with a useful publication being that of the U.S. Geological Survey in which detailed variations in bed materials and form are illustrated by means of photographs alongside their n values (Barnes, 1967). For example, some *approximate* values for n related to different channel floor materials are: sand $n = 0·02$; gravel $n = 0·03$; cobbles

$n = 0.04$; boulders $n = 0.05$. Mountain streams with rocky beds have values of n of between 0.04 and 0.05.

The movement of material along the bed of a river channel can be studied through a series of empirical relations which revolve around an assessment of *Shield's entrainment function* (F_s) and the *particle Reynold's number* (R^*_e). The latter indicates the relation between the forces of inertia and those of viscosity.

Shield's entrainment function (F_s) is given by:

$$F_s = \frac{V^{*2}}{gD(S_s-1)} \tag{4.3}$$

where

g = acceleration due to gravity;
D = material (particle) diameter;
S_s = sediment specific gravity;
V^* = shear velocity (i.e. (shear stress on channel bed $\div$ fluid density)$^{\frac{1}{2}}$).

F_s can also be expressed as:

$$F_s = \frac{RS}{D(S_s-1)}. \tag{4.4}$$

For any one portion of a river channel all of the items on the right-hand side of the equation can be measured and F_s calculated.

The particle Reynold's number (R^*_e) can also be related to measurable properties in that:

$$R^*_e = \frac{(\gamma RS)^{0.5} D\rho_f^{0.5}}{\mu} \tag{4.5}$$

where ρ_f = fluid density and μ = fluid viscosity. These two values (F_s and R^*_e) are of greatest interest when they are plotted against each other on double log axes (Fig. 4.2). This shows that the threshold of sediment transport occurs at the minimum values of F_s and when R^*_e increases beyond a value of 6. By the time $R^*_e > 400$ and $F_s = 0.056$ turbulent flow is fully developed and material is removed from the channel bed. Since R^*_e must exceed 400 for full turbulence to exist a reordering of Equation 4.5 allows (γRS) to be stated as:

$$(\gamma RS)^{0.5} = \frac{400\mu}{D\rho_f^{0.5}}. \tag{4.6}$$

Since γ, μ, and ρ_f are constants for water at defined temperatures and both D and S are known for the channel concerned, R can be calculated, and related to channel cross-sectional area in order to establish the likely wetted perimeter (Fig. 4.1) for the channel when the movement takes place of sediment of a specific size.

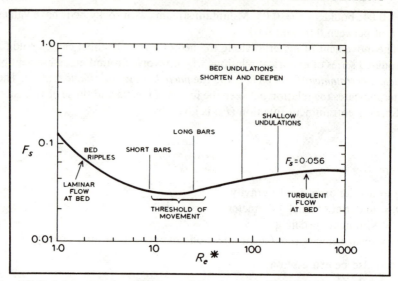

FIG. 4.2. Bed characteristics and flow conditions in terms of Shield's entrainment function (F_s) and the particle Reynold's number (R^*_e)

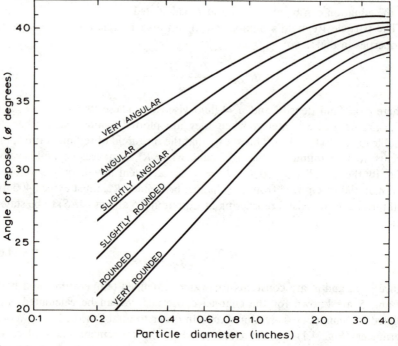

FIG. 4.3. Angle of repose of non-cohesive materials in terms of particle size and shape

(b) Channel-bank erosion

Of the various contributing parts of a river basin to the sediment load carried by the river, a high proportion is derived from the channel walls. As with any inclined slope, the material of which the channel walls are composed (assuming that it is not bedrock but sediment or regolith) in the short term will remain stable as long as the angle of slope (θ) does not exceed the angle of repose (ϕ) of the material. The latter is related both to particle size and shape (Fig. 4.3). On a river channel bank, unlike a hill slope, there is not only the gravitational force ($W \sin \theta$) exerted on each particle (where W is the weight of the particle) but also the lateral force generated by the flowing water. This force is proportional to τ_o, the shearing force on the stream bed. The resistance to movement is provided by:

$$\text{Resistance} = W \cos \theta \tan \phi. \tag{4.7}$$

In this situation τ_o is related to τ_c (the critical shear stress required to move a particle) according to the relationship:

$$\frac{\tau_o}{\tau_c} = \sqrt{1 - \frac{\sin^2\theta}{\sin \phi}}. \tag{4.8}$$

On a side slope:

$$\frac{\tau_o}{\tau_c} = 0.75\gamma ds \tag{4.9}$$

where

γ = specific weight of the fluid;
d = vertical depth of flow;
s = channel slope.

The sediment from the channel banks combines with that washed in from the valley slopes and eroded from the channel bed to give the total sediment carried by the river. However, its actual release from the channel banks is a function of the size and shape of the material, its packing characteristics, the steepness of the bank, the depth of river flow, and the channel gradient.

(c) Channel-bed deposition

The empirical equations used to describe various states of channel-bed erosion can also be used to detect the likelihood of deposition taking place. The converse of erosion threshold values not being passed is that deposition will occur. If for engineering purposes it is necessary to induce the deposition of material above a certain size (e.g. to avoid damage to machinery) then this can be governed by channel shape and gradient characteristics, so that, for example, R^*_e (Fig. 4.2) is kept well below 400.

Alluvial channel deposits have been found to adopt specific forms according to described criteria (Simons and Richardson, 1961). A sequence of six channel-bed forms has been recognized, each related to different stream velocities (Fig. 4.4). Linked to each there is also a value for the Manning roughness coefficient (n) as used in Equation 4.2. Observation of the alluvial bed forms can therefore lead to an estimate of the likely extreme values of n, and these, for a channel having prescribed R and S values (Equation 4.2), can be used to estimate the likely range of stream velocities before the bed forms will change. Reference is also made in Figure 4.4 to a Froude number (Fr), where:

$$Fr = \frac{V}{(gd)^{0.5}} \qquad (4.10)$$

in which

V = mean river velocity;
g = acceleration due to gravity;
d = vertical depth of water flow.

If figures 4.4. and 4.2 are compared it will be seen that ripples occur at $R*_e$ values well below those required for transport under turbulent conditions. At high velocities (Fig. 4.4) ripples give way to dunes and bars (Fig. 4.2) as turbulence begins. The upstream migration of anti-dunes implies sediment movement and is probably little short of the $R*_e$ values required for erosive conditions. Field observation of bed forms in alluvial channels can sometimes be used, therefore, to estimate values employed in hydraulic engineering.

Channel floors are irregular in cross-section with the greater depths normally occurring on the outside of meander bends. The inside bend of a meander forms the site for most deposition on the channel floor, though sometimes longitudinal bars can occur in the river bed some distance out into the channel. Along straight channels the belt of fastest river flow, and of deepest water, seldom follows a straight line down the middle of the channel, but meanders within the channel. In practical terms, such as when considering the navigability of a channel, it is the shallowest sections that determine the passage of a vessel. On meanders the width of the deeper parts has to be enough to allow the vessel to pass, and predictable enough to enable a navigability line to be maintained.

4.3. Materials in the River System

(a) Types of river load

The nature and size of river-bed materials is very variable. Many headwater tributary streams flow on bedrock surfaces, while others are cut into a cover of superficial deposits, such as glacial drifts, and have bed materials that reflect the nature of these deposits. Thus, boulders washed out of boulder clay may introduce discontinuous sections of very rough channels adjacent to sections of smoother character. Lower down the system rivers tend to flow in their own alluvial material.

Whether or not a river has the competence to move the materials that it encounters depends on the processes referred to in Section 4.2. A river may in fact carry material as bed load, saltation load, suspended load, or in a dissolved form. *Bed load* is composed of material too large to be carried upwards in turbulent eddies, but in times of high discharge it may nevertheless be moved along the river bed. *Saltation load* is that part of the bed

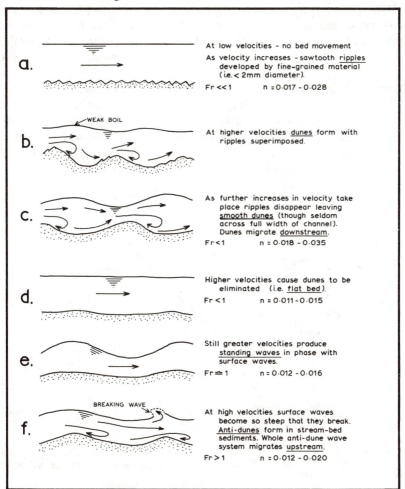

FIG. 4.4. Stream-bed forms in alluvial channels, their roughness (*n*) and relation to the Froude number (*Fr*) (modified from Simons and Richardson, 1961)

load which may in periods of very high-energy conditions in the river, be bounced along the channel floor. The material is never able to stay in suspension for long. The mechanism is similar to that described for saltating wind-blown sand particles (Chap. 3).

In hydraulic engineering terms bed load may be said to include the saltation load as defined above (see Einstein *et al.*, 1940), since the saltating particles (unlike the suspended load) from time to time supplement the bed load. *Suspended load* is maintained by turbulence, and is composed of the finer particles. These may constitute up to 90 per cent of the mechanical load of the river. Einstein *et al.* (1940) called this the *wash load*. *Dissolved load* may be high in areas where solution is an important weathering process. In many industrial areas, including areas being heavily fertilized for agricultural purposes, a high proportion of the dissolved load may be directly attributable to pollution resulting from a human introduction of chemicals into the system.

(b) Relationship between load and other channel characteristics

The shear stress on a river-bed, τ_o, can be related to the size of material that can just be moved. If it is assumed that the specific gravity of the sediment (S_s) is 2·65 and that the kinematic viscosity (i.e. fluid viscosity ÷ fluid density, v) is $1\cdot2\times10^{-5}$ ft²/sec, then Figure 4.5 illustrates the theoretical relationship between τ_o/γ, as represented by RS, and material size (D).

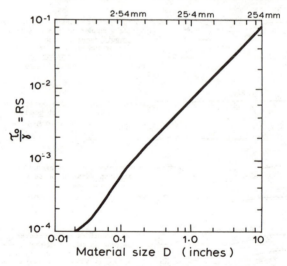

FIG. 4.5. The size of bed-load material mobile in a river channel for specific combinations of hydraulic radius and channel slope. τ_o = unit tractive force; γ = specific weight of water (modified from Henderson, 1966)

From this diagram it is therefore possible to arrive at a theoretical measure of the size of material that should be mobile for a given hydraulic mean radius

(R) and channel slope (S). Conversely, and as important, is the possibility of measuring the diameter of channel material in order to discover what the minimum value of R must have been to bring the material to that point. This is on the assumption that there will have been no appreciable change in S since the material was deposited. If it is desirable for practical purposes to ensure that the bed load in question does not move again then enough water has to be diverted from the channel to keep R below a specified maximum value, namely that value necessary to move the bed load for the given value of S. The appropriate maximum value of R can be estimated from Figure 4.5.

Other equations have also been defined for the relationship between sediments and channel characteristics. For example, Rubey (1952) suggested that when the form (width–depth) ratio of a channel is constant:

$$SF = \frac{kL^a \bar{D}^b}{Q^c} \qquad (4.11)$$

where

S = graded slope;
F = optimum form ratio (the depth–width ratio which gives to a stream its greatest capacity for traction);
L = amount of load (through any cross-section per unit time);
$\bar{D}$ = average diameter of bed load;
Q = discharge (through any cross-section per unit time).

Morisawa (1968) modified this relationship by including a product (n) on the left side of the equation, where n is channel-bed roughness. In effect the equation implies that changes in either the load (L and/or D) or of discharge (Q) will lead to changes in either slope and/or channel form, as well as influencing bed roughness. This illustrates the dangers of considering any one, or even any pair of channel parameters in isolation, for each has to be assessed in the context of the whole system in which it operates.

Experience in hydraulic engineering has made it possible to isolate those parameters upon which design decisions can be based. For example Lacey (1929–30) based the design of irrigation channels on a regime theory expressed by:

$$\frac{V^2}{R} = 1 \cdot 324 f_{VR} \qquad (4.12)$$

$$R^{\frac{1}{4}} S^{2/3} = 0 \cdot 0052 f_{RS} \qquad (4.13)$$

$$P = 2 \cdot 67 Q^{\frac{1}{2}} \qquad (4.14)$$

where

V = mean velocity of flow;
R = hydraulic radius of channel;
S = channel slope;
P = wetted perimeter;
Q = discharge;
f_{VR} and f_{RS} are silt (sediment) factors.

From these predictive equations the known parameters can be used to estimate the others, or channel characteristics can be designed to cope with
prescribed silt factors. However, as in many similar design situations, constraints upon the use of these equations exist. For example, they only apply
to channels that have a bed of loose material of the same type as that being
moved along the bed. In addition the channels should also be in equilibrium;
and these equations may be applicable only to those areas in India and
Pakistan for which they were developed.

One of the problems in designing artificial channels lies in the decision
over the width, depth, and shape of the channel required, for a given bed
material, which will allow the input of water and sediment to be effectively
discharged. One assumption here is that bed material is constant, another
that friction at the channel bank can be neglected. On this basis a family of
design curves has been compiled (Chien, 1956) for the determination of
channel slope and water depth which will cope with prescribed quantities
of water discharge and total sediment load over beds of (a) fine sand, (b)
coarse sand, and (c) gravel respectively (Fig. 4.6). From a knowledge of the
likely water and sediment input the most appropriate depth and slope design
criteria can be estimated. In this, as in other similar cases, the empirical
formulae upon which the design relationships are based are derived from the
behaviour of alluvial river channels, laboratory models, and artificial channels. Local conditions can always override the general case, and in the application of design models (such as that in Fig. 4.6) caution always needs to
be exercised.

A study of natural river channels has indicated relationships between
parameters of channel size, shape, flow, and materials which have subsequently been applied in hydraulic engineering. The most relevant investigations have been on the nature of channels flowing in alluvial materials. Here
rivers tend to meander, with the meanders themselves introducing local
channel variations. Along parts of their courses rivers may also be braided
(i.e. split up into channels which divide and rejoin), as is often the case at the
outflow from a glacier snout or in areas of seasonal flood levels where at
the dry stage braided channels may occur within the major flood channel.
Meandering channels usually occur on the lower portion of the river's
course, near the river mouth. On the geological time-scale meanders are

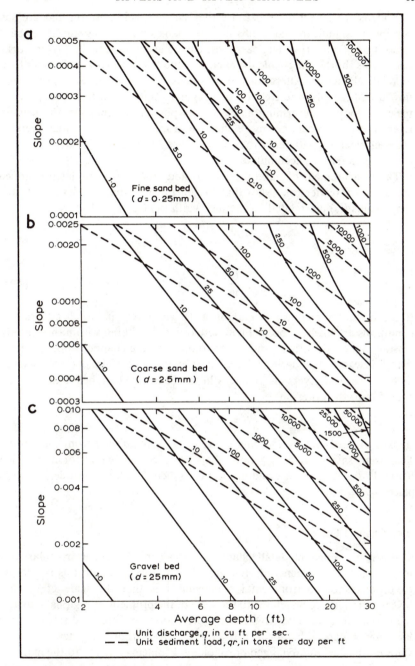

FIG. 4.6. Alluvial channel dimensions required to conduct specific discharge and sediment load (from Bruun and Lackey, 1962; after Chien, 1956)

considered to be unstable because they change their position and are susceptible to significant variations in outer-bank undercutting and inner-bank deposition. To the river engineer these can be accommodated by river works, and on his shorter time-scale the engineer tends to consider the meander form as relatively stable. Alterations in meanders for the purposes of controlling flow usually involve the creation of a regular gentle downstream sequence of meander loops, conforming as closely as possible to those in the original river. These meanders are more effective in inducing currents that keep the channel clear of sediments than is the case in straight channels in which central bars tend to accumulate.

The measurement of natural channels and laboratory simulated channels has shown (Henderson, 1966) that in general meander wavelength (λ) and loop radius (r_c) are related to channel width (w) (see Fig. 4.1) by:

$$\frac{\lambda}{w} = 7 \text{ to } 11 \tag{4.15}$$

$$\frac{r_c}{w} = 2 \text{ to } 3. \tag{4.16}$$

If the hydraulic engineer is making a design system in which equilibrium conditions are likely to occur then these ratios should fall within these limits.

Braiding, on the other hand, tends to occur where channel slope is steeper than it is where meandering takes place. Leopold and Wolman (1957) showed that for natural rivers in coarse alluvial beds (median bed material size >6·35 mm) the criterion:

$$S = 0.06Q^{-0.44} \tag{4.17}$$

enables braided and meandering rivers to be distinguished from each other. That is to say when S is greater than the predicted value for a specified bankfull discharge (Q), braiding occurs. When it is less, meandering takes place.

4.4. Channel Forms

When examined in detail channels display much greater irregularity in long profile than is generally recognized. One way in which this is relevant to problems of environmental management is that the scale of channel investigation has to be related to the scale of the problem. A regional management scheme (e.g. involving general predictions of stream discharge over large areas) can be based on predictions derived from the general relationships established in fluvial geomorphology. Problems relating to the management of specific stretches of a river (e.g. in channel control for irrigation or navigation) will have to take account of local variations based on local measurements. The following part of this account therefore attempts to do two things. It sets out to indicate the type of general relationships concerning river channels, their discharge, and their sediments; and, by taking a specific example, it also sets out the type of local variation and behaviour that may occur despite the general trends. Such a discussion cannot be comprehensive, but it does indicate how necessary it is for each problem to be considered bearing in mind the likelihood of unpredictable detailed changes in channel characteristics both in time and in space.

(a) The channel in long profile

In general a river displays a concave profile from source to mouth. This profile may be interrupted by steps at rejuvenation knickpoints, which frequently coincide with the outcrop of rocks that are comparatively resistant. Between these steps channel slope is approximately inversely proportional to river discharge. The concave profile tends to approach a graded state at which there is a balance between the forces of water and sediment movement within the system and the geometry of the river channels. At any one moment in time there may be a steady state along any one portion of a river channel when input by way of water and sediments equals output over a prescribed period of time. Such a steady state is time-independent in that at any time in the river's history and at any locality along its profile steady-state conditions may occur. However, unless a change in climate or geology intervenes, as grade is approached so more and more of the long profile of a river will approach a steady state.

Relationships between drainage-basin morphometry and fluvial processes, and between channel forms and fluvial processes are based on the concept that dependent relationships exist between form, process, and materials, so much so that under steady-state conditions empirical predictive relationships can be established. A section of a channel in steady state will lose its fine balance between input and output if major controlling components (e.g. climate, sediment supply, or human structures) change.

The general downstream decrease in channel gradient is not only related to discharge, but also according to Rubey's equation (4.11) to the amount and size of load. In general the calibre of sediment load decreases downstream.

A detailed examination of channels in Virginia and Maryland by Hack (1957) supported the general findings of Rubey, but some additional points emerged. Over limited sections of a long profile the calibre of the sediment load may increase, decrease, or remain constant in a downstream direction according to local geological conditions. Hack found greater variations in material size across channels than along their long profiles. In the channels examined by Hack bars of fine gravel commonly occurred on the inside of each bend and boulders tended to be concentrated in the deepest parts.

Channel slope was found by Hack to be directly proportional to the 0·6 power of the median size of material for any given drainage area. That is to say:

$$S = 18 \left(\frac{M}{A}\right)^{0\cdot 6} \tag{4.18}$$

where

S = channel slope (ft per mile);
M = median particle size of bed material (mm);
A = drainage area (sq miles).

The constant, 18, is determined by the mixture of measuring units used. i.e.

$$S \propto \left(\frac{M}{A}\right)^{0\cdot 6} \tag{4.19}$$

According to Hack (1957) there is a consistent relationship between channel slope (S) and channel length (L), where the latter is the distance in miles from a locality on the stream to the drainage divide at the head of the longest stream above it:

$$S = kL^n \tag{4.20}$$

However, values of k and n vary according to bedrock lithology. The close relationship between lithology and long profile provides an interesting approach to landform analysis, for it supplies the link between form and materials as a result of fluvial processes. In an applied sense this is important in that downstream conditions cannot be thought of in isolation from properties and events upstream. For example, where rivers flow through areas of differing lithologies the downstream areas must carry away the erosion products of upstream portions of the basin. If these are resistant they determine the minimum channel slopes required to carry them away and thus affect the rate of downcutting of the main river. This in turn will control the profiles and form of tributary valleys for which the main river acts as the local base level. A downstream site investigation should not be undertaken in isolation from a consideration of upstream conditions.

Channel slope can be related directly to stream-flow characteristics in that for many river channels the Manning equation (Equation 4.2) can be simplified. Since for most river channels the hydraulic radius (R) is approximately equal to the mean channel depth ($\bar{d}$) and the roughness coefficient (n) has a mean value ($\bar{n}$) of 0·03, the Manning equation can be generalized to:

$$V = 50 . \bar{d}^{2/3} . S^{\frac{1}{2}} \tag{4.21}$$

$$\text{or } S = \left(\frac{V}{50\bar{d}^{2/3}}\right)^{1/2} . \tag{4.22}$$

Values of S can be calculated from Equation 4.22 for specified values of V and d, and on a plot of V against d a family of curves can be drawn to represent different values of S (Fig. 4.7) Actual river measurements of velocity and depth along the length of the channel can be plotted on Figure 4.7 so as both to judge the general changes in its slope and to see how its behaviour compares with that predicted from the Manning equation. The data for the Kansas–Missouri–Mississippi river system plotted on Figure 4.7 and referred to again in the case study on river navigability (Sect. 4.7a), show river slope to decrease from over 0·005 (4·9 m per km) to less than 0·00005 (4·7 cm per km).

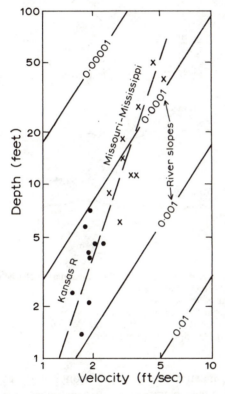

FIG. 4.7. River slope in relation to channel depth and flow velocity (modified from Langbein, 1962)

(b) *Channel patterns*

When seen in plan river channels are seldom straight but tend to meander. Meanders can occur in bedrock or in alluvial material. This discussion is confined to the latter, as it is here that most river control schemes operate and that information about the relationship between channel patterns, river flow, and sediment discharge is most relevant.

Schumm (1963) classified the pattern of channels developed on alluvial material as tortuous, irregular, regular, transitional, and straight (Fig. 4.8).

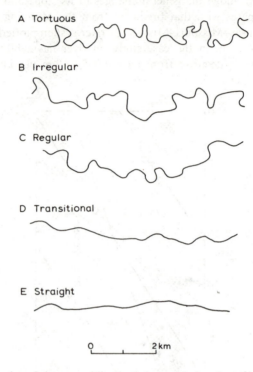

FIG. 4.8. Classification of rivers according to their meandering characteristics.
(a) White River near Whitney, Nebraska
(b) Solomon River near Niles, Kansas
(c) South Loup River near St. Michael, Nebraska
(d) North Fork Republican River near Benkleman, Nebraska
(e) Niobrara River near Hay Springs, Nebraska
(from Schumm, 1963)

This is only a classification for analytical convenience as in reality there is a continuum of types between the two extremes. For the most tortuous channels meander bends are deformed, and the smoothness typical of the ideal meander curve is absent. Irregular meanders are irregular only with respect to the smoothly curved regular meander pattern, and may appear to consist of a meander pattern of low amplitude and wavelength superimposed on a larger pattern. Hjülstrom (1949) suggested that in such cases the smaller meanders may be related to periods of low perennial flow, while the larger occur in response to higher flows related perhaps to the mean annual flood.

The regular pattern is the one most amenable to quantitative analysis because it approximates to a regular wave form in plan. The transitional pattern is characterized by very flat curves which may also approach a regular wave-like oscillation in plan. The straight pattern is almost unknown, but for this classification it is the channel which has only minor bends showing no regularity.

From his study of such channels on the Great Plains Schumm (1963) was able to summarize their main form, material, and discharge properties. This shows that as sinuosity decreases the width–depth ratio decreases, as does the percentage of silt and clay in the channel banks and the channel perimeter. Conversely, as channels become straighter the mean annual discharge increases. Other parameters such as the median grain size of channel sediments, and the gradients of both the channel and the valley floor, show no progressive change with decreasing sinuosity. Schumm goes on to show that sinuosity appears to be determined by the proportions of wash load (suspended sediment) and bed-material load (bed load and saltation load). A relatively wide and shallow channel is associated with the movement of a high proportion of bed-material load, whereas a narrow deep channel is associated with the transport of a sediment load predominantly maintained in suspension.

Empirical relationships have been suggested between discharge (Q) and the measurable geometry of alluvial channels. The latter include meander wave length (λ), meander amplitude (A_m), and channel width (w) (Fig. 4.1). For example Albertson and Simons (1964) reported that:

$$\lambda = k_1 Q^{\frac{1}{2}} \tag{4.23}$$

$$A_m = k_2 Q^{\frac{1}{2}} \tag{4.24}$$

$$w = k_3 Q^{\frac{1}{2}} \tag{4.25}$$

where $k_1 = 29 \cdot 6$, $k_2 = 84 \cdot 7$, and $k_3 = 4 \cdot 88$, but each of these may vary from locality to locality. These, however, only define general tendencies and may not be applicable to specific situations. More complex relationships can be arrived at by including more varied information about the meandering channels. For example Schumm (1967) showed that:

$$\lambda = \frac{1890 Q_m^{0 \cdot 34}}{M^{0 \cdot 74}} \tag{4.26}$$

where

λ = meander wavelength;

Q_m = mean annual discharge;

M = percentage of silt and clay in channel perimeter (which is representative of the river load).

This illustrates the point that channel meanders should not be considered independent of their material properties. This is still further supported by the discovery (Schumm, 1969) that channel width also is related not so much to Q_m but to the ratio of Q_m and M:

$$w = 2 \cdot 3 \, \frac{Q_m^{0 \cdot 38}}{M^{0 \cdot 39}} \qquad (4.27)$$

and likewise:

$$d = 0 \cdot 6 M^{0 \cdot 34} Q_m^{0 \cdot 29} \qquad (4.28)$$

where d is channel depth. Variations in channel dimensions with constant discharge are probably attributable to changes in the calibre of the sediment load. For example, the width–depth ratio of alluvial channels appears to be determined by the nature of the sediment transported through the channel. A high width–depth ratio is associated with large bed-material load, and vice versa.

(c) The channel in cross-section

Any consideration of river channels in cross-section has to take account of the following facts: (i) the velocity of water flow is constant neither in depth nor across the channel width; (ii) the position of the zones of maximum turbulence in a channel can vary with its cross-section, thus influencing the position of maximum scour of the river-bed; (iii) channel cross-sectional form can change very rapidly; (iv) channels differ in cross-section over very short distances.

Short-term changes in cross-section are more likely in channels cut in alluvial material rather than those in bedrock. For example, large changes were observed with the passing of a flood on the San Juan River (Fig. 4.9). The arrival of the flood, with an increase in discharge from 635 cfs (cubic feet per second) to 6,560 cfs, caused a rise in channel-bed level, but scouring soon began. This both deepened the channel and changed its cross-profile. As the flow decreased from 59,600 cfs to 18,000 cfs so sedimentation was renewed and the channel cross-profile was once again modified (Leopold, Wolman, and Miller, 1964).

Measurements of both channel characteristics and stream-flow characteristics are thus time-dependent. Stream flow also varies with the position in the channel at any one cross-section. This variability must be recognized in the field measurement of stream and channel properties and in making predictions based upon these concerning flow or sediment discharge. Data acquired in the past about channel conditions at a point may not remain constant for future site development plans.

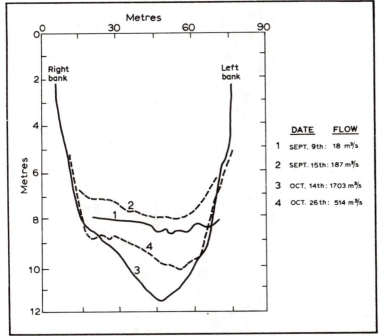

FIG. 4.9. Changes in channel cross-profile during the passing of a flood, Sept.–Dec. 1941, San Juan River, near Bluff, Utah (after Leopold, Wolman, and Miller, 1964, © 1964 W. H. Freeman & Co.)

(d) Channel classification

River channels are principally of two types, *bedrock-controlled* channels and *alluvial* channels. The former are confined between outcrops of rock and the material forming both bed and banks determines channel morphology. Alluvial channels on the other hand acquire their dimensions, shape, pattern, and gradient as a response to the hydraulic characteristics of the river. Schumm (1971) produces a classification of alluvial channels according to their stability characteristics. Stability, or the lack of it, can be thought of in terms of the total sediment load delivered to the channel. Deposition occurs when the calibre of sediment exceeds the power of the river to transport it, erosion occurs when stream power rises and enables material to be moved from channel banks or bed, and between these two extremes is the stable channel. In stable channels there is no progressive change in gradient, dimensions, or shape; it is a channel in steady state.

When this threefold subdivision is further subdivided according to the predominant mode of sediment transport then a total of nine classes are generated (Table 4.1). Stable, depositing, and eroding rivers can be classified according to whether the predominant type of load is in suspension (0–3 per cent bed load), is bed load (3–11 per cent bed load, including saltating

TABLE 4.1
Classification of alluvial channels

Mode of sediment transport and type of channel	Channel sediment (M) (percentage)	Bed load (percentage of total load)	Channel stability		
			Stable (graded stream)	Depositing (excess load)	Eroding (deficiency of load)
Suspended Load	20	3	Stable suspended-load channel. Width–depth ratio less than 10; sinuosity usually greater than 2·0; gradient relatively gentle.	Depositing suspended-load channel. Major deposition on banks causes narrowing of channel; initial stream-bed deposition minor.	Eroding suspended-load channel. Stream-bed erosion predominant; initial channel widening minor.
Mixed load	5–20	3–11	Stable mixed-load channel. Width–depth ratio greater than 10, less than 40; sinuosity usually less than 2·0 and greater than 1·3; gradient moderate.	Depositing mixed-load channel. Initial major deposition on banks followed by stream-bed deposition.	Eroding mixed-load channel. Initial stream-bed erosion followed by channel widening.
Bed load	5	11	Stable bed-load channel. Width–depth ratio greater than 40; sinuosity usually less then 1·3; gradient relatively steep.	Depositing bed-load channel. Stream-bed deposition and island formation	Eroding bed-load channel. Little stream-bed erosion; widening channel predominant.

N.B. Sinuosity is the ratio of channel length to valley length.
Source: Schumm (1971).

particles), or is a mixture of the two (transporting 3–11 per cent bed load). In general the bed load equates with the sand-size fraction, while the suspended load consists of clay- and silt-sized particles. The use of sediment load in this way is justified on the grounds that the manner of material transport is a major factor in determining the character of a stream channel.

4.5. River Discharge

Knowledge as to the amount of water discharging from a drainage basin is important in many situations. The planning of reservoir construction is based on a knowledge of the expected rate of reservoir filling, and on the anticipated balance between supply (water discharging into the reservoir) and demand. An inventory of the land resources of an area needs to include at least a good estimate of the available water resources, including river discharge. The amount of water required from a river channel for human use depends on the agricultural, industrial, and population demands of an area. In this section consideration is given to methods of estimating the likely amounts of discharge when no gauging stations are present. Elsewhere (Chap. 5) the equally important factors relating to the frequency and duration of specific flows are discussed in relation to flooding.

River discharge has figured in several of the equations given in the preceding sections. Each of these allow discharge to be estimated from other more readily measured characteristics. To these others can be added.

Leopold and Maddock (1953) demonstrated simple relations between channel dimensions, velocity of flow, and discharge:

$$w \propto Q^{0.5} \tag{4.29}$$

$$d \propto Q^{0.4} \tag{4.30}$$

$$V \propto Q^{0.1} \tag{4.31}$$

where

w = channel width;
d = channel depth;
V = velocity of flow;
Q = discharge.

(Note that the powers sum to 1·0, as the product of w, d, and V is Q.) Figure 4.10 illustrates the relations between Q and each of w, d, and V for the Missouri–Mississippi Rivers, and for the Kansas River basin. In these cases d increases faster than w with Q along the Missouri–Mississippi, but vice versa on the Kansas River. As is shown in Section 4.7a, this is of some importance when considering the navigability of these two river systems.

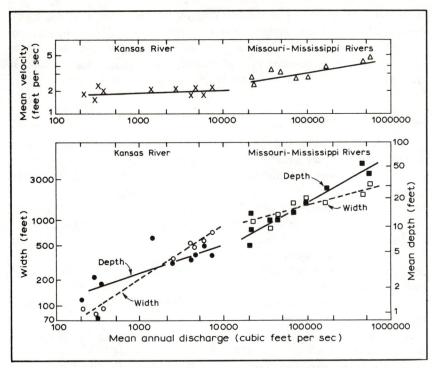

FIG. 4.10. Relation between discharge and channel width, depth, and river velocity for the Kansas River and the Missouri and Mississippi Rivers (after Langbein, 1962)

Working on the premise that mean annual discharge (Q_m) is related to channel slope (S) by a power law:

$$S \propto Q_m{}^a. \tag{4.32}$$

Carlston (1968) found that the value of a varied significantly with the character of the channel. On sections which were graded (i.e. in steady state), and where the channel was cut into an alluvial bed, a could be defined and the correlation between S and Q_m was good. In ungraded sections there was no correlation between S and Q_m. Thus on the graded portions of Red River, Tennessee River, and Delaware River a varied between -0.55 and -0.93. For the mountain reaches, or sections of a valley along which bedrock is highly variable, channel slope and mean annual discharge are uncorrelated. Recognizing this significant distinction between graded and ungraded portions of a river, it is possible from the relationship given above to assess the influence of S on Q_m in particular cases.

An alternative approach to the estimating of discharge is through examining the morphometry of the drainage basin concerned rather than concentrating on channel characteristics. For example, discharge may be thought of as being proportional to some function of basin area. Leopold and Miller (1956) tested the general equation:

$$Q = jA^m \qquad (4.33)$$

for parts of central New Mexico and found that the flood discharge equalled or exceeded in 2·3 years ($Q_{2.3}$) was closely estimated by:

$$Q_{2.3} = 12A^{0.79} \qquad (4.34)$$

(see also Chap. 5.3b iv). Values for j and m vary from one area to another, although m tends to remain in the range 0·5–1·0. Topographically similar areas nevertheless probably give rise to very similar j and m coefficient values.

Discharge can be more closely predicted by including additional measures of basin morphometry, and also by including a climatic parameter. Patterson (1970) analysed the available data for Arkansas and found that:

$$Q = 0.082A^{1.02}P^{0.75}E^{0.06} \qquad (4.35)$$

where

Q = mean annual discharge (cu ft/sec);
A = basin area (sq miles);
P = mean annual precipitation (in inches) minus 30;
E = mean elevation of the basin above sea-level.

The omission of E from the equation increases the standard error of estimated values of Q by only 0·5 per cent, thus indicating the dual importance of both the amount of precipitation and the size of the catchment.

4.6. Sediment Discharge

Unwanted sediment in a river can damage industrial machinery through which water has to pass, cause the siltation of reservoirs, and choke irrigation canals. Much of this may derive from water erosion on valley-sides, and not least from areas disturbed by man (Chap. 2). Although sediment load can be measured directly, it may also be necessary to estimate likely sediment loads where no gauging sites exist. Lane (1955) suggested that bed-material load (Q_s) and sediment size (D) are related to water discharge (Q_w) and channel gradient (S):

$$Q_s D \rightleftharpoons Q_w S \qquad (4.36)$$

In many cases, however, water and sediment discharges can be independent of each other. For example, in natural rivers, climatic fluctuations, changes in land use, river regulation, and diversions can significantly modify the balance between water discharge and sediment load.

Schumm (1969) reconsidered Lane's suggestion and concluded that:

$$Q_s \simeq \frac{w\lambda s}{dP} \qquad (4.37)$$

where

Q_s = bed-material load;
w = bankfull width;
λ = meander wavelength;
s = gradient;
d = maximum channel depth;
P = sinuosity (ratio of channel length to valley length).

Schumm indicated how this relationship can be used to predict man-induced changes. For example, a change in bed-material load at constant mean annual discharge causes a change in channel dimensions (w and d), wavelength (λ), slope (s), and sinuosity (P). Deforestation or clearing land for housing development can significantly increase Q_s, and thus induce downstream changes in these other parameters.

Continuing with this analysis Schumm (1969) showed that:

$$Q_w Q_t \simeq \frac{w\lambda F}{P} sd \qquad (4.38)$$

where

w, λ, P, s, and d are as in Equation 4.37;
Q_w = water discharge;
Q_t = percentage of total sediment load that is bed load;
F = width–depth ratio;

thus linking water and sediment discharge in one empirical statement which can be applied to situations such as that of dam construction whereby water and sediment are cut off so that Q_w and Q_t are decreased for all channels below the dam. This results in downstream decreases in channel width, wavelength, and width–depth ratio, and an increase in sinuosity. Channel gradient would probably decrease with the increase in sinuosity and depths might also increase.

The quantitative prediction of sediment yields was considered in Chapter 2, but these relate to large areas and to time spans of a year or so. It is difficult to make accurate predictions of the sediment yield of a river as a result of a particular rainfall, for much depends on antecedent conditions. For example, in a small catchment a previous rainfall may have flushed through the system most of the sediment available for transport. In a large catchment the consequences of a rainfall will depend upon the location of maximum rainfall intensity. Thus in estimating sediment yields it is possible to define general relationships applicable over long time-periods or large areas. Particular predictions are made with less accuracy and are no substitute for specific site analysis.

4.7. Case Studies in the Human Use of Rivers

Two examples will serve to illustrate some of the problems attendant upon man's use of rivers. There are of course many other examples. These include the redirecting of unwanted sediment away from irrigation canals or machinery (see Blench, 1969; Inglis, 1947; 1949). Another worldwide problem is that of the sedimentation of reservoirs, thereby decreasing their storage capacity. The problem is exacerbated by the fact that this sedimentation occurs as the demands upon the reservoirs are increasing (see Stall *et al.*, 1949; United States, 1957; Gottschalk, 1964; and Dendy, 1968). In recent years the ecological consequences of man's interference with river systems has become better documented (e.g. Blench, 1972).

The first of the two case studies selected relates to the use of water channels for navigation, while the second concerns the fate of salmon in rivers modified by man.

(a) River channel navigability

The potential use of a channel for navigation depends on the physical characteristics of both the channel and the vessels which are to use it. For the channel, knowledge is required as to the fluctuations which are likely to occur in water level (river stage), the velocity of river flow, sediment load and movement (e.g. bed load movement over shoals), changes in depth (both in time and space), channel width, and meandering properties. In particular navigation is concerned with the depth and velocity at shallow sections. The most difficult reach can determine the value of the whole river as a route for commercial traffic.

Relevant vessel characteristics include its proposed speed of travel, draft, and its specific tractive force (T_s) where (in Imperial units):

$$T_s = \frac{\text{Horse power} \times 550}{\text{Speed in ft per sec} \times \text{displacement in tons} \times 2,240} \tag{4.39}$$

which shows T_s to be a measure of the horse-power–hours expended per ton–mile of transport (i.e. energy consumption per unit transported). Some examples of T_s values are given in Table 4.2, and these are related to both design speed and draft in Figure 4.11. A comparison of T_s values for river barges (0·0026), rail freight transport (0·01), lorries and trucks (0·04), and aircraft (0·1 to 0·2) shows that barges have a low specific tractive force which puts them in a competitive position for the movement of bulk goods (Langbein, 1962). The minimum specific tractive force required to sustain upstream

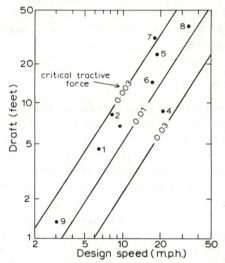

FIG. 4.11. Vessel design speeds, drafts, and critical tractive force for selected vessels
1. Mississippi river boats and barges (630 hp); 2. Mississippi river boats and barges
(3,200 hp); 3. Seine river barges; 4. Fast river steamers; 5. Great Lakes cargo ships; 6. Car
ferry; 7. Tankers; 8. S.S. Queen Elizabeth; 9. Canal barge (after Langbein, 1962)

navigation can be expressed in terms of channel depth and river velocity
(Fig. 4.12). When curves are superimposed on this diagram to represent
specific rivers then indications as to their navigability begin to emerge. For
example, as Langbein (1962) showed, the minimum tractive force which would
be required for navigation on the San Juan River near Bluff, Utah, is of the
order of 0·02, which is greater than that for many commercial vessels. At
Red River Landing, on the Mississippi, however, the specific tractive force
for upstream movement is about 0·00015 as compared with the 0·004
normally available to commercial craft on this river. The Kansas River at
Bonner Springs lies between these extremes with a minimum T_s value for
upstream navigation of about 0·002.

TABLE 4.2

Some characteristics of commercial vessels

Vessel	Horse-power	Displacement (long tons)	Design speed (mph)	Specific tractive force
1. Mississippi R. towboats and barges	630	3,500	6·5	0·0046
2. Mississippi R. towboats and barges	3,200	15,000	8·1	0·0044
3. Seine R. barges	200	400	9·5	0·009
4. Fast river steamers	4,500	1,500	21·0	0·024

Source: Langbein, 1962.

By this method, therefore, Langbein (1962) developed a basis for comparing rivers as to their navigability potential, one which he suggests could be used in legal examinations. He also concluded that in practice rivers with specific tractive forces above 0·002 are not used for navigation. To navigate rivers that require tractive forces greater than this amount would necessitate most of the energy developed by the vessel being expended on overcoming the river current rather than on actually carrying cargo. Within the range 0·002 to 0·001 navigation is usually limited to ferry or short-run operations. Major navigation appears to be associated with rivers that require tractive forces less than 0·001.

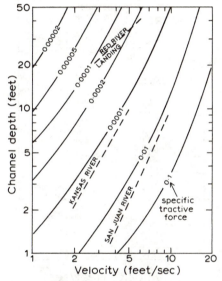

Fig. 4.12. Minimum tractive force required to sustain upstream navigation in terms of river velocity and depth (modified from Langbein, 1962)

Langbein (1962) also defined a river transport-capacity index (C) as:

$$C = D^3 \left(\frac{1 - V_w^2/10D^{4/3}}{f}\right)^{1/2} \tag{4.40}$$

where

D = channel depth;
V_w = river current velocity;
f = the ratio of the resistance against vessel movement at a certain speed in shallow water to that at the same speed in deep water.

Values of C can also be plotted as a family of curves in terms of river velocity and channel depth (Fig. 4.13) for upstream navigation, and clearly channel depth is the more important control on the value of *C*. If the limit of suitability for navigation, in terms of the specific tractive force limit of 0·002, is added to this diagram, then rivers which plot to the left (i.e. in the upper left corner of Fig. 4.13) are suitable for navigation, while those to the right (with higher velocities and shallower channels) are generally not suitable.

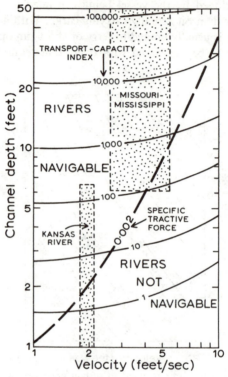

FIG. 4.13. River navigability classification based on the specific tractive force and a transport-capacity index (defined in terms of velocity and channel depth) (based on Langbein, 1962)

In earlier sections (4.5 and 4.6) reference was made to the Kansas River and the Missouri–Mississippi river system. In particular Figure 4.10 shows that for the Kansas River velocity values range from 1·8 to 2·1 ft/sec, and depth values from 0·9 to 6·5 feet. The Missouri–Mississippi rivers on the other hand have a range of velocity values of 2·5 to 5·5 ft/sec, and depths of 6·0 to 50·0 feet. Transferring these data to Figure 4.13 shows that the Kansas River would sometimes be classed as non-navigable. This is rarely the case for the Missouri–Mississippi systems.

Langbein (1962) compared the transport-capacity index with recorded amounts of commercial river traffic and showed that no traffic is recorded for the Kansas River in 1957, while as the transport-capacity index rises from approximately 400 on the Missouri to over 10,000 downstream of New Orleans, Louisiana, so traffic density rises from 0·27 to 50 million ton–miles per mile.

The relation between actual traffic density and the transport-capacity index of navigable rivers in the United States is so close that its use as a comparative index of navigability appears to be justified.

(b) Dam construction and salmon

In their annual search for spawning grounds salmon swim from the seas up-river, and frequently find their passage upset by man-induced changes to the river. Alternatively the spawning grounds may have been rendered unsuitable by man. Salmon are very sensitive to both the depth and the velocity of the water in which they are swimming. Indeed, their whole life habits are geared through centuries of evolution to the seasonal variations of their flowing-water environment. For example, along the McClinton Creek, British Columbia, Pritchard (1936) found a high positive correlation between the numbers of pink salmon migrating each day from the sea into the stream and the maximum daily water height. Reduced flows, as can be caused either by upstream storage or by extraction along the course of the river, can cause serious delays in salmon migration. In addition, reduced discharge along the river can cause points of difficult passage, or even generate complete barriers to further migration, despite the presence of adequate flows for passage in most of the stream.

Salmon can also become confused if there are artificial changes in flow direction. For example, irrigation pumping on the San Joaquin River delta, California, caused a local reversal in stream flow and effectively reduced the number of salmon successfully continuing upstream. Salmonids migrating downstream from the spawning grounds are especially sensitive to current flow orientation.

At the spawning stage velocity and depth are again the most important features of discharge. Optimum discharge, within the tolerance limits, depends on the configuration and gradient of the stream bed. The basic necessity for effective spawning, however, is a suitable bed of gravel. Suitable conditions decrease with excessively high or low discharge. Too high a discharge can make the gravel mobile, thus reducing the chances of egg survival, and too low a discharge can leave the gravel banks exposed above water level and out of reach of the salmon. In addition, low flow can increase the predation on downstream migrant juvenile salmonids.

Normal high flows, before the salmon reach the gravel beds, as occurs with snow melt on mountain streams, tend to flush any accumulation of fines out of the gravel, thus leaving space for egg-laying.

The building of dams has generally had two distinct effects on salmon spawning. It has disturbed the natural drift of food organisms, and may even 'starve' the salmonids by trapping the organisms behind the dam. Secondly, discharge from the dam has either reduced downstream flow too much or it does not flush the fines out of the river gravels. The latter has been noted for the gravels below the reservoir on the Trinity River in California. Instead sediment from tributary streams, below the dam, has caused the spawning grounds to become covered by unsuitable finer deposits (Fraser, 1972).

In order to preserve the migrant spawning cycle of the salmon it thus becomes necessary to maintain an adequate downstream discharge, to create or maintain suitable gravel spawning grounds, and to allow for the passage of a sufficient quantity of food organisms. In some instances it is possible to build salmon runs, as is done alongside some of the dams in Scotland, so as to allow the fish to bypass the obstruction. Einstein (1972) suggested that the natural spawning grounds could be replaced with artificially prepared deposits of gravel, preferably within diversion canals so that a flow is maintained over the gravel which continues into the same stream in which the historic spawning grounds are located. These canals can be made to carry an amount of suspended load, including a food supply, equal to that which the river used to carry. Flood flows through the gravel can also be generated artificially.

4.8. Conclusion

Rivers form an essential element in man's use of the environment, and man's way of life may be geared to the use of river waters for drinking, fishing, navigation, or as a means of waste disposal. Rivers are dynamic systems, however, in which fluvial processes are governed not only by external factors such as the amount of rain falling in a drainage basin, but also by internal conditions such as channel morphology and sediment supply. Between these there are empirically definable relationships which make possible the prediction of consequences resulting from changes in any one of the member parameters in the system. In all cases the management of rivers has to allow for the possibility that local influences may cause significant deviations from the general situation implied by the established empirical statements.

General estimates can be made from more easily measured parameters of such things as river discharge. A knowledge of the behaviour of river systems and the interrelationships between form, process, and material characteristics can provide important information concerning water supply and river behaviour in areas not being monitored by gauging equipment.

The consequences of river modification are seldom confined to the spot where these modifications are taking place. They may extend for great distances downstream influencing further uses of that channel, and often dramatically affecting the ecological territories of wild life and fish.

5

FLOODPLAINS, FANS, AND FLOODING

5.1. Dimensions of a Hazard

Discharge of water and sediment in rivers varies greatly in space and time. Discharge is normally confined below the banks of channels, but occasionally the channels are unable to contain the discharge, and water and sediment spill on to and move across the adjacent surfaces. Adjacent to perennial rivers these surfaces are usually (but not always) *alluvial floodplains*, which are created by the fluvial system specifically to accommodate the larger, less frequent flows. In certain areas where flow is ephemeral, floods often spread across the surfaces of *alluvial fans*. As floods commonly pose a hazard to man and his activities it is important to emphasize at the outset that they are a natural phenomenon and are only a problem where man has elected to use areas susceptible to flooding and where he has induced flooding that would otherwise not have occurred.

Yet flooding is undoubtedly a serious hazard. Table 5.1 lists the average

TABLE 5.1

Estimates of average annual loss of life and property from selected hazards in the United States

Hazard	Annual loss (persons)	Annual loss (property) $
Tuberculosis	11,456	—
Veneral disease	3,069	—
Lightning	600	100 million
Tornadoes	204·3	45 million
Hurricanes	84·8	100 million
Floods	83·4	350 million–1,000 million
Weeds	—	4,000 million

Source: Burton and Kates (1964).

annual losses in terms of life and money caused by selected natural hazards in the United States. These data are rather crude estimates but they do indicate the approximate order of importance of various hazards. The approximate nature of the data is indicated by the fact that annual flood damages in the United States have been estimated at as little as $350 million by the U.S. Weather Bureau, at over $900 million by the U.S. Corps of Engineers, and at $1,200 million by the U.S. Department of Agriculture (Burton and Kates, 1964).

Flooding is not only a relatively serious environmental problem; there is evidence to suggest that the damage caused by floods, despite extensive and expensive efforts to reduce it, is actually increasing in some areas (Holmes in White, 1961; Hoyt and Langbein, 1955). For example, in the United States, floods today cost the nation $1,000 million per annum, a figure twice as great as that when the Flood Control Act was passed in 1936 (Hanke, 1972). It is no coincidence that some of the worst flood disasters in the United States were also some of the most recent—such as those in South Dakota and in the eastern United States in 1972, and along the Mississippi in 1973. Many towns and cities in North America still have serious flood problems (see Sewell, 1965; and White, 1961).

Because flooding is a natural phenomenon that often becomes a problem to man, it is important to ask why man continues to live in, and to exploit, flood zones. The question may be answered in a variety of ways. Firstly, the advantages of living and working in flood zones must clearly outweigh the disadvantages. Secondly, the advantages are numerous, and frequently complex. In agricultural areas, the increment of sediment from floods, the irrigation of water meadows, or the fertility of alluvial soil may be amongst the benefits of using a flood zone. Access to water—in places a right associated with riparian law—for transport, industrial consumption, or waste disposal may figure prominently in industrial decisions favouring a flood-zone location. Or the hazard of flooding may be reflected in relatively low land values and prices, thus making flood-zone land more desirable for certain activities. Or again, transport planners may find flood zones to be convenient, topographically subdued avenues for the location of communications. A third answer may lie in the many ways in which man perceives the hazard. Some people may simply be ignorant of the hazard; others may know of it but not expect to be personally affected by a future flood; there may be those who expect a flood, but do not expect to be at a loss as a result of it; some may expect to bear a loss, but not a serious one; and a few may expect to take a serious loss, are worried about it, and plan to take action to reduce such losses (Kates, 1962). It is the last group that is most concerned to understand and control floods.

Flood analysis, prediction, and control are generally in the hands of national, regional river or water-administration authorities. These normally operate in the context of major drainage basins (catchments), for flood administration, like river administration, is essentially a catchment-contained problem. Dissonance between drainage basins and administrative areas has at times been a cause of serious management difficulties.

The general study of flooding and flood-hazard management is examined in several important books. A comprehensive American review is that by Hoyt and Langbein, *Floods* (1955), while an interesting survey focused on management problems in the United States is Leopold and Maddock's *The Flood*

Control Controversy (1954). Most of the hydrological matters relating to flooding are examined conveniently in Chow's *Handbook of Applied Hydrology* (1964). Some British experience is presented in the Institution of Civil Engineers' (1966) *River Flood Hydrology* and Smith's *Water in Britain* (1972). Coastal flooding as a hazard is discussed briefly in Chapter 8, and is described by Burton, Kates, and Snead in *The Human Ecology of Coastal Flood Hazard in Megalopolis* (1969). This volume is one of a series of important geographical monographs on aspects of flooding published as the University of Chicago Department of Geography Research Papers (White, 1945; White *et al.*, 1958; Murphy, 1958; Sheaffer, 1960; White 1961; Kates, 1962; White, 1964; Kates 1965; and Sewell, 1965).

5.2. Floodplains and Fans

(a) Floodplains

A river floodplain results from the storage of sediment within and adjacent to the river channel. Two principal processes are involved. The first is the accumulation of sediment, often coarser sediment, within the shifting river channel. Sediment is commonly deposited, for example, on the slip-off slopes on the inside of meander bends to produce point-bars. As the river migrates in the direction of the outside of the bend, the point-bar grows and the floodplain deposit is augmented. Much of the sediment is only temporarily stored in a point-bar and it may be moved further downstream from time to time. This type of within-channel accumulation, which can occur at any point along the valley, is mainly associated with below-bankfull discharges. Secondly, suspended sediment carried by overbank discharges across the valley floor may settle and provide a further increment of floodplain sediment, either generally over the flooded surface or, occasionally, locally along the channel margins to form levees. Where floodplain sediments comprise both coarse and fine material, most of the coarse fraction is the result of deposition by lateral accretion within the channels, and some of the fine material may result from overbank accretion; where the floodplain sediment is composed largely of fine material it is likely that most of it will be deposited within a channel (Wolman and Leopold, 1957; Leopold, Wolman, and Miller, 1964).

The frequency with which bankfull discharge and flooding occur is fairly similar for different rivers, even though they may be located in contrasted environments. The recurrence interval for bankfull flow is in the general range of 0·5–2 years, with 1·5 years a common value (Leopold, Wolman, and Miller, 1964; Nixon, 1959). The similarity of the recurrence interval in contrasted environments and the fact that channels normally do not become progressively deeper as floodplain deposition continues, strongly imply that channels are adjusted to accommodate discharge generated within the watersheds for much of the time, that the floodplain is adjusted to transmit larger flows for the remainder of the time, and that the two features are functionally related to one another.

Floodplains may be defined in various ways. To the geomorphologist, the floodplain is an area characterized by a distinctive suite of forms and deposits; to the hydrologist it is the area inundated by flood events of particular magnitudes and frequencies; and to planners and lawyers it may be an area defined by statute (Kates and White, in White, 1961; Fig. 5.1). A particularly

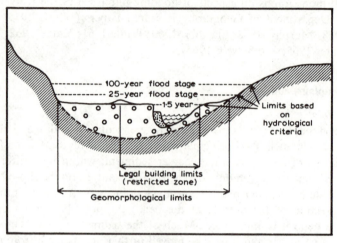

FIG. 5.1. Various definitions of a floodplain

useful distinction on many floodplains, from the point of view of hazard problems, is that between the *floodway* and the *pondage area*. The floodway is the area of the floodplain, usually marginal to the main channel, in which land filling or drainage concentration as the result of construction would significantly increase water levels (Lee, 1972). The pondage area lies adjacent to the floodway and is where water is stored during flooding.

(b) Alluvial fans

Alluvial fans are fairly common landforms in arid and semi-arid areas and they are usually found where ephemeral flows from mountains spread out on to adjacent plains (e.g. Bull, 1968; Cooke and Warren, 1973). Individual fans vary enormously in their dimensions, but they are normally cone-shaped in plan, focusing on an apex near the mountain front. Fans frequently coalesce to form composite alluvial slopes. Deposition on fans arises from changes in the hydraulic geometry of flows as they leave the major feeder channels from the mountains—the flows increase in width, decrease in depth and velocity, and often lose water by infiltration into permeable alluvial deposits, thus causing deposition. Flooding on alluvial fans can occur in two principal locations: along the margins of the main supply channels, and in the depositonal zones beyond the ends of the supply channels. A general model of flooding danger on fans which reflects this statement is shown in Figure 5.2.

There are several perplexing problems associated with such flood zones. Firstly, because flow on fans is ephemeral, and at most times there is no flow, the likelihood of flooding is often minimized or ignored by man.

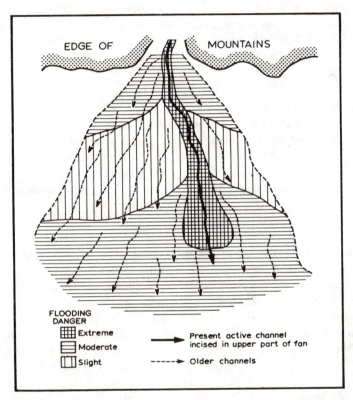

FIG. 5.2. Relative flooding danger on a typical alluvial fan in the western United States (after Kesseli and Beaty, from Schick, 1971)

Secondly, flow in alluvial-fan systems may vary from 'river' flows with low sediment concentration to highly viscous debris flows capable of moving large boulders. Thirdly, the channels followed by floods can change from time to time—for example when a channel becomes blocked with debris—and thus the loci of entrenchment and deposition and hence the hazard zones may vary in space and time. The nature of alluvial-fan flooding is therefore of great interest to farmers working agricultural land on fans, to those concerned with building and maintaining lines of communication across them, and to residents in communities built on them.

5.3. Flooding: Characteristics and Controlling Variables

A number of physical characteristics of floods are important in considering the impact of flooding on man: the frequency of flooding; peak flow (magnitude); total flood runoff volume (the discharge above an arbitrary base-flow); the rate of discharge increase or decline; the lag time (the time between the centre of mass of a rainstorm and the centre of mass of associated runoff, see Fig. 5.3b) or the flood-to-peak interval (time difference between flood elevation at which damage begins and flood peak); the area inundated; the velocity of flow; the duration of inundation; the depth of water; the sediment load of the flood; and the time of year when the flood occurs (seasonality).

(a) Graphical descriptions

Many of these characteristics can be described in two fundamental documents: the *flood-frequency* curve, and the *flood hydrograph*. A typical flood-frequency curve is shown in Figure 5.3a. The recurrence interval, I, the period of time in which a given event is *likely* to be equalled or exceeded once, is calculated using the simple formula:

$$I = \frac{n+1}{m} \tag{5.1}$$

where n = number of years of discharge record, and m = the rank number of an individual item in an array. The rank values are flood magnitudes, such as annual peaks (measured in terms of discharge or depth of flow), or number of flood peaks above a chosen base. A major problem in some areas is that appropriate historical records may not be available. Each recurrence interval is plotted against the appropriate measure of magnitude on semi-logarithmic (or probability) graph paper. From such graphs, statements relating to the *statistical probability* of flood events can be made. For example, the mean of a series based on annual peaks—the *mean annual flood* ($Q_{2.33}$)—commonly has a recurrence interval of 2·33 years. Or it can be said that a flood of given magnitude is likely to occur, on average, once in every n years. This does not mean, of course, that such a flood can be forecast for a particular year. Put another way, a flood with a 10-year recurrence interval has a 10 per cent chance of recurring in any year. It should be noted that the technique described here is only one of several ways of arriving at flood-frequency curves (e.g. Chow, 1964; White, 1964).

Because discharge is a product of mean depth, width, and velocity of flow, and because its relations to characteristics such as area of inundation and sediment load can be precisely described, curves can be drawn relating these variables to their frequency of occurrence. Magnitude–frequency graphs are useful for many reasons, but perhaps the most important is that they provide a yardstick by which the hazard and the cost of preventing it can be assessed. The flood hydrograph describes the change of discharge during a flood for a single station. An example (Fig. 5.3b) shows several important flood characteristics, notably peak flow, total runoff, and the rate of discharge rise and fall (rising and falling limbs); when the storm rainfall is plotted on the same graph, lag time can be determined; and when flood

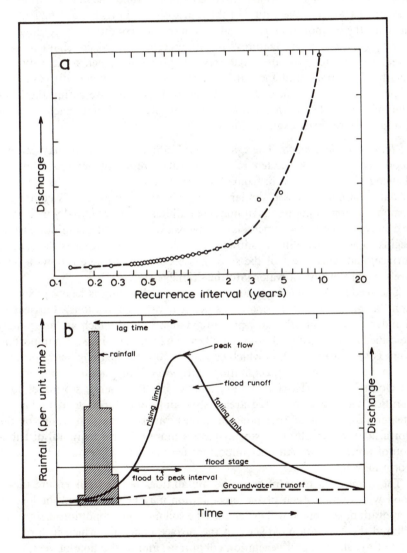

FIG. 5.3. a. A typical flood-frequency curve
b. A typical flood hydrograph, showing several important flood characteristics

stage is defined, the flood-to-peak interval is also clear. As with the frequency–magnitude plots, it is possible to construct hydrographs relating many specific flood characteristics to time—depth and area of inundation, for example.

(b) Factors controlling flood characteristics

Each of the important flood characteristics can be explained in terms of interrelated factors, of which three groups are fundamental: (i) transient phenomena, such as the nature of the rainstorm, and other changing features such as the evaporation rate and soil moisture conditions preceding the storm; (ii) permanent features, including especially basin characteristics (area, shape, etc.), the drainage network properties (such as density and length of streams), and the nature of the drainage channels (for example, slope, roughness, width, and depth), and (iii) the land use within the basin that may be either permanent or transient (Rodda, 1969). Each of these groups requires brief examination.

(i) *Transient phenomena.* There are several circumstances that may cause a flood. The first is a sudden release of water, resulting perhaps from the thawing of a snow cover, or from blockage of flow by ice jams. (The breakage of dams, of course, has a similar effect.) The thawing process is commonly responsible for spring floods in middle and high latitudes: initially meltwater in a snowfield freezes as it percolates downwards; as melting continues, water begins to accumulate in the snow; eventually meltwater reaches the ground surface, and when most of the snow has become slush there follows a relatively rapid flushing of water into the drainage system.

The second, and the most common, cause of flooding is heavy precipitation, a phenomenon normally associated with depressions, hurricanes, thunderstorms, and other low-pressure systems, all of which are adequately described in standard textbooks (e.g. Barry and Chorley, 1968). An important point is that three variables which contribute towards defining heavy precipitation, namely depth of precipitation, duration of fall, and area covered, are all closely related (Hoyt and Langbein, 1955). In general, storms of long duration tend to cover large areas, short-duration storms tend to have a greater proportion of precipitation concentrated in the first hours of the storm, and longer-duration storms have a more uniform temporal distribution of precipitation. Another important feature is the direction and rate of storm movement.

The mere occurrence of a heavy rainfall does not in itself ensure that a flood will follow. Much of the precipitation may be intercepted, or lost by evaporation, and much will depend on the soil moisture conditions before the fall and the infiltration capacity of the surface materials, which in turn will relate to geological and vegetation conditions, and to the time of year.

(ii) *Basin characteristics.* Basin, drainage network, and channel characteristics are extremely important in determining the nature of the flood hydrograph, given a certain input of water. Channel geometry is also important. As discharge is a function of width, depth, and velocity, and as velocity depends on hydraulic radius, channel slope, and bed roughness (see Chap. 4), clearly the nature of these variables will influence the nature of discharge. For example, velocity of flow, all other things being equal, will be directly proportional to the square root of channel slope, and therefore channel slope will directly influence lag times.

Many other factors are also significant. For instance, peak discharge is related to the area of the drainage basin and also is reduced by basin storage (e.g. ponds, meander cutoffs, and the channels themselves). Perhaps it should be emphasized in this context that just as the basin characteristics help to explain differences of flood characteristics between drainage basins, they also influence flood characteristics at different stations in the same drainage basin.

(*iii*) *Land use*. The importance of land-use characteristics can be illustrated by reference to the relations between urbanization and changes in the unit hydrograph (Leopold, 1968). The unit hydrograph is the average time-distribution graph of discharge from a *unit* or standard storm, such as a storm which produces one centimetre of runoff. In Figure 5.4a hypothetical unit hydrographs are shown for an area before and after urbanization. The building of the urban area normally has several consequences. Water runs off more quickly from the relatively impermeable surfaces of streets and roofs and flows through drains and sewers more efficiently than across 'natural' vegetated surfaces. This reduces lag time. As the time a given amount of water takes to run off shortens, so flood peak increases and the rising and falling limbs steepen. In general, the greater the area served by sewers and the greater the impervious area, the greater will be the ratio of peak discharge to discharge before urbanization. This relationship is shown in Figure 5.4b, where discharge is defined as the mean annual flood (see above) for a one-square-mile (2·59 km²) drainage area, and the data are taken from various studies in the United States (Leopold, 1968). Similarly, the frequency of overbank discharge might be expected to increase as a result of urbanization. Figure 5.4c shows the relationship between extent of urbanization and the ratio of number of floods to the number of floods preceding urbanization. This graph is also based on a one-square-mile (2·59 km²) drainage area, and employs data from various sources (Leopold, 1968). Thus in some places urbanization might be expected to increase the height and frequency of floods, to reduce lag time, and to increase rate of flood rise and fall.

(*iv*) *Flood description and forecasting*. The previous discussion suggests that many flood characteristics should be precisely describable in terms of their numerous controlling variables. As the number of variables is so great, much effort has been directed towards identifying the most important variables that

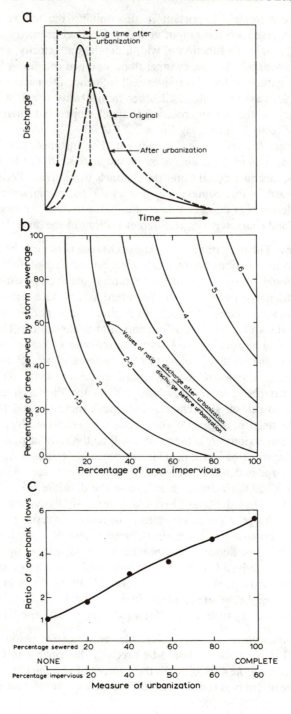

are relatively simply derived. Geomorphological variables figure prominently among the variables used. For example, Rodda (1967), using data (in Imperial units) from twenty-six small drainage basins in the United Kingdom, expressed the relationship between mean annual flood, $Q_{2\cdot33}$, and other variables as follows:

$$Q_{2\cdot33} = 1\cdot08A^{0\cdot77}R_{2\cdot33}^{2\cdot92}D^{0\cdot81} \tag{5.2}$$

where

A = drainage area (sq miles);
$R_{2\cdot33}$ = mean annual daily maximum rainfall (in);
D = drainage density (miles per sq mile).

The value of the multiple correlation coefficient for this study was 0·90, and the factorial standard error of the estimate was 1·58.

A similar study of 164 drainage basins in New England (United States) by Benson (1962) demonstrated the following relationship:

$$Q_{2\cdot33} = 0\cdot4 + 1\cdot0 \log A + 0\cdot3 \log Sl -$$
$$0\cdot3 \log St + 0\cdot4 \log F + 0\cdot8 \log O \tag{5.3}$$

where

$Q_{2\cdot33}$ = mean annual flood;
A = drainage area (sq miles);
Sl = main channel slope (feet per mile);
St = percentage of surface storage area plus 0·5 per cent;
F = average January degrees below freezing (°F);
O = an orographic factor.

Closely allied to the notions of flood description used in these studies and of flood probabilities discussed earlier, is the shorter-term problem of flood forecasting (Chow, 1964). The importance of forecasting, of course, lies in the necessity of giving advance warning of a flood so that emergency action can be taken in potentially vulnerable areas. There are several prerequisites for an efficient forecasting system. Of great importance are adequate data on the history of rainfall and runoff in the basin. Much of this information is ideally described in terms of graphs showing rainfall-runoff relations, unit hydrographs, routing methods (i.e. the ways of determining the timing and

---

FIG. 5.4. a. Hypothetical unit hydrographs for an area before and after urbanization (after Leopold, 1968)
b. Effect of urbanization on mean annual flood for a one-square-mile drainage area (after Leopold, 1968)
c. Increase in number of flows per year equal to or exceeding channel capacity (for a one-square-mile drainage area), as ratio to number of overbank flows before urbanization (after Leopold, 1968)

magnitude of the flood wave at successive points along a river), and stage–discharge relations. Even more important is the creation of an information collection and dissemination network that can handle all the data relating to a particular flood event. Telephone, telegraph, or radio communications are often used, sometimes based on data provided by completely automatic recording stations. The earliest forecasts may be based on predictions of discharge from rainfall information; when a flood crest is initiated, the forecast can then be related to routing methods, using information on gauge heights, channel storage, and rates of travel.

5.4. Perception and Responses

(a) Perception of flooding

How man responds to flooding depends, in the first instance, on how he perceives the hazard. The study of flood perception is fraught with difficulties and work on this subject suffers to some extent from the problem of deriving generalizations other than those that are immediately obvious. Nevertheless, it is of interest here to identify a few useful generalizations.

Firstly, perception will in part be related to some or all of the physical characteristics of the hazard itself. For example, Burton, Kates and White (1968) showed that perception and consequent adjustment to flooding are directly related to flood frequency.

Secondly, perception of the hazard varies significantly between different cultural groups. For instance, the problem of adjustment to flooding is widely perceived as being important in the technically advanced societies of western Europe and the United States, but no such problem exists for technically primitive floodwater-farming groups in southern Arizona and northern Mexico. Here the difference is one of degree, for the Indians have successfully come to terms with flooding, whereas the technically advanced groups have only partially adjusted to the phenomenon. Equally, some groups may aspire to controlling floods, whereas others may see them as Acts of God to be endured.

Thirdly, within any particular cultural milieu there are often pertinent differences within and between groups, such as scientific personnel, resource users, and the general public. Scientists often differ in their perception of the flood hazard. In part such differences relate to inadequate data and the complexities of the problem, as the controversy over calculating flood insurance rates reveals (e.g. Langbein, 1953; Kunreuther and Sheaffer, 1970). Similar variations occur between resource users. Burton and Kates (1964) suggested that such differences might arise, for example, from variations in the nature and degree of personal experience and in the variety of ways that flooding affects different users of flood-hazard zones. For example, to one factory manager a few centimetres of flooding for a short time on rare occasions may present no problem; to another manager, faced with frequent, deeper flooding, there may be a serious problem. But if the second manager is new to his job he may fail to appreciate the problem, especially if he has transferred from an area where there was no flooding hazard. Or again, the second manager might have serious labour-relations and production difficulties that distract his attention from the problem of flooding until the event occurs. And in any case, the managers' perception of flooding is likely to be quite different from that of neighbouring farmers or shopkeepers who are affected in different ways, and perceive the hazard in different economic and social contexts.

There are also frequently conflicts between the attitudes of scientists, resource users, and the general public. Burton and Kates (1964) cited the example of Fairfield, Connecticut where users of waterfront property opposed the construction of a flood-protection dike mainly on the grounds that such protection would seriously interfere with their view and result in loss of breeze.

(b) Responses to flooding

Individuals or groups can adjust to flood hazards in a number of ways. Research at Chicago by White and his students has led to the identification of a suite of possible responses. These are: bearing the loss; public relief; flood abatement and control; land elevation; emergency evacuation and rescheduling; structural adjustment; land-use change and regulation; flood insurance. Although this range of adjustments may theoretically be available to an individual or group, it by no means follows that any particular individual is free to select the most appropriate adjustments, for much will depend on the perception of the problem, and certain adjustments may be precluded for various technical, social, or economic reasons. The characteristics of the flood, the success with which it has been forecast, and the degree to which floods generally have been predicted are all important considerations here. Thus in examining responses to floods it is useful to bear in mind two lists of variables, and to examine the links between them in terms of individual and group possibilities. The two lists are shown in Table 5.2. Some of these relations are considered briefly below.

TABLE 5.2

Adjustments to floods	Critical flood characteristics
1. Bearing the loss	1. Depth
2. Public relief	2. Duration
3. Abatement and control	3. Area inundated
4. Land elevation	4. Velocity of flow
5. Emergency action	5. Frequency-recurrence relations
6. Structural adjustment	6. Lag time (or flood-to-peak interval)
7. Land-use regulation	7. Seasonality
8. Flood insurance	8. Peak flow
	9. Rate of discharge increase and decline
	10. Sediment load

(*i*) *Bearing the loss.* Bearing the loss is more often an involuntary than a voluntary act, especially in areas where flood forecasting and drainage-basin management are poorly developed, or where such facilities fail to provide sufficient warning or prevent flooding. The ways in which the loss can be carried vary greatly, of course, according to the individual or group situation. A good example of a guide to accepting loss from flooding is the U.S. Department of Agriculture's *First Aid for Flooded Homes and Farms* (1970a).

(*ii*) *Public relief*—in forms ranging from gifts from friends to government and international aid—is a common response to a disaster. Such community help may ease immediate distress, but it may also encourage the persistence of inappropriate occupation of flood-hazard zones.

(*iii*) *Emergency action.* The effectiveness of emergency action, which normally includes the removal of people and property, various flood-fighting techniques (e.g. sandbag defences), and rescheduling of activities, depends on advance warning of a flood and certain flood characteristics. Advance warning is related mainly to the availability of flood-forecasting techniques, and to lag time (or flood-to-peak interval). For example, Sheaffer (1960) suggested that where flood-to-peak interval is less than a day, only minimum emergency action is possible; if the interval is one to three days, responses will be governed by the effectiveness of pre-flooding planning by individuals and groups; if the interval is over three days, there should be sufficient time for most emergency adjustments. In addition, emergency action is generally most effective where flood duration is short, where velocity is low, and where frequency of flooding is high.

(*iv*) *Flood-proofing.* The main forms of structural adjustments—sometimes referred to as flood-proofing—include control of seepage (for example, by sealing walls); sewer adjustment (by the use of valves, for instance); permanent closure of unnecessary openings, like windows; protection of openings (such as flood-control gates at entrances to underground stations); protective coverings to buildings, machinery, etc. to prevent such problems as rusting, and elevation of structures above flood levels (Sheaffer, 1960). These measures and other similar ones, are designed to reduce damage to structures and goods within hazard zones. Some are permanent, some are contingent upon action being taken on receipt of a flood warning, and others are used only in emergencies. Some require structural changes. Many are restricted by hydrological limitations: neither depth of water nor hydrostatic pressure should be too high; velocity over two feet per second might begin to cause damage; duration should be relatively short; and generally flood-proofing is more applicable to pondage rather than floodway zones. The measures are therefore particularly beneficial in areas where flood control measures are absent, where depth, velocity, and duration are short, and where forecasts are available. A study of flood-proofing is provided by Sheaffer (1960).

(*v*) *Land use*. The regulation of land use in flood-hazard zones and the maintenance of adequate floodways are further ways of limiting the damage produced by floods. In an American survey, Murphy (1958) pointed to the use in various places of channel-encroachment statutes, flood-plain zoning ordinances, subdivision regulations, building codes, and various other devices such as permanent evacuation. Critical to the successful adoption of such measures is the precise analysis of the physical geography of the flood-plain and the appraisal of development potential in terms of such analysis. Inevitably, as Murphy discovered, there are wide differences of interpretation that are reflected, for instance, in methods of determining channel-encroach-ment lines (i.e. limits for development). Kates and White (in White, 1961) identified several problems related to land-use regulation including, for example, difficulties of estimating the damage and use potential of land, the danger of protection works giving a false sense of security and encouraging risky development, and confusion of terminology.

(*vi*) *Flood insurance*. Flood insurance has been seen for a long time as a useful alternative in floodplain management, but its lack of availability and a heritage of bankrupt companies testify to the difficulties of creating realistic policies. Among the more intractable problems are lack of essential data, such as those on frequency and magnitude, the formidable difficulties arising from creating policies that will attract floodplain subscribers and yet cover potential claims for a hazard that strikes erratically, and the possible en-couragement to floodplain development by groups able to afford the pre-miums (Hoyt and Langbein, 1955; Langbein, 1953; Renshaw, in White, 1961). In the United States these difficulties have been met by a series of methods for determining flood insurance premiums. One was based on the applicant determining the amount of insurance cover he required in the light of information on the probability of floods reaching given heights. Another sought to allocate a single rate to a specific type of property within a major river basin regardless of location. A recent method arises from The Flood Insurance Act of 1968 (U.S. Congress, 1968) and is based on the calculation of rates, some of which are federally subsidized, for different classes of struc-ture within specified hazard zones in the light of 100-year flood, stage-frequency, and depth-damage information (Kunreuther and Sheaffer, 1970).

(*vii*) *Abatement and control*. Perhaps the most effective ways in which man intervenes to modify flooding in fluvial systems are through the alteration of land use within watersheds ('flood abatement') and through protective measures along flood channels ('flood control'). These two approaches are not mutually exclusive, although in some countries they have tended to be the responsibility of different authorities. In the United States different ap-proaches and administrative rivalry have been reflected in the 'upstream-downstream' flood-control controversy (Leopold and Maddock, 1954). The principal protagonists were the Department of Agriculture and the Corps of Engineers, and a major issue was the relative merits of small 'headwater' dams (often associated with land-use management) and large, 'mainstream' dams—both types being designed to store floodwaters and reduce flood peaks. The large dams, some argued, flooded extensive areas of productive land and displaced people and settlements (e.g. Kollmorgen, 1953). On the other hand, it has been argued that numerous headwater dams—although they use less valuable land, affect no single community greatly, and are likely to cause relatively little damage if they fail—do not control larger, rarer floods downstream, and the cost per unit of storage increases rapidly as reservoir size declines. There are many other issues in the controversy, ranging from flood-routing and sedimentation problems to the recreational and other uses of lakes.

Land-use practices, such as those discussed in the context of soil conservation in Chapter 2, may or may not be specifically designed to modify floods, but they often serve to reduce and delay surface runoff to river channels. Unfortunately, the precise consequences of these practices on flooding are not always clear. One unequivocal experiment was carried out by the Tennessee Valley Authority in a 6·9 km² watershed in eastern Tennessee (Hoyt and Langbein, 1955). The land in the basin had at various times been cleared and cultivated, the soils were severely eroded, and about two-thirds of the watershed was composed of burned and grazed woodland. Extensive abatement measures were introduced in 1934–5, tree planting was carried out, and the area was retired from use and protected from fire. The effect was clear: runoff data showed that, although the volume of floodwater was not affected, the shape of the annual hydrograph was significantly modified (Fig. 5.5). Peak discharges, especially summer flood peaks, for comparable storms 15 years after treatment averaged only 15 per cent of those before treatment, and lag time increased from 1·5 to 8 hours. Computations by the

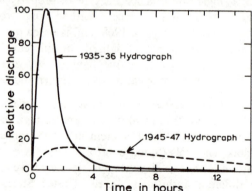

FIG. 5.5. Effect of land management on flood runoff from White Hollow Basin, east Tennessee (after Hoyt and Langbein, 1955, © 1955 Princeton U.P.)

U.S. Department of Agriculture show that, in general, land-treatment measures reduce peak flow from small storms giving a few centimetres of runoff, but have little effect on large floods (Leopold and Maddock, 1954).

Flood control by *channel change* has come to be focused on a relatively small range of engineering devices: structures to confine floods to the floodway, by embankments, etc.; modifications to natural channels to improve their efficiency and/or capacity by, for example, straightening, steepening, widening, deepening, or 'smoothing'; creating diversionary routes for use in times of flood; and construction of reservoirs to store floodwater and regulate the passage of water downstream. Many of these techniques are illustrated in Figure 5.6, which shows various solutions commonly adopted in the United Kingdom to the problem of flooding in urban areas (Nixon, 1963). Solution A involves protective embankments, with sluice gates or a pumping station at the confluence of mainstream and tributary. The main problem with this solution is that it may increase flood levels upstream and downstream. An example is the Nottingham Flood Protection Scheme on the River Trent in England. Solution B simply requires enlargement of channels to accommodate larger discharges. One problem with such schemes is that as the enlarged channel is only rarely fully used by flowing water, weed growth may increase and riparian land may become overdrained. The River Don Improvement Scheme in Yorkshire adopted this solution in overcoming problems of land subsidence.

In Solution C a flood-relief channel removes excessive flows. This solution is often appropriate where it is difficult to modify the original channel, and it does not have the disadvantages of Solution B, but it tends to be rather expensive. An example is provided by the River Welland Improvement Scheme in which the town of Spalding is bypassed by the Coronation Channel. Intercepting or cut-off channels are used in Solution D. These channels normally divert part of the flow away from the area subject to flooding, leaving some flow for riparian users in the protected area. The Great Ouse Protection Scheme in the Fenland is an example. In Solution E, flood-storage reservoirs are used. The arguments relating to reservoirs are admirably explored by Leopold and Maddock (1954); suffice it to say here that this solution is widely used especially as many reservoirs created for water-supply purposes may have a secondary flood-control role. An example of a multi-purpose reservoir project in Britain is the Bala Lake Scheme. The final solution, F, the removal of settlements threatened by flooding, is rarely used. Naturally, in many circumstances a combination of flood-control techniques is used to overcome a flooding hazard.

(*viii*) *A note on economic considerations.* Almost all economic aspects of flooding are concerned in some way with determining flood losses and the costs and benefits associated with abatement and control efforts. Various

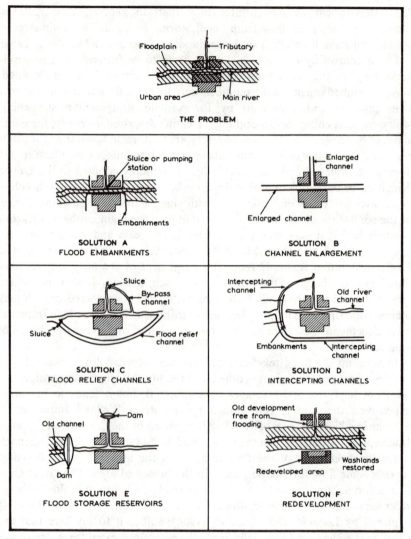

FIG. 5.6. The main methods of flood control in the United Kingdom (after Nixon, 1963)

methods have been used to put a value on these gains and losses, and most have been seriously criticized. Cost–benefit analysis, the most widely used technique, is an example (e.g. White, 1964).

A U.S. Corps of Engineers procedure for calculating damage-frequency curves is shown in Figure 5.7. The flood characteristics used are flood depth (elevation), discharge, and frequency—other features such as sediment load are not directly involved. Calculations are made for both natural (initial) and improved (projected) conditions, with an assessment of frequency–discharge relations (*c*) first being combined with the description of stage–discharge relations (*a*) and then applied to the stage–damage curves (*b*), to give an average annual estimate of flood damage (e.g. Kates, 1965). (Specific

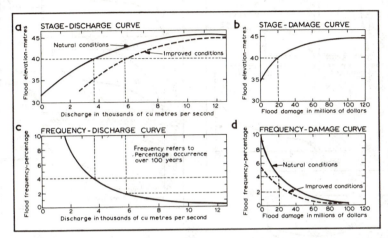

FIG. 5.7. Stages in the calculation of flood-damage frequency curves, based on U.S. Corps of Engineers procedure (after Kates, 1965)

examples of stage–damage curves for some establishments in La Follette, Tennessee, are shown in Figure 5.8a.) The average annual estimate can be derived by measuring the area under the frequency–damage curve and dividing it by the number of years of record, or by multiplying the probability of each stage by the corresponding damage and summing the expected values. In this way, flood-control benefits are expressed as the difference between the 'natural' and 'improved' estimates.

There are, of course, numerous problems in making these calculations. Firstly, not all flood characteristics are considered. But it is certainly possible to take additional variables into account in calculating damages, as White (1964) has shown. Figure 5.8b, for example, illustrates generalized relations of depth, duration, and velocity to urban flood damage for a representative commercial establishment. Secondly, while obvious features of structural damage may be easily costed, there is a host of incidental effects that may be difficult to identify and to cost. To take one example, what is the value of unanticipated leisure time, time that is possibly partly offset by sacrificing leisure at a later date (Renshaw, in White, 1961)? Further difficulties arise in projecting estimates into the future, when general economic conditions (interest rates, inflation, economic growth, etc.) and the period over which capital investment in flood-protection work is written off have to be considered.

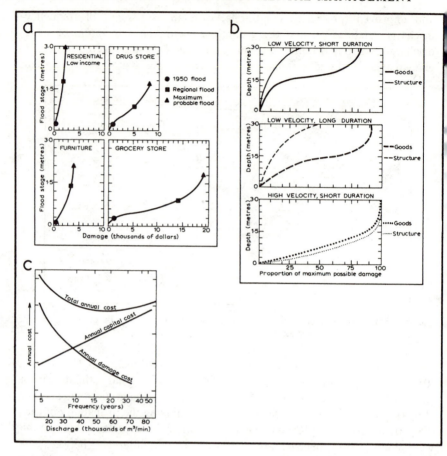

FIG. 5.8. a. Stage-damage curves for representative establishments in La Follette, Tennessee (after White, 1964)

b. Generalized relations between depth, duration, and velocity and urban flood damage for a representative commercial establishment suffering losses of goods and damage to structure (after White, 1964)

c. Hypothetical relations between discharge and frequency and annual costs of damage and capital investment in a flood-protection project. The most economical structure is where the total annual cost curve is lowest

In deciding the design of flood-control works, economics have to be examined in the context of the hydrological situation. Usually it will be economically impossible to provide protection against the maximum probable flood, but if some smaller 'design flood' is selected, risks of damage will be greater. In short, damage costs must be offset against capital (construction and protection) costs, calculated on an annual basis, and the most economical structure is that where the total annual costs are least (Fig. 5.8c).

5.5. Case Study: Alluvial-Fan Flooding in Southern Israel and Jordan

Detailed studies of desert floods are scarce because they rarely affect man and settlements, and research stations monitoring desert floods are few. An exception is the review by Schick (1971) of a storm and flood in the southern Negev Desert and adjacent areas on 10–11 March 1966. Figure 5.9 is a map of the region which largely comprises a north–south trending depression some 180 km long.

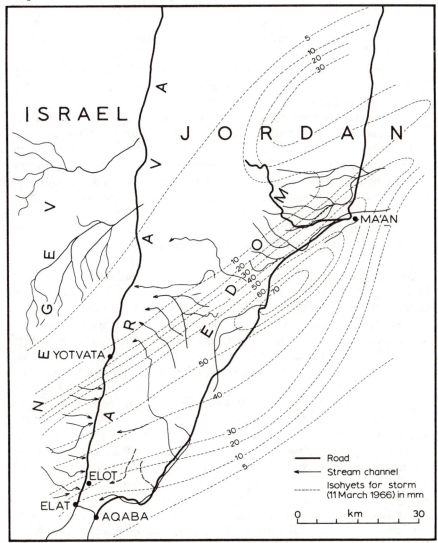

FIG. 5.9. The area in southern Israel and Jordan affected by the storm of March 1966, showing the pattern of rainfall (after Schick, 1971)

A torrential rain fell in the basin on 10–11 March 1966. The rainfall was caused by cold northerly air that penetrated southwards along an upper air trough associated with a Mediterranean depression moving eastwards across the Sinai Peninsula. The pattern of precipitation is shown on Figure 5.9 which indicates that over 50 mm was recorded in places, with the highest rainfall occurring south-west of Ma'an. Over most of the storm belt, precipitation exceeded the annual mean, and rainfall amounts varied greatly from place to place. The storm was possibly the second largest in the area for twenty years.

Analysis of rainfall-intensity and infiltration-capacity data shows that some 50 per cent of all rainfall in the area normally occurs in intensities exceeding 2·5 mm/hr, a rate equivalent to a reasonable infiltration capacity in the area, and that 30 per cent exceeds 5 mm/hr. For this reason, a considerable proportion of rainfall can be expected to form runoff. Such was the case in the storm of 1966. Floods began in mountain tributaries and were fed into the main valley along alluvial-fan channels. The flows accumulated huge quantities of sediment and deposited much of it within the basin. A rough estimate of sediment load suggested that 1·57 million tons of sediment reached the southern Arava, about half entering the Yotvata Basin, and some 0·73 million tons the southernmost area; 0·12 million tons are assumed to have reached the Gulf of Aqaba. In all, 1·45 million tons were added to the alluvial fans and playas of the southern Arava, approximately the equivalent of 1–2 mm of aggradation over the depositional area.

Damage in the Negev amounted to about 1 million Israeli pounds, and included partial flooding of the Timna open-pit and underground copper mines; disruption of traffic on the road north of Elat; deposition on roads and erosion of roads, embankments, etc.; filling of drainage ditches with sediment on farmland at Elat; cutting of power and telephone lines; isolation of Elat for 40 hours; and the flushing into the sea of 5,000 tons of salt prepared for shipment. In Jordan, on the eastern side of the basin, the damage was even more serious. Some 70 people were killed and over 250 were injured, mostly in the town of Ma'an, where the artificial dam to an apparently natural overflow channel passing through the centre of the town was breached. Many houses built on flood channels after the dam was constructed were destroyed at about 6.30 a.m., before the inhabitants had left for their work. Buildings on the banks of the channel were also demolished as the result of lateral scour by the floodwaters. Damage in Jordan was estimated at between 1·3 and 2·7 million dollars.

The fact that this heavy but not wholly exceptional storm did so much damage arose for several reasons, all of which are of general significance. Firstly, the spottiness of rainfall in the desert, to say nothing of scarce rainfall-recording stations, makes forecasting difficult. Secondly, the flood-to-peak interval for desert floods is rather short, so that insufficient advance warning exacerbates the hazard. Thirdly, design values for desert structures are meagre and indeed many engineering works are often semi-improvised, despite the enormous erosive force of some floods. Fourthly, urban planners and highway engineers may be ignorant of or may become insensitive to the potential dangers of dry river beds and closed-basin lakes in dry lands. And finally, alluvial fans represent preferred sites in terms of local topography, water supply, and accessibility.

5.6. Conclusion

Flooding has seriously affected man for millennia, and despite prodigious efforts to reduce flood damage and loss of life in some areas, the problem remains serious. To the geomorphologist and the hydrologist, flooding is a natural process fundamental to the transfer of water and sediment in drainage basins and important in the modification of valley-floor and alluvial-fan landforms. The characteristics of floods of significance to the natural scientist, such as their episodic nature and discharge characteristics, are also of direct interest to those concerned with the consequences of floods on man and his use of flood-prone land. This survey has ranged freely beyond the limits of any narrowly defined geomorphological interest in floods, but in doing so it may have illuminated some fundamental issues in environmental management; it has emphasized how important it is to study a wide range of environmental variables (in this case within a drainage-basin context) if a physical phenomenon is to be understood, predicted, and managed successfully; and it has stressed the role played in environmental management by problem perception and economic analysis.

6

LANDSLIDING

6.1. Introduction

Almost every year a major disaster is caused by landsliding. On 2 September 1806 during a heavy rainstorm a landslide at Goldau in Switzerland destroyed a village, killing 457 people. In 1881 a landslide at Elm also in Switzerland, which lasted only 50 seconds, killed 115 people (Heim, 1932). In 1966 the village of Aberfan in Wales lost many of its children in a tragic slide on the flank of a coal tip-heap (H.M.S.O., 1969). These dramatic events attract public attention. In addition there are also hundreds of smaller, less newsworthy cases of slope instability which at the very least are a nuisance, and many of which cost money both in terms of the damage they do and in their effective control.

A disaster is said to occur only if landsliding takes place in an area where man lives or works, though there are many landslides in uninhabited areas. Particularly serious disasters occur in areas of high population density. Among these are areas of dense settlement on the footslopes of mountains or of steep hills, and especially at the mouths of confined mountain valleys. For example, the twin cities of Victoria and Kowloon (Hong Kong), sited on a mountain-front footslope, were heavily hit by landslides descending from the slopes above after the severe rainstorms of June 1966 (So, 1971). Similarly sited is Yungay, Peru, where a debris avalanche descending a mountain valley overran settlements on the flatter country below, killing up to 25,000 people (Fig. 6.1; Cooke and Townshend, 1970; Clapperton and Hamilton, 1971). The list of landslide and other damage resulting from slope instability is a long one, and it is increasing. New slides are occurring not least because man's power to change a hillside has been transformed by technological developments. Excavations are going deeper, man-made structures are larger, and areas are being used for civil engineering sites which are at best marginally suitable and which had previously been avoided while more favourable alternatives remained available. In Brazil legislation was introduced in order to prevent this type of expansion. The Forest Law of 1959 made illegal construction work above a specified level in order to preserve the stabilizing influence of forests and to prevent construction across springs. A second law, the Law for Licence of Construction in Uneven Terrain, passed in 1967, regulated construction on steep slopes or wherever instability was possible. It demanded that the contractor should obtain proof of slope stability before building construction was allowed to commence (Barata, 1969).

Publications concerned with landsliding occur in many fields traditionally thought of as geomorphology, engineering geology, soil mechanics, and civil

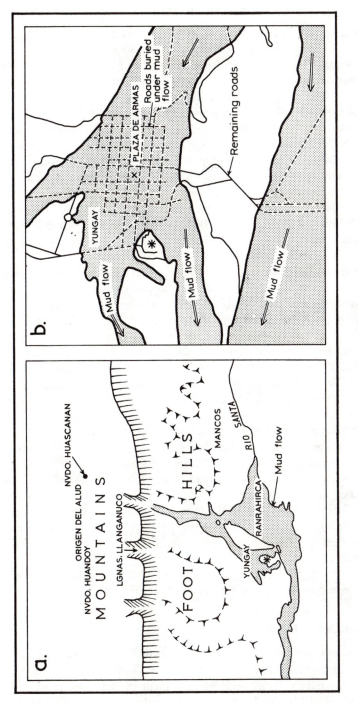

FIG. 6.1. Landsliding at Yungay, Peru

a. General situation
b. Flow details at the site of Yungay
   (compiled from aerial photographs)

engineering. This also reflects the breadth of active interest in the subject. Thus Sharpe's classic study of 1938 deals mainly with the geological and geomorphological aspects of landslides, while Zaruba and Mencl (1969) include greater reference to engineering aspects of landslide studies. Terzaghi (1950) reviewed the mechanisms of landsliding, and saw the roles of the geologist and engineer as being complementary to each other. The stability of slopes composed of coherent, massive, bedrock are discussed in many publications concerned with mining and quarrying, a good example being that edited by van Rensburg on *Planning Open Pit Mines* (1970). The analysis of sliding in less coherent beds (e.g. clays) and soils is described in many texts, including the frequently cited work by Terzaghi and Peck, first published in 1948 but revised in 1967, and more recent books such as Lambe and Whitman (1969), and Wilun and Starzewski (1972). The more important papers include those by Hutchinson (1967, 1969) and Peck (1967), together with those mentioned in the rest of this chapter.

Specialist volumes include that edited by Eckel (1958) on *Landslides and Engineering Practice*, which contains many useful case studies from the United States. In addition there are a number of conference reports, such as those of the International Conference on Soil Mechanics and Foundation Engineering; and the regional conferences of the International Society for Soil Mechanics and Foundation Engineering. Of the latter those held in Europe have included the subjects of the stability of earth slopes, published in *Géotechnique*, Vol. 5 for 1955; pore pressures and suction in soils, with a report volume published by Butterworths in 1961; and a conference on the shear strength properties of natural soils and rocks held in Oslo in 1967, with the proceedings published in two volumes.

Important journals in this field include *Géotechnique*, the *Quarterly Journal of Engineering Geology*, and that of the Soil Mechanics and Foundations Division of the American Society of Civil Engineers, who published a special volume of the papers presented at the Research Conference on the Shear Strength of Cohesive Soils held in Boulder, Colorado, in 1960. The papers delivered at the Speciality Conference on the Stability of Slopes and Embankments held in Berkeley, California in 1966, were printed in the *Proc. ASCE 93*, No. *SM4*, July 1967.

A comprehensive bibliography on slope stability, compiled by Hoek, appears at the end of the volume by van Rensburg (1970).

6.2. Forces Producing Instability

There is seldom only one cause of slope instability. Instability usually results from a sequence of events that ends with downhill movement. Landslides occur because the forces creating movement exceed those resisting it (shear strength of materials). In general it is important to distinguish those factors that contribute to increasing shear stress (disturbing forces) (Table 6.1) from those that contribute to low shear strengths within the materials

TABLE 6.1

Factors leading to an increase in shear stress

1. Removal of lateral or underlying support
   —undercutting by water (e.g. river, waves), or glacier ice
   —weathering of weaker strata at the toe of the slope
   —washing out of granular material by seepage erosion
   —man-made cuts and excavations, draining of lakes or reservoirs
2. Increased disturbing forces
   —natural accumulations of water, snow, talus
   —man-made pressures (e.g. stock-piles of ore, tip-heaps, rubbish dumps, or buildings)
3. Transitory earth stresses
   —earthquakes
   —continual passing of heavy traffic
4. Increased internal pressure
   —build up of pore-water pressures (e.g. in joints and cracks, especially in the tension crack zone at the rear of the slide).

(resisting properties) (Table 6.2). Evaluating each of these in any one situation may then lead to both a valid assessment of specific causes and enable the relative balance between shear stress and shear strength to be defined (Eckel, 1958).

TABLE 6.2

Factors leading to a decrease in shearing resistance

1. Materials
   —beds which decrease in shear strength if water content increases (clays, shale, mica, schist, talc, serpentine) (e.g. when local water table is artificially increased in height by reservoir construction, or as a result of stress release (vertical and/or horizontal) following slope formation
   —low internal cohesion (e.g. consolidated clays, sands, porous organic matter)
   —in bedrock: faults, bedding planes, joints, foliation in schists, cleavage, brecciated zones, and pre-existing shears
2. Weathering changes
   —weathering reduces effective cohesion (c'), and to a lesser extent the angle of shearing resistance (ϕ')
   —absorption of water leading to changes in the fabric of clays (e.g. loss of bonds between particles or the formation of fissures)
3. Pore-water pressure increase
   —higher groundwater table as a result of increased precipitation or as a result of human interference (e.g. dam construction) (see 1 above)

In general terms the stability of a slope may be defined by a safety factor F_s, where:

$$F_s = \frac{\Sigma \text{ forces resisting slope failure}}{\Sigma \text{ disturbing forces}}.$$

The following discussion illustrates how difficult it may be to define and measure these adequately. If $F_s > 1 \cdot 0$ stability is likely, but if $F_s < 1 \cdot 0$ instability exists. The calculation of F_s depends upon the on-site measurement of the critical engineering parameters of the materials. This stage is only required, however, after good reasons have been found for suspecting that mass movement is likely to occur, or if an engineering project requires specific knowledge about slope stability conditions at a site.

Predicting the likelihood of mass movement taking place is usually based on an analysis of the landforms, geology, and geomorphological processes to be found in the area, and this includes past processes. To emphasize this last point some examples may be given. For instance, it may be important to recognize areas of previous slope steepening by glacial erosion, perhaps followed by glacial deposition, as in the case of the landslides in Snoqualinie Pass, Washington State, in August 1953. During the Pleistocene a glacier cut a steep slope into highly-fractured graywacké. The glacier snout then retreated and a landslide began, the toe of which was subsequently removed by a readvance of the glacier later in the Pleistocene. With the next and final retreat of the ice the slide was partially buried and left supported by a 3 m thickness of glacial till. The resulting slope remained stable until excavation when road works in the foot of the old slump removed much of the supporting material. This reactivated the instability and severely damaged the construction scheme (Ritchie, 1958).

Another legacy from past processes which has caused frequent engineering difficulties is that of solifluction features. Under periglacial activity, solifluction is a common form of slope instability with the seasonal thawing and saturation of the surface debris layers causing mass movement (Chap. 9). Movement can take place in one of three ways: (i) as viscous flow, when the soil is extremely wet and greatly exceeds its liquid limit (Sect. 6.6biv); (ii) by sliding on a shear plane, the water content of the soil being up to or even above the liquid limit; (iii) freeze-thaw movements bringing about the down-slope creep of unsorted materials, and which can operate on slopes of as little as 2°. Stability returns both as the periglacial climate is replaced by warmer climates with no seasonal freeze–thaw and less extreme groundwater conditions, and as the toes of the solifluction lobes or sheets come to rest at the foot of slopes. Reactivation of the sliding type of movement (where a shear surface of low shear strength already exists) may begin if excavations decrease support from below by re-steepening any part of the lobe (especially the toe), or by loading the slope to increase the pressure exerted on the solifluction material. Considerable trouble was caused during the construction of the Sevenoaks bypass in south-east England because fossil solifluction forms were reactivated by loading with a bank and then by excavations into the lobe (Weeks, 1969; Higginbottom and Fookes, 1970).

Around Bath, England, Lias clays are overlain by limestones. In Pleistocene times cambering, valley bulging, and mass movement by sliding took place under periglacial conditions; the cambering taking place on the interfluves where the more coherent beds now curve downslope more steeply than the general dip, and the bulging of less coherent materials (e.g. clays) occurring in the valley floors where periglacial activity caused these materials to be thrust upwards and contorted. As later urban development occurred some movement was reactivated, and now old slips are stabilized (mainly by drainage) before being used as building sites. Slopes exceeding 15 ° generally remain unstable, and have been designated as public open spaces within the city boundary (Kellaway and Taylor, 1968). Shallow movements still take place on the slopes around Bath, as they did during the heavy rains of July 1968.

It is imperative, therefore, that in evaluating any area for proposed engineering works due consideration should be given to the recognition not only of signs of present instability but also of instability in the past.

6.3. Materials and Landslides

Strata composed of cohesive rock are not prone to landsliding in the same way as less cohesive beds and regolith. In the former the presence of weak elements, such as lineaments (including joints, faults, or bedding planes) or bands of weaker material are critical in determining the overall ability of the bedrock to resist landsliding. Less cohesive materials (e.g. dry sands, clays) tend to respond to instability conditions by shearing or deformation within themselves.

(a) Bedrock materials

Landsliding in coherent predominantly unweathered bedrock is not caused by the lack of any inherent strength in the rock itself but is due to the presence of internal weaknesses, such as joint planes and faults. Whatever the causes of a failure in bedrock the actual plane of movement will very largely be determined by the position of the weakest zones (e.g. joints).

One of the important controls of bedrock instability is the angle at which the major joints intersect the ground surface. Some examples are shown in Figure 6.2, and illustrate the importance of the relation between the angle of friction (at which under gravity sliding would occur) along the joint plane (ϕ_f) and the inclination of the joints (α). Even if $\alpha < \phi_f$ the slope could stand at any angle up to 90 ° (Fig. 6.2a), but if $\alpha > \phi_f$ then gravity would induce movement along the joint plane and the hillside would not exceed the critical angle (ϕ_c) which would have a maximum value equal to the inclination of the joint planes (case b). It is seldom that bedrock displays only one set of joints. Examples c–e in Figure 6.2 illustrate the relations between ϕ_f, α, and ϕ_c when a secondary set of joints is also present.

The angle of friction is determined by a number of different characteristics along the discontinuities (here referred to under the term joints, but this could apply also to faults, fissures, bedding planes, and shear planes). Other contributing characteristics include the continuity of the joints within the rock

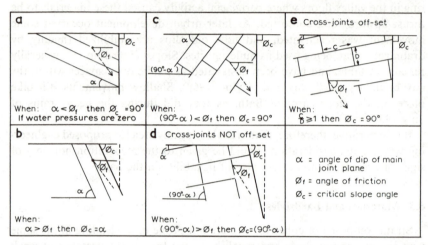

FIG. 6.2. Some possible relationships between joint planes in bedrock and slope stability (based on Terzaghi, 1962)

mass, the roughness of the joint planes, and their undulations (waviness). Roughness and waviness can be defined as shown in Figure 6.3 (but note that a slickensided plane (roughness category 1) implies that it has been polished by previous shearing). In some instances the most important control on the stability of the bedrock is not so much the joints themselves but the weathering products (or in the case of a fault, the breccia) which may fill them. This gouge material may separate two bedrock surfaces from each other, and slope stability will depend on the shear strength of the gouge rather than on the plane to plane contact of two bedrock surfaces. In some cases the gouge/bedrock interface may have even lower shearing strengths than the gouge itself. With respect to the presence of gouge material on failure, four situations are possible (Piteau, 1970): (i) sliding plane passes entirely through gouge, and shear strength is dependent upon gouge material only; (ii) sliding plane passes partly through gouge, partly through joint wall, and shear strength a complex made up of contributions from both; (iii) gouge is very thin, and only marginally modifies the shear resistance of the joint plane; (iv) no gouge, and shear strength is entirely that of the joint plane.

During a field survey of joint spacing, direction, and inclination, and the nature and thickness of any gouge materials, it is also worth recording the hardness of both the bedrock and the gouge material. This can be done on a qualitative scale such as that given in Table 6.3, and is likely to be related not only to the lithological nature of the material but also to its degree of weathering (see Chap. 11 and the examples in Chandler, 1969, 1972). This qualitative scale is interpreted by Robertson (1970) in terms of the minimum likely unconfined compressive strength of the materials, which is a measure of its resistance to internal shearing (Fig. 6.4). In areas which a field survey shows to be of limited stability samples of these materials can be taken for a laboratory analysis of their strength, mineral, and structural properties. The analysis of the data collected on such a field survey is described by Robertson (1970).

TABLE 6.3

A scale for the field assessment of material hardness

S1 *Very soft*—easily moulded in fingers; shows distinct heelmarks
S2 *Soft*—moulds in fingers with strong pressure; faint heelmarks
S3 *Firm*—very difficult to mould in fingers; difficult to cut with a hand spade
S4 *Stiff*—cannot be moulded in fingers; cannot be cut with hand spade and requires hand-picking for excavation
S5 *Very stiff*—very tough and difficult to move with handpick; requires pneumatic spade for excavation
R1 *Very soft rock*—material crumbles under firm blows with a sharp end of a geological pick and can be peeled off with a knife; it is too hard to cut a triaxial sample by hand
R2 *Soft rock*—can just be scraped and peeled with a knife; indentations 1·5 mm to 3 mm show in the specimen with firm blows of the pick point
R3 *Hard rock*—cannot be scraped or peeled with a knife; hand-held specimen can be broken with a hammer end of a geological pick with a single firm blow
R4 *Very hard rock*—hand-held specimen breaks with hammer end of pick under more than one blow
R5 *Very, very hard rock*—specimen requires many blows with geological pick to break through intact material

Source: Piteau (1970).

The orientation of joints is also important in determining the passage of water through the bedrock. This movement of water is related to the degree to which joints are continuous or form an interlocking network. Free water flow prevents the build-up of locally excessive water pressures, and thus decreases the likelihood of instability. The underground continuity of joints is usually difficult to determine. However, discontinuous joints do not normally increase the tendency towards instability.

Within joints it is not the quantity of water but the water pressures which are important, for as the latter increase so does the shear resistance decrease. Since bedrock frequently contains joints running in several directions, providing predetermined routes for water movement, they exhibit anisotropic and non-homogeneous flow properties. In this respect they differ from soils, unless the joint spacing is close and there is a random hydraulic connection between the joints (Lyell, 1970).

The shear strength of a joint plane is related also to bedrock mineral

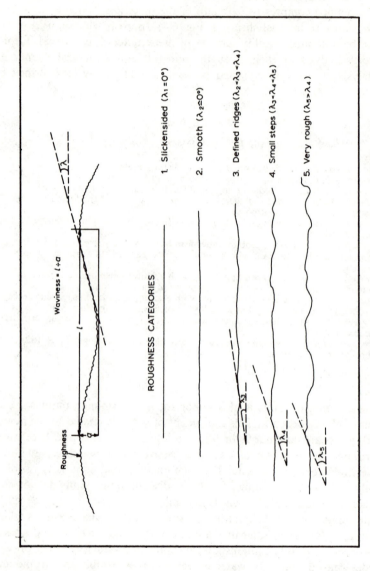

Roughness

Waviness = $i \div a$

ROUGHNESS CATEGORIES

1. Slickensided ($\lambda_1 = 0°$)

2. Smooth ($\lambda_2 \simeq 0°$)

3. Defined ridges ($\lambda_2 < \lambda_3 < \lambda_4$)

4. Small steps ($\lambda_3 < \lambda_4 < \lambda_5$)

5. Very rough ($\lambda_5 > \lambda_4$)

FIG. 6.3. The roughness and waviness characteristics of a joint plane (after Piteau, 1970)

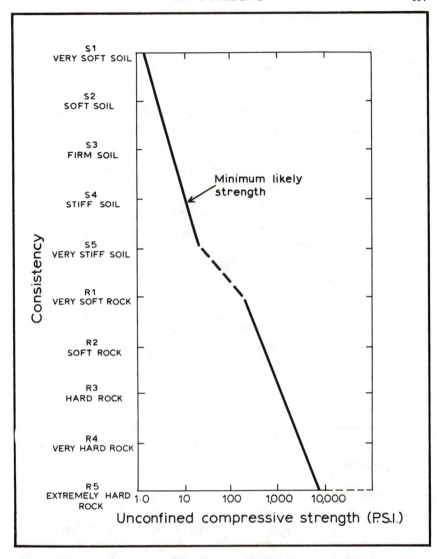

FIG. 6.4. Relationship between material hardness (see Table 6.3) and its minimum unconfined compressive strength (based on Robertson, 1970)

composition. For example, quartz and calcite increase their resistance to
sliding if they become moist, but talc-like minerals are less resistant to sliding
when they are wet. Weathering along the joint plane can have important
effects. Montmorillonite (a clay mineral) has a low shear strength, and
swells exceptionally on wetting, making for potentially unstable conditions.
Soluble minerals, such as rock salt, gypsum, limestone, or dolomite can be

dissolved and cause a decrease in shear resistance with time as they are removed along the joint plane (Piteau, 1970). Conversely they increase shear strengths if they are deposited within joints.

Three categories of rock slope failure are illustrated in Figure 6.5. These show how joints (or any other physical discontinuity) can control the shape of both local (case a) and large-scale (case b) failures. If these discontinuities are very close together they allow deep and intense bedrock weathering to

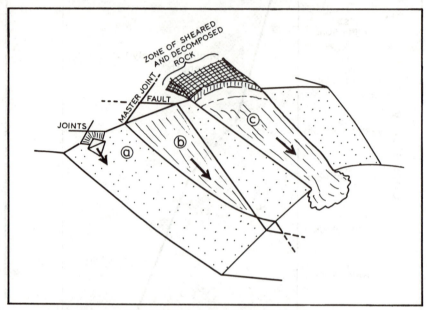

FIG. 6.5. Three types of bedrock failure
a. Local failure of joint wedge
b. Large-scale wedge failure
c. Mass movement within a zone of low shear strength (based on Patton, 1970)

occur, degrading the original bedrock strength, changing its deformability properties and permeability characteristics, and producing a complex of (i) residual soil, (ii) weathered rock, and (iii) unweathered rock. Such zones have a major influence on groundwater flow, and may cause excess pore-water pressures to build up, thus leading to instability (case c). Once movement has taken place, in any type of bedrock failure, the residual strengths will be much lower than the initial resistance to shearing, and subsequent movements occur with less inducement. The observation of an initial small displacement is thus critical in the assessment of bedrock slope stability.

Some materials that are generally classified as bedrock, such as clay beds, may behave in a similar manner to soils, and their stability may depend on criteria similar to those examined below for soils. Clays tend to be distinctive both in their physical properties and their behaviour on sliding. For example, stiff fissured clays may have a shearing resistance of between 10 and 20 tons (metric) per square metre, which is their peak strength as measured by laboratory tests on small samples. However, if a cut is made into the clay (e.g. by excavation, or by an incising stream) this value may fall to as little as 3 tons/m². This is because, as the cut develops, stresses in the clays are relaxed and joints open up into which rainwater can pass. This in turn increases the pore-water pressures in the clay, which can build up to high internal values over the years, independent of the season of the year or the dampness of the clays at the surface. In addition the water can cause swelling, clay blocks to break up, and shearing resistance to decrease. As soon as it becomes equal to the average shearing stress on a potential surface of sliding, the slope fails (Terzaghi, 1950). Such failures can occur long after the cut was made. For example, the banks of a railway cutting through stiff, fissured, Weald Clay at Sevenoaks, Kent, suffered sliding seventy years after the cut had been excavated.

In an engineering situation, such as in an excavation for a cutting, it is important to distinguish between 'short-term', 'intermediate', and 'long-term' failures. After excavation has taken place in a clay slope a new reduced state of stresses applies to the slope materials. Over time this enables them to increase their water content and a swelling of the clay will occur. 'Short-term' failures take place soon after excavation and before the water content has changed. 'Intermediate' and 'long-term' failures occur largely as a result of the swelling that follows excavation and the reduction in stress. Because of the low permeability of many clays, though not those containing silt and sand layers, the swelling of the clay towards a new equilibrium pore pressure is a comparatively long process that could well occupy the seventy years of the Sevenoaks Weald Clay example. As swelling and water content increase so shear strength decreases (see below). If sliding occurs during the swelling process it is classified as an 'intermediate' failure; if it occurs at or near the end of the swelling period it is a 'long-term' failure. Laboratory measurement of the shear strength of materials in this situation over-estimates actual shear strength values. This is largely because the small laboratory sample does not adequately represent the physical controls of failure (e.g. fissures running through the material) in the slope itself.

The time-lags between slope exposure and subsequent failure in clays appear also to be related to slope steepness. Skempton (1948) found that in the highly colloidal, stiff, fissured London clay, a vertical slope with a height of 6 metres may stand for several weeks before sliding takes place. A slope of 1 in 2 (25°), but of the same height, tends to fail after 10 to 20 years. Progressively gentler slopes take longer to fail, so that a slope of 1 in 3 (18 °) is likely to remain stable for up to 50 years. The steepest natural slopes in London clay are of the order of 1 in 6 (8 °), at which relative stability is achieved; these being failed slopes which have reached equilibrium.

(b) Soils

The shear strength of a soil is the maximum available resistance that it has to movement induced by shear stress. This shear strength (s) is a function of the friction at grain-to-grain contact, and is related to the amount of grain interlocking, which increases with the angularity of the material, and the density of packing (consolidation) of the grains. These control an important parameter in engineering studies of soils, known as the angle of shearing resistance (ϕ'). In addition, shear strength is related to the effective pressure transmitted between particles (σ') and their effective cohesion (c').

For *normally consolidated* soils, with no internal cohesion:

$$s = \sigma' \tan \phi' \qquad (6.1)$$

For *overconsolidated* soils (i.e. subject to over-burden pressure) and displaying some effective cohesion:

$$s = c' + \sigma' \tan \phi' \qquad (6.2)$$

where s is the shear strength at the potential slip surface.

These values are illustrated in Figure 6.6. If water, under pressure u, is present in the pores of the soil then σ will be reduced (i.e. $\sigma' = \sigma - u$) for a *saturated* soil (Fig. 6.6c); the relationships are more complex if the soil is only partly saturated and either air or gas bubbles are present.

When shear stress rises to the value of shear strength then displacement occurs between two parts of the soil body, usually along a well-defined rupture plane. Once movement has taken place along a slip surface there is a significant decrease in the shear strength of the material at the slip surface, for a given effective pressure, and subsequent movements will take place with lesser induced stress. Figure 6.6d illustrates how, after movement, the residual values of ϕ' (ϕ'_r) are lower in both cohesionless and cohesive soils, while in the latter the effective cohesion is reduced to a residual value (c'_r). Any further stability calculations using Equations 6.1 and 6.2 thus need to use ϕ'_r and c'_r values after movement has taken place. The difference between peak (before displacement) and residual values for these two parameters varies with the texture of the material, generally increasing with the clay content (Skempton, 1964). Once movement has taken place the clay particles become reoriented at the slip surface, tending to parallel each other and the direction of movement, thus making further movement much easier. In addition stiff clays may show an increase in water content associated with a dilation of the soil within the zone of shearing.

It was suggested above that stratified clays behave like soils rather than like more coherent bedrock. Figure 6.6c is modified from a study of clay slopes by Skempton (1964). Using the parameters defined in this diagram

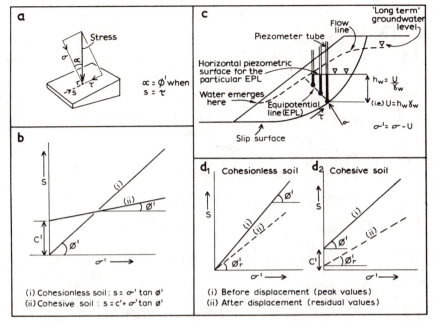

FIG. 6.6. Some fundamental principles of soil mechanics which affect slope stability analysis

a. Stress at a point can be divided into two components, the pressure normal to the slope (σ) and the shear stress (τ) which operates in the same plane but in the opposite direction to the shear strength (s). When $s = \tau$, the angle α = angle of shearing resistance (ϕ')

b. Shear strength (s) against effective pressure (σ') for cohesionless and cohesive soils (ϕ' may or may not be greater for the one than the other according to local conditions)

c. Relationships on a natural clay slope in hydrostatic equilibrium with the groundwater (based on Skempton, 1964)

(N.B. h_w = piezometric head
γ_w = specific weight of water)

d. The relation between peak and residual values of ϕ' and c'
In (d_1), $\phi'r$ is due only to dilation (which involves a decrease in density and an increase in water content) relative to density in the ϕ' state. In (d_2), $\phi'r$ is due to dilation plus a loss of strength resulting from the reorientation of platey clay particles

Skempton found that on a stable slope the resistance to sliding offered by the clay along the slip surface (i.e. its shear strength s) is given by:

$$s = \bar{c}' + (\sigma - u) \tan \bar{\phi}' \qquad (6.3)$$

where $\bar{c}'$ = cohesion intercept (see Fig. 6.6b) and $\bar{\phi}'$ = angle of shearing resistance. $\bar{c}'$ and $\bar{\phi}'$ are average values of c' and ϕ' around the slip surface and both of these are expressed in terms of the effective stress. Based on this relationship the average shear stress ($\bar{\tau}$) may be defined as:

$$\bar{\tau} = \frac{\bar{c}'}{F} + (\sigma - u)\frac{\tan \bar{\phi}'}{F} \qquad (6.4)$$

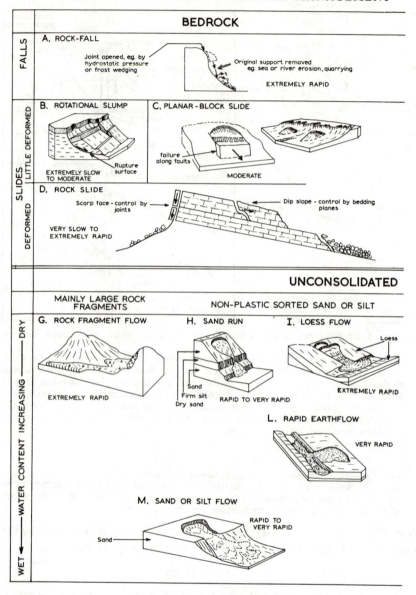

FIG. 6.7. A classification of landslides (after Varnes, 1958)

where F is a factor of safety. When a slip occurs $F = 1\cdot0$ and the total shear force equals the total shear strength.

This type of approach to the study of mass movement through the engineering properties of materials is illustrated by a case study in Section 6.7b. It is

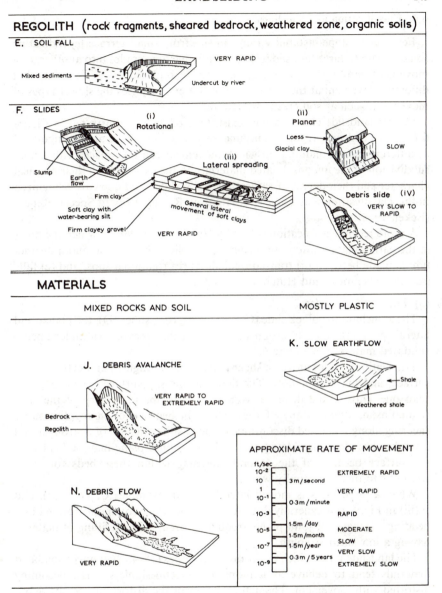

REGOLITH (rock fragments, sheared bedrock, weathered zone, organic soils)

E. SOIL FALL

VERY RAPID

Mixed sediments

Undercut by river

F. SLIDES

(i)
Rotational

(ii)
Planar

Loess

Glacial clay

SLOW

(iii)
Lateral spreading

Slump

Earth flow

Firm clay

Soft clay with water-bearing silt

Firm clayey gravel

General lateral movement of soft clays

Debris slide   (iv)

VERY SLOW TO RAPID

VERY RAPID

MATERIALS

MIXED ROCKS AND SOIL

MOSTLY PLASTIC

K. SLOW EARTHFLOW

J.   DEBRIS AVALANCHE

VERY RAPID TO EXTREMELY RAPID

Shale

Bedrock

Regolith

Weathered shale

APPROXIMATE RATE OF MOVEMENT

N. DEBRIS FLOW

ft/sec		
10^{-2}		EXTREMELY RAPID
10	3 m / second	
1		VERY RAPID
10^{-1}	0.3 m / minute	
10^{-3}		RAPID
10^{-5}	1.5m /day	MODERATE
	1.5m /month	SLOW
10^{-7}	1.5m /year	VERY SLOW
10^{-9}	0.3 m / 5 years	EXTREMELY SLOW

VERY RAPID

an approach which is critical at the site analysis stage of an engineering project. For example, it provides valuable information about the physical characteristics of materials liable to landsliding. Nevertheless it is a time-consuming approach which is most efficiently employed after an analysis of the ground has indicated the sites of potential or actual instability.

6.4. Classification of Landslides

Features of slope instability range in size from small terracettes indicating soil creep up to large landslides. Frequently one type grades into another, or a movement can start in one way only to be replaced by a secondary but different movement at the same site. For present purposes the slower types of movement, such as soil creep, are ignored.

Several landslide classifications exist. One of these is illustrated in Figure 6.7. In this bedrock, regolith, and unconsolidated materials are distinguished as different parent materials in which movement can take place, with a finer subdivision of the unconsolidated materials on the basis of both their physical character and their degree of saturation. Falls, slides, and flows are recognized in this classification made in a U.S. Highway Research Board Special Report (Eckel, 1958).

A more recent classification is that by Hutchinson (1968) in which the flows in the earlier classification are recognized as slide phenomena, and a distinction is made between (a) translational slides (b) rotational slips, and (c) falls (see also Skempton and Hutchinson, 1969).

(a) Translational slides

These include the block slide (C in Fig. 6.7), rock slides (D), the planar and lateral spreading slides of group F, and all of the types shown under unconsolidated materials in Figure 6.7.

Translational slides move on shear planes which roughly parallel the ground surface, and may be shallow. The depth to the slip surface is generally less than one-tenth of the distance from the toe to the rear scarp of the slide (Hutchinson, 1968). Many of these slides in bedrock or in regolith suffer shearing along a marked discontinuity, such as the junction of a stronger and a weaker bed. In consolidated materials the failure plane may be kept near the surface because of the general tendency within these beds for shear strengths to increase rapidly with depth.

When a large mass is moved laterally (as in examples C, D, F (ii), and F (iii) in Fig. 6.7) it generally does so on a slip surface commonly formed by a bedding, cleavage, or joint plane, frequently occupied by a filling of material having a low shear strength.

The landslide types which occur on steep slopes (15–40 °) in unconsolidated materials tend to behave as a more or less cohesionless mass, becoming distorted with movement. The actual details of the distortion and the depth of the slide (mostly referred to as flows in the earlier classification illustrated in Fig. 6.7) depend on the nature of the slide debris. Sand runs involve dry, cohesionless granular material experiencing only shallow movement, whereas clays, being more cohesive, tend to move in less distorted units. The trigger mechanism for the movement of slides in the unconsolidated materials frequently appears to be a heavy rainfall which creates high pore-water pressures in the slide material. In the earthquake zones of the world the mechanical vibration induced by the shock waves of an earthquake can also bring about movement (Simonett, 1967).

Many of the movements in unconsolidated materials tend to be sited within valley-side depressions, where weathered materials have accumulated, or where the concentration of ground water has locally increased pore-water pressures. When slides of this type debouch onto inhabited plains or mountain footslopes they can be especially disastrous (Fig. 6.1).

(b) Rotational slips

Rotational slips occur with a concave-upwards curved failure plane (rupture surface) passing through a thick and relatively homogeneous bed of clay or shale (Hutchinson, 1968). They are more deep-seated than translational slides. Movement tends to be rotational on this slip surface and thereby induces a backward tilt which is shown by an inward-sloping head to the moving mass (e.g. example B in Fig. 6.7) and heaving at its toe. The main features of a rotational slide are shown in Figure 6.8.

Although the rupture surface tends to be semi-circular this form is seldom perfectly achieved since its shape is influenced by original structures such as faults, bedding planes, and joints. The back-tilting at the head of the mobile mass forms hollows in which water can collect. This tends to drain down the rupture surface and maintain permanently high pore pressures. By this means it may to some extent be self-perpetuating as an area of instability.

(c) Falls

Rock-falls may occur where a steeply-sloping rock-face is characterized by well-jointed rock. Rock-falls tend to be generated in three stages. The first sees the creation of the cracks from which the block eventually breaks. These cracks become enlarged and the fall takes place because support has been removed from the base of the rock. The crack enlargement process arises mainly from pressures related to the freezing of water or the growth of plant roots, or directly as a result of gravity. Undercutting at the base may be the result of erosion by the sea, rivers, differential weathering of a softer layer that lies beneath the capping rock or human interference with the slope. Rock-fall debris may be removed by processes working at the base of the slope, in which case further falls may occur, or it may accumulate to form a talus footslope.

Soil falls can occur wherever a soil bank or terrace face is undercut. The unstable unit is initially separated from the parent cliff by tension cracks, before its abrupt collapse as shear failure occurs at the base of this unit.

All falls tend to be confined to the surface zones in which the effects of pressure release and of pressure build-up through water freezing can be most

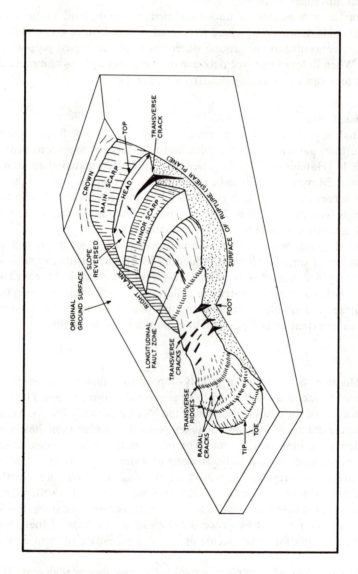

FIG. 6.8. The main features of a rotational slide (after Varnes, 1958)

effective. In addition it is here that water pressures can be built up behind a veneer of ice on the surface of the slope.

6.5. A Morphometric Approach to the Study of Landslides

The illustration in Figure 6.7 suggests that the nature of landslide movement is indicated by the forms which result. This being the case it should be possible to measure the form characteristics of a landslide and arrive at some conclusions about the nature of the mechanisms at work. Such evidence would be supplementary to a geotechnical examination of the mechanical properties of selected slides, but it would allow a preliminary classification to be made of the landslides of an area. This would make it possible to detect quickly if significantly different modes of failure had occurred amongst these slides.

Skempton (1953) indicates a distinction between 'surface slips', 'deep rotational slips', and 'slumps' on the basis of the ratio between the depth of the slide and its overall length. Crozier (1973) takes this principle much further. He defines a number of unique indices based on the morphometry of a landslide (Fig. 6.9). Of these he identifies the ratio of depth (D) to length (L) as the critical ratio indicative of the primary process responsible for a particular landslide.

Crozier acquired data for sixty-six landslides in New Zealand which he classified according to the five groups which are listed in Table 6.4 (although the 'flows' would have been more precisely described if they had been called 'slides'). There is a statistically significant difference (at the 99 per cent confidence level) for each type of landslide, between the mean values of the form (D/L), flowage, displacement, and tenuity indices (Fig. 6.9).

TABLE 6.4

The process groups for landsliding used by Crozier (1973)

Process Group	Class of movement
Fluid flow (F)	mud flows, debris flows, debris avalanches
Viscous flow (V)	earth flow, bouldery earth flow
Slide flow (S)	slump/flow
Planar slide (P)	turf glide, debris slide, rock slide
Rotational slide (R)	earth and rock slump

Source: Crozier (1973).

It was suggested by Crozier that selected pairs of indices might be used to discriminate between the landslide types defined in Table 6.5. To test this hypothesis Figure 6.10 has been compiled from Crozier's data to show the relationship between the values for D/L and the flowage, displacement, and tenuity indices in turn. The rotational and planar slides do not provide data

a

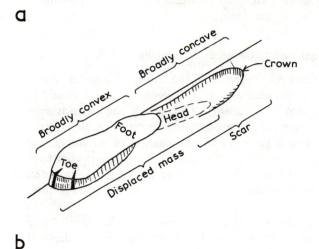

b

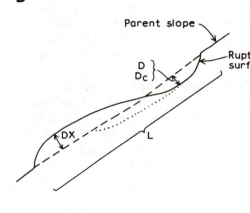

FIG. 6.9. Morphometric indices for landslides (after Crozier, 1973)

a. Main classification index $= \dfrac{D}{L} \times 100\%$

   Dilation index $= \dfrac{Wx}{Wc}$

b. Flowage index $= \left(\dfrac{Wx}{Wc} - 1\right) \dfrac{Lm}{Lc} \times 10$

   Displacement index $= \dfrac{Lr}{Lc}$

c. Viscous flow index $= \dfrac{Lf}{Dc}$

   Tenuity index $= \dfrac{Lm}{Lc}$

c

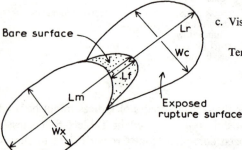

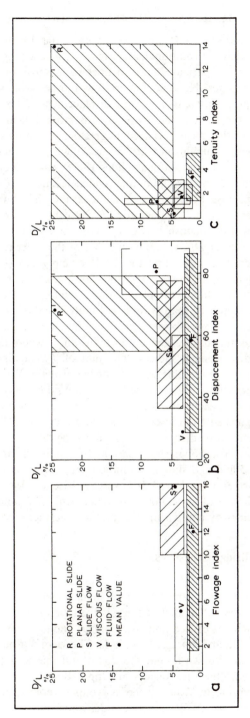

FIG. 6.10. Pair-wise discrimination between landslides on the basis of selected morpho-metric indices
The mean values of the indices are shown by a dot. The enclosed areas define the possible range of values lying within one standard deviation from the mean
(data from Crozier, 1973)

for the flowage index and so do not appear on Figure 6.10a. From 6.10b and c, however, it is clear that the rotational slides tend to have a range of index values which for the most part are quite distinct from those of the other landslide types. The only possible area of overlap with other groups is that of low D/L values coupled with either a high displacement index or a low tenuity index. For all the other types, however, it is more difficult to justify their discrimination by this means. Indeed their close grouping on Figure 6.10 is in keeping with Hutchinson's classification whereby these are all considered to be forms of translational slides (Sect. 6.4a). Nevertheless, within this general group some discriminating tendencies may be discernible. For example, a comparison between the D/L index and the flowage index suggests that in general the data values for 66 per cent (= two standard deviations) of cases will distinguish between viscous and slide flows. Fluid flows will tend to have lower D/L values than either of the other two groups. A comparison between D/L values and the displacement index (Fig. 6.10b) suggests that in general slide flows and fluid flows may be successfully discriminated from each other, with planar slides tending to have higher pairs of values for these indices than would be found in either of the other two groups. However, viscous flows would be difficult to separate from fluid and slide flows on this basis except in those instances where the viscous flows had very low displacement indices. The plot of D/L against tenuity index (Fig. 6.10c) shows considerable overlap for the probable range of data values for all but the rotational slides. Slide flows and fluid flows could probably be distinguished from each other on this basis, but neither are likely to be successfully distinguished from viscous flows, and only about half of the planar flows would tend to be distinguishable from slide flows.

In general, therefore, rotational slides should be readily distinguishable from the other types. Fluid and slide flows may, for the most part, be separated because the former have lower D/L values. Viscous flows are best distinguished from fluid and slide flows by comparing D/L and flowage indices; and planar slides by comparing D/L and displacement indices.

6.6. Managing (Stabilizing) Landslides

(a) Introduction

As far as their effects upon environmental management are concerned landslides are of three types. The first of these is the type of slide which was not predicted and which has caused immense damage and grief. Two of these were the slides at Aberfan and Yungay (Sect. 6.1). The second group consists of those slides which are known about, the threat and the need to control them being appreciated. One such occurs at Upper Boat, about 2 km up-valley of South Pontypridd in Wales, where a group of comparatively new houses stands just below the toe of an active slide, whose movement is being controlled by drainage pipes releasing the high internal pore-water pressures. There are many examples of this type of landslide around the world. The third type important in environmental management is that which occurs either along a proposed transport route (road, railway, or canal) or which potentially influences the site of a projected development (e.g. reservoir dam, office block, housing estate, or industrial development site).

The geomorphological contribution to the management of each of these three groups lies in (i) predicting which areas are susceptible to landsliding, and (ii) identifying the controlling environmental characteristics of known landslides. The former involves both the prediction of landsliding of the catastrophic type (i.e. the first group) and those likely to be encountered during land development. The identification of important environmental characteristics (cf. Tables 6.1 and 6.2) relates especially to the second and third groups, but must of necessity form the basis of a search for landslides of the first group. The physical control of landslides is an engineering problem whose solution lies in the correct identification of the causes of movement.

(b) Identifying landslide-prone areas

In order to identify landslide-prone areas it is relevant to distinguish between situation and site. Those situations where landslides may be expected to occur include mountain fronts, steep-sided mountain valleys, and steep hill-slopes and scarps. By contrast, situations not susceptible to large amounts of mass movement include relatively flat areas without deep river incision. Site, on the other hand, refers to the location of a particular landslide, for example along a specified steep slope. General relief categories (e.g. intensely dissected mountains) are sufficient if it is necessary only to identify those areas within which sliding is likely to occur. These closely approximate to the land systems to be defined and described in Chapter 13, and the land-systems approach as a whole may be validly employed in situation mapping. One of the advantages of this approach is that land-systems boundaries can be defined from aerial photographs, thus saving much time and expense in field mapping.

In many countries advance knowledge already exists about the areas which are susceptible to sliding, but it is frequently not appreciated which sites are potentially liable to mass movement. For such a site investigation full use can be made of a checklist (e.g. Table 6.5) whereby each separate and discrete slope unit is classified according to a stability rating on a scale from stable through degrees of potential instability to those slopes which have already failed. In the context of Section 6.2 it is clear that attention also has to be paid to old mass movements which could become reactivated. On the other hand those slopes which are geomorphologically speaking 'old' and which show no signs of instability are unlikely to be the site of landsliding in the near future. The only way in which they can become so is through

TABLE 6.5

A check list for sites liable to large-scale instability

RELIEF

Valley depth	Small ☐	Moderate ☐	Large ☐	Very large ☐
Slope steepness	Low ☐	Moderate ☐	Steep ☐	Very steep ☐
Cliffs	Absent ☐			Present ☐
Height difference between different valleys	Small ☐	Moderate ☐	Large ☐	Very large ☐
Valley-side shape	Spur ☐	Straight ☐	Shallow cove ☐	Deep cove ☐

DRAINAGE

Drainage density	Low ☐	Moderate ☐	High ☐	Very high ☐
River gradient	Gentle ☐	Moderate ☐	Steep ☐	Very steep ☐
Slope undercutting	None ☐	Moderate ☐	Severe ☐	Very severe ☐
Concentrated seepage flow	Absent ☐			Present ☐
Standing water	Absent ☐	Present at local base level ☐	Present slowly draining ☐	Present rapidly draining ☐
Recent incision	Absent ☐	Small ☐	Moderate ☐	Large ☐
Pore-water pressure	Low ☐	Moderate ☐	High ☐	Very high ☐

BEDROCK

Jointing density	Low ☐	Moderate ☐	High ☐	Very high ☐
Direction of major joints (faults or bedding planes) with respect to steepest slopes	Away ☐		Normal ☐	Towards ☐
Amount of dip (or steepness of joint and/or fault planes)	Horizontal ☐	Small ☐	Moderate ☐	Large ☐
Strong beds over weak beds	Absent ☐			Present ☐
Degree of weathering	None ☐	Small ☐	Moderate ☐	Large ☐
Compressive strength	High ☐	Moderate ☐	Low ☐	Very low ☐
Coherence (particularly of lower beds)	High ☐	Moderate ☐	Low ☐	Very low ☐

SOILS(incl. drift materials)

Site	Valley floor ☐	Gentle slopes ☐	Moderate slopes ☐	Steep slope ☐
Coherent over incoherent beds	Absent ☐			Present ☐
Angle of rest	Low ☐	Moderate ☐	Steep ☐	Very steep ☐
Depth	Small ☐	Moderate ☐	Large ☐	Very large ☐
Shear strength	High ☐	Moderate ☐	Low ☐	Very low ☐
Liquidity index	Low ☐	Moderate ☐	High ☐	Very high ☐

EARTHQUAKE ZONE

Tremors felt	Never ☐	Seldom ☐	Some ☐	Many ☐

LEGACIES FROM THE PAST

Fossil solifluction lobes and sheets	Absent ☐	Rare ☐	Some ☐	Many ☐
Previous landslides	Absent ☐	Rare ☐	Some ☐	Many ☐
Deep weathering	None ☐	Slight ☐	Moderate ☐	Much ☐

MAN-MADE FEATURES

Excavations-depth	None ☐	Small ☐	Moderate ☐	Large ☐
Excavations-position	Hillcrest ☐	High valley ☐	Low valley ☐	Bottom valley ☐
Reservoir	Absent ☐	Small ☐	Moderately deep ☐	Very deep ☐
Drainage diversion across hillside	Absent ☐			Present ☐
Lowering of reservoir level	None ☐	Small ☐	Moderate ☐	Large ☐
Loading of upper valley side	None ☐	Some ☐	Moderate ☐	Large ☐

a gradual decrease in the cohesive properties of the slope-forming materials, assuming that external influences do not change.

The advantage of having a checklist, either in a field investigation or during air-photo interpretation, is that each of the main categories of influencing and controlling parameters is systematically examined. By carefully assessing the state of the relief, drainage, bedrock, and regolith, the incidence of earthquakes, legacies from past processes, and man-made features the likelihood of landsliding can be assessed. If checking from this list reveals gaps in the available information then any further work can be directed at filling in these gaps.

The check list (Table 6.5) has been designed so that by placing an X in the appropriate places not only are the main causes of instability considered in turn, but the more the X signs align themselves to the right the more the slope concerned is approaching an unstable state. Such a list has the advantage that if only one or two X's occur to the left side of the check list, this identifies those parameters which should remain undisturbed if disaster is to be avoided. It is not realistic to supply on such a checklist absolute values for each category since the significance of a particular value depends on its local context. A 9 ° slope in clays may be potentially unstable while a slope three times as steep in a more coherent material (e.g. sandstone) may be extremely stable.

Each category on this list is now considered in turn in the context of the preceding discussions on soil and rock mechanics, and in terms of the techniques which can be used for measuring the appropriate properties. Techniques well documented elsewhere are not described here, but appropriate references are given. In many instances, such as in the examination of relief, drainage, structure (e.g. jointing), lithology, and the search for signs of past conditions of instability, much can be gained from a study of aerial photographs. This can save a considerable amount of expensive field time not only by directly providing information but also by identifying those sites worthy of the most detailed examination in the field. For the use of aerial photographs see American Society of Photogrammetry (1960), Lueder (1959) and V. C. Miller (1961).

(*i*) *Relief.* The steeper a slope the more liable it is to be unstable. Steep slopes tend to predominate in areas of deep valleys, as do bedrock cliffs, which are potential rock-fall sites. Slope steepness may be obtained from contour maps, together with the other measures of relief, though as a data source such maps can be notoriously unreliable. However, if a general guide as to steepness is required the scales for determining slope from contoured maps provided by Thrower and Cooke (1968) are particularly useful. A quick assessment of the mean steepness of a slope can be obtained from a measure of valley width (W) and valley depth (D), since $\tan^{-1} D/0{\cdot}5W = \bar{\theta}$, the mean angle of the valley-side slope. If the slope is not straight, however, then its steepest portions will be in excess of this value $\bar{\theta}$.

The significance of recording the height difference between adjacent valleys lies in the fact that one of the most likely locations for landslides is the lower end of a spur between a higher and a lower stream. Groundwater seepage may take place laterally through the spur especially if the bedrock is fractured or jointed. This leads to instability on the valley side of the lower stream. Similarly the recognition of coves, or embayments, in the valley side, as seen in plan, is significant in that higher pore-water pressures are likely at such sites since they are water receiving sites from the slopes around.

(*ii*) *Drainage*. High drainage densities are a sign of such things as impervious strata, high rainfall, little vegetation (including deforestation), and active stream incision, all of which may be associated with mass movement. Likewise steep river gradients generally indicate a phase of active and rapid incision, unless the river is in a phase of steady state (Chap. 1.2). Rock falls, earth falls, and sliding can arise directly from such incision, but they can also occur because of undercutting by a stream whose predominant forces are directed laterally rather than vertically downwards. Lateral undercutting can be achieved by rivers of low gradient in floodplain channels, as well as by mountain torrents.

Seepage from a hillside (e.g. along a spring line) can produce seepage erosion in fine sands and silts by a drag effect which takes with it individual soil particles, or a seepage pressure may be generated within the ground materials (especially in cohesionless water-bearing beds) so that particles are carried outwards from the slope. Eventually undermining results and slope stability is lost (Terzaghi, 1950; Hutchinson, 1968). Seepage sites may be identified through the recognition of the associated features of belts of damp ground, incipient gullies, spring-sapped hollows, or the collapse of natural underground pipes. In clays, however, it is unlikely that water will be seen at the surface as the rate of evaporation tends to be greater than the rate of flow out of the slope.

Groundwater pressures (including pore-water pressures and water pressures in joints) are critical in many slope stability problems, as much of the above discussion has shown. These are usually measured with piezometers (Lambe and Whitman, 1969), by which a piezometric water surface is defined for use in stability calculations such as that in Equation 6.3 (see also Fig. 6.6c).

(*iii*) *Bedrock*. Joint density, or that of other lineaments such as bedding planes or faults, will determine the size of falling blocks. Joint directions and inclinations with respect to the orientation and steepness of the ground surface have a direct bearing on the overall stability of the slope (Sect. 6.3a). These all need to be measured in the field, though sometimes aerial photographs can provide valuable indications of actual conditions. One approach to effective field recording is through the identification of a structural region within which joint spacing is more or less homogeneous, but which differs from that of the neighbouring parts. Methods for the field description of joints are given by Piteau (1970), Robertson (1970), and Ruth Terzaghi (1965). Other important data to be collected on a field survey include the nature of bedrock lithology, and the coherence, compressive strength, and degree of weathering (Chap. 11) of the beds. Compressive strength is the load per unit area under which a rock fails by shearing or splitting, and can be measured by laboratory shear-testing equipment. If clays (weak beds) underly more coherent beds (strong beds), then the more clay that is exposed the more likely it is that instability will arise.

(*iv*) *Regolith* (*soils and drift materials*). The site occupied by regolith materials, or the outcrop position of non-coherent beds such as clays, has an important bearing on stability conditions, the steepness of the slope being an important factor in predicting the likelihood of landslides. However, a great deal also depends on their particular relationship to other beds, and to rivers, slope drainage, and human interference.

The susceptibility of materials to failure is in part also related to their physical condition, for this very largely determines their shear strength. Relevant properties include texture, structure, coherence, grain shape, relative density, permeability, the porosity of granular soils, and the void ratio in cohesive soils. In engineering terms the Atterberg limits (liquid and plastic limits) may need to be known. These limits are based on the observation that a fine-grained soil can exist in any one of four states depending upon its water content. A soil is *solid* when it is dry, but on the addition of water the soil passes through the *semisolid*, *plastic*, and *liquid* states as the water content increases. The water content at the stage when the soil is passing from the semisolid to the plastic state is known as the *plastic limit*, while the water content between the plastic and liquid states defines the *liquid limit*.

$$L_I = \frac{w_n - PL}{LL - PL} \qquad (6.5)$$

where

w_n = natural water content;
PL = plastic limit;
LL = liquid limit.

L_I expresses the water content of a soil in a dimensionless way. Each of these physical parameters and its measurement or derivation is discussed in soil mechanics volumes such as Terzaghi and Peck (1967), Lambe and Whitman (1969), and Wilun and Starzewski (1972).

One other important factor is the stress history of the materials, and this may be intimately related to their geological history. For example, London Clay has been overconsolidated by the pressure of overlying deposits which, together with large amounts of the London Clay itself, have been removed by erosion. Overconsolidation leaves a clay with a much lower water content, at a specified depth, than in a normally consolidated clay at the same depth. In a normally consolidated clay c' of Equation 6.3 equals zero, but in an overconsolidated clay c' has a finite value (see Fig. $6.6d_2$) which is a result of the overconsolidation. In addition overconsolidation leaves the clay with much higher lateral stresses than occur in normally consolidated clays. These lateral stresses are not relieved to any great extent by the removal of the overburden. Relief may only be found by slope failure (progressive failure) when these lateral stresses exceed the shear strength of the clay.

(v) *Earthquakes.* Landslides are more likely if the area occurs within an earth-tremor belt.

(vi) *Legacies from the past.* Evidence of previous mass movement, such as solifluction lobes and sheets, and old landslides, indicates that slope failure could occur again. In particular nothing should be done to load the top of the slope or to take away support from the bottom of the slope. If an engineering structure is to cross the old landslide then very careful slope stability analyses will need to be made, and appropriate precautionary measures taken during construction. Where deep weathering has taken place the soils may be particularly prone to movement, especially if they lie on steep slopes, or have been subject to deforestation. Bore-hole records, or advantageous sections (e.g. in cuttings) may be the only sources of information about the location and depth of weathering products.

(vii) *Man-made features.* Each of the items listed has been discussed above in its appropriate context. If a large number of X symbols have been entered in or near the right-hand column in the check list (Table 6.5) then every endeavour should be made to ensure that man's modification of the land surface keeps all X's in this section well to the left, or instability could result.

In addition to using a checklist, such as that in Table 6.5, for evaluating particular sites, it is important that the relationships between the physical characteristics should be identified across the area. Their spatial dimension needs to be mapped. The technique most relevant for this purpose is geomorphological mapping, or at least a geotechnical or soil engineering derivative from it (see Chap. 14). At a less involved level, however, a technique of morphological mapping, by which landforms are identified by specific symbols (Chap. 14), can be used to identify and delimit the extent of instability. This was done by Brunsden and Jones (1972) for the unstable valley-sides of the Char Valley in Dorset, England. Figure 6.11a illustrates the nature of the

FIG. 6.11. Morphological mapping and the delimitation of landslides
a. Morphological map (for key to symbols see Fig. 14.3)
b. Landslides (p. 159)

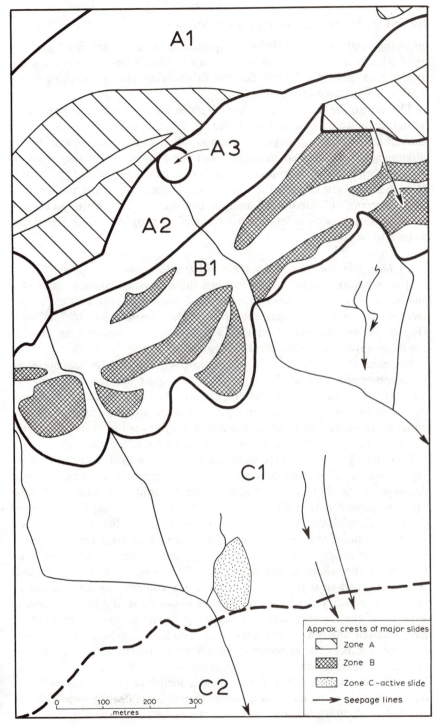

A1–3 degradational slopes in Lower Greensand B1, C1–2 accumulation slopes
(based on Brunsden and Jones, 1972)

mapping, and Figure 6.11b its interpretation. A more definitive geomorphological map of a landslide on the coast of Dorset, also by Brunsden but reproduced from Geological Society Engineering Group Working Party (1972), is illustrated in Figure 6.12.

The advantage of having identified those slopes which are potentially unstable, or which have already failed, is that this knowledge can be considered at the planning stage, for example of an engineering project. Route alignment can be decided so as to avoid these slopes, or, if they must be crossed then remedial or controlling methods can both be planned and costed. If such slopes are identified at the reconnaissance stage of a project then field sampling and laboratory analysis can concentrate on the potentially unstable slopes, with a less intensive sampling programme for the slopes known to be stable.

(c) Identifying and controlling the cause of a landslide

It is important to distinguish between those physical characteristics of a site which make landsliding possible and the actual cause (the 'trigger mechanism') which initiates movement. The classification of landslides (Fig. 6.7) shows the former, which is also the category that includes most of the items appearing on the checklist (Table 6.5).

For most slides (translational or rotational), given that the site possesses the appropriate physical characteristics, the trigger mechanism for movement is an increase in pore-water pressure. As has already been shown, most hillslopes which are potentially unstable are also particularly sensitive either to an increase in the load which they bear at the top of the slope, or a decrease in the amount of support which they have at the toe.

Recognizing which of these causative factors is in operation can in itself suggest the cure. Excessive pore-water pressures can be relieved by adequate drainage of the slide, with the appropriate removal of the cause of these high pressures if it is at all possible. In addition to draining the slide itself diversion channels can be placed above the head of the slide to prevent surface runoff onto the slide itself. Care has to be taken, of course, over the disposal of the water caught in these channels so as not to generate the conditions for yet another slide on the slope where this water is being discharged. Where overloading at the head of the slide is a contributing cause of movement it is necessary to relieve that load by removing it. If greater support is required at the toe then adequate retaining structures have to be built. If river undercutting is a contributory cause then its channel must be diverted. In some instances the engineering answer to a problem slide has been to remove it by excavation and then to regrade the slope down to a stable angle. For slides involving less than 2 million cubic metres of moving materials this may even prove to be the most economical method of dealing with it (Baker and Marshall, 1958). Several other control measures exist which are described

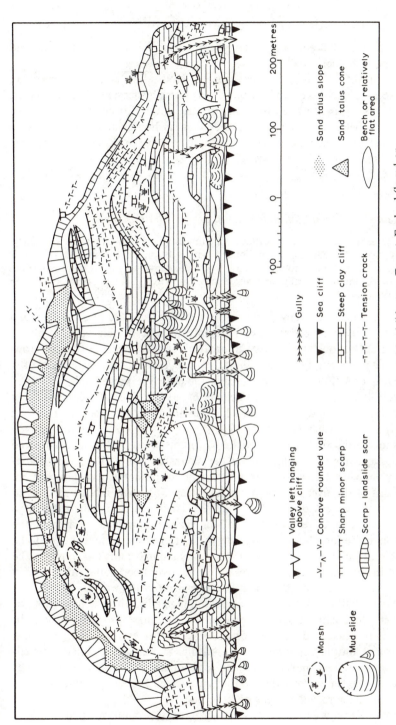

Fig. 6.12. Geomorphological map of a coastal landslide area, Dorset, England (based on Geological Society Engineering Group Working Party, 1972; original mapping by Brunsden)

Legend:

- Valley left hanging above cliff
- -V-∧-V- Concave rounded vale
- Sharp minor scarp
- Scarp - landslide scar
- >>>>> Gully
- Sea cliff
- Steep clay cliff
- -T-T-T-T- Tension crack
- Sand talus slope
- Sand talus cone
- Bench or relatively flat area
- Marsh
- Mud slide

200 metres

in appropriate civil engineering texts (e.g. Terzaghi, 1950; Eckel, 1958; and Zaruba and Mencl, 1969).

There is yet one other primary cause of landsliding over which man can exercise no control, and that is the shock waves generated by earthquakes. In such cases it is necessary to recognize that in the earthquake zones of the world it is better for human structures such as houses and dams to avoid sites which are either themselves potentially unstable or which have unstable slopes immediately above them.

6.7. Case Studies

(a) Landslides in a 'flat, monotonous, prairie' State, Illinois

As most people associated landslides with mountainous, hilly or rugged country, and as most of them also seem imbued with the misconception that Illinois is nothing but a flat, monotonous prairie, the statement that in this state landslides are very common generally occasions some surprise. (Ekblaw, 1931, quoted in DuMontelle *et al.*, 1971.)

Illinois is similar to many parts of the world in that 'despite the fact that landslides are common and do frequently present troublesome problems, they are not serious and possess no great potential dangers in Illinois. However, the prevalent false idea that they are non-existent engenders utter disregard, and this in turn does pave the way for preventable disasters.' (Ekblaw, 1931, quoted in DuMontelle *et al.*, 1971). This false conception of the environment and the 'it can't happen to me' philosophy is a perfect prescription for trouble.

In the vicinity of La Salle and Peru, Illinois (Fig. 6.13) much industrial development has taken place along the navigable Illinois River and consequently there has been a considerable demand for residential accommodation. The bluffs of the river are comparatively steep with short incised sections to the tributary valleys of the Illinois River. These bluffs and incised valleys break the comparative monotony of the plains above and because of the common desire for a home with a view these steeper slopes have proved to be desirable as building sites. Among the most favoured plots are those on the crest of the bluff some 30–45 metres above the river floodplain. This situation and mental attitude has many parallels elsewhere in the world.

Heavy rainfall in early 1970 induced slumping and earth flows in portions of the river bluff, causing considerable anxiety amongst property owners. In reviewing this case study it is important first of all to establish the geological background and the topographic relationships as well as the climate prevailing in the area. This makes it possible to recognize the cause of landsliding; it may also be shown how the landsliding affected particular homes with an indication of the remedial measures adopted.

The stratigraphy of the area is shown in cross-section in Figure 6.13. A superficial spread of gravels, glacial tills, soil, and loessic deposits overlies interbedded shales and limestone. Of these the shale beds are the thickest and form the widest outcrops. The shales contain more than 50 per cent of the expandable types of clay minerals and the shear resistance of the shales when weathered is decreased by either freeze–thaw, or alternating wet and dry conditions. The limestones, because they are thin, are generally unable to prevent movement on the steep slopes. An exception to this is provided by the La Salle Limestone member of the series. The limestone is well jointed thus making available blocks of limestone which themselves slide down-slope. The scarp slopes are generally inclined at between 20 and 25 ° with springs appearing at various horizons along the face of this slope. Rapid erosion

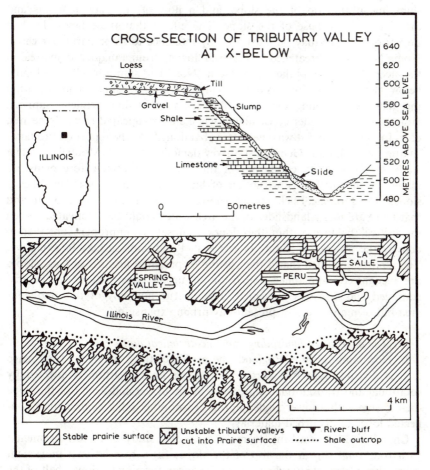

FIG. 6.13. Geology and situation of the unstable scarp slopes of the Illinois River banks (based on du Montelle *et al.*, 1971)

occurs at the toe after heavy rains have fallen. Climatically this area commonly experiences extreme conditions of temperature and precipitation. Records show that in five years prior to the 1970 landslides there was a consistent cycle of alternating freezing and thawing conditions as well as a period with high-intensity storms followed by dry conditions. The frequency of this alternating wetting and drying was of the order of less than one month. This period culminated in an excess of 1,140 mm of rain falling during the spring and summer of 1970, with more than 254 mm during both the months of May and September. This heavy rain reduced the support at the toe of the slope by erosion and increased the weight of the weathered shales, while at the same time building up pore-water pressure and thus decreasing their shearing resistance. The types of movement that followed consisted of slumping at the top of the slope, and a flow of the clay materials and the shales at the base of the slope. The effects that these landslides had on various homes may be illustrated by reference to three particular cases. Home No. 1 was threatened by material flowing from a major slump area on the steep slopes behind the house. Home No. 2 had slump blocks developing below it and stabilization has been attempted through sealing the joints within this material, draining the front blocks, and providing artificial support. Home No. 3 was under construction when cracks appeared through the area of its foundation and it was thereafter abandoned. As the report on this landsliding by the Illinois Geological Survey demonstrated (DuMontelle, 1971) a study of the stability conditions of this area is essential before plans are made for the further construction of any houses on or near these scarp slopes. The geological field studies showed that sliding will occur again and that there are many landslides in the area which could be reactivated. When the critical balance within the slopes is upset by improper construction activity the process of sliding is accelerated and damage may be extensive. The report concluded that if the existing or potential hazards of these slopes are recognized, unplanned expansion of urban construction can be discouraged and disasters prevented. This is true not only of this particular area but of many areas of present-day urban expansion.

(b) The analysis of landsliding using geotechnical data

In Northamptonshire, England, the profile of the Upper Lias Clay scarp face is divisible into two parts. The steeper upper portion is affected by rotational and translational slips, while at about the mid-height of the slope these merge with a lower-angled apron of smoother relief composed of connected lobes of disturbed material.

Chandler (1971) was able to make a detailed study of three movements: (i) a rotational slip at the top of the slope; (ii) a translational movement below the crest; (iii) reworked clay in the apron forming the lower half of the slope. Stability analysis required the measurement of (i) the effective-strength parameters c' and ϕ' (Sect. 6.3b) applicable to the slip surfaces; (ii) the highest likely position of the groundwater level; (iii) the density of the soil within the slip. The values used for the three slips are listed in Table 6.6.

TABLE 6.6

Data relating to stability analysis

Slip	Soil density (kg/m^3)	Water content (%)	ϕ' when $F = 1$ and $c' = 0$	F For given values of c' and ϕ'
1. Rotational	1920	32	16·7 °	1·28 (0; 21 °)
2. Translational	1890	34·5	21·0 °	1·00 (0; 21 °)
3. Reworked clay	1860	37	14·3 °	1·32 (0; 18·5 °)

Source: Chandler (1971).

The stability analysis adopted follows that of Morgenstern and Price (1965), which is equally applicable to circular and non-circular slip surfaces. In this a factor of safety F is defined which is the value by which the shear-strength parameters must be reduced in order to bring the slip into a condition of limiting equilibrium, as given in Equation 6.4.

Table 6.6 shows that slips 1 and 3 are stable under present conditions, with $F > 1$ in each case. Chandler deduced that movement in slip 1 is only possible with higher pore-water pressures, such as might be obtained with colder conditions and large quantities of unfrozen groundwater. Slip 3 also requires wetter and cooler conditions to bring F below unity, such as would enable the groundwater to rise to the level of the present ground surface. Slip 2, on the other hand, with an F value of 1·00, is only marginally stable and could move under present-day groundwater conditions. The texture of the material of slip 2 is suggestive of a mudflow of well-comminuted soil without a vegetation cover, such as might occur beneath a melting snow bank. The present-day actual stability of this quite small slip probably derives from root growth across the slip surface into the stable material beneath. This effectively anchors the slide.

This study illustrates the use of a stability analysis approach to unstable hillsides. It also illustrates, once again, the importance of past conditions in terms of the legacy of instability forms to be found on present-day hillsides.

6.8. Conclusion

Preventing or controlling landslides is not always an economically viable proposition, but where such control is necessary it is of fundamental importance that the nature of slope instability should be fully understood. The engineer approaches the problem through the subjects of soil and rock mechanics by which a quantitative assessment is made of the conditions of slope stability and of the necessary controlling structures. The geomorphologist comes to the problem from a somewhat different view: his investigation is concerned with the situation and site of the unstable slopes; and it involves a consideration of the relationship between slope form and slope stability together with the geological and geomorphological processes (including man) that influence slope stability. The two approaches need to be, and more and more frequently are, fully integrated in order to achieve a proper understanding of mass-movement phenomena.

7
GROUND-SURFACE SUBSIDENCE

7.1. Introduction

Visitors to Venice over a period of years are aware that parts of the historic city are being flooded with increasing frequency; residents of many coal-fields are familiar with the problems of structural failures developing in buildings and roads; and near Long Beach, California, railroad tracks have been buckled and oil-well casings sheared. These examples illustrate the widespread and diverse problems that may arise from ground-surface subsidence associated with human activity. Subsidence can occur as the result of changes in which man has played no part (Allen, 1969). In limestone areas, subsidence may be a consequence of the subterranean solution of carbonate rocks (Sparks, 1971); several areas of Alaska subsided as a result of the violent earthquake of March 1964 (U.S. Geological Survey) and, less spectacularly, the North Sea Basin has been subsiding for many years in response to large-scale crustal forces (Edelman, 1954; Smalley, 1967). At times subsidence from natural causes may seriously influence human activity. An example is subsidence of the surface over dolomites in certain areas of South Africa (Jennings, 1966). Nevertheless, subsidence is most significant where man has so altered surface or subsurface conditions that the existing surface level can no longer be maintained (Lofgren, 1969; Poland and Davis, 1969; Prokopovich, 1972; UNESCO, 1969).

On a world scale, subsidence resulting from man's activities is highly localized. It is most extensive in California (Fig. 7.1) where hundreds of square kilometres are affected; in the San Joaquin Valley alone over 900,000 hectares of irrigable land have subsided by more than 0·3 m. Elsewhere, areas of subsidence normally coincide with the occurrence of intensively exploited major subterranean mineral resources such as oil, coal, and salt, and with large urban-industrial complexes that depend heavily on local groundwater supplies.

Subsidence is often a subtle phenomenon that may pass unnoticed. Where ground level is falling at a slow rate over long periods of time precise techniques are required to monitor changes in surface elevations. The most widely used method is that of repeatedly surveying by levelling a series of precisely located benchmarks or other points of known relative altitude. A second method, usually used in conjunction with the first, is the direct measurement of compaction of deposits by means of a compaction recorder (Fig. 7.2) (Lofgren, 1961). The device consists of a heavy anchor weight that is set in the rocks below the well casing and is attached to a cable counterweighted at the surface to maintain constant tension. As compaction occurs so the length of the cable at the surface increases and this change is constantly recorded. Several recorders may be used in one locality with their anchors emplaced at different levels in order to monitor the extent of compaction in different parts of the sedimentary column. Measured compaction equals subsidence if the recorder spans the total thickness of compaction deposits.

The subsidence system may be described very simply. The ground surface is maintained at an equilibrium level which represents a balance among the various stresses that are applied to the materials of the earth's crust. The equilibrium condition can be disturbed so that the surface subsides as the result of a relative increase in downward directed stresses, a decrease in the strength of crustal materials, or a combination of both. For example, the imposition of a heavy load, such as a large reservoir, on a section of the crust may cause subsidence of the initial surface (e.g. Crittenden, 1963). Equally,

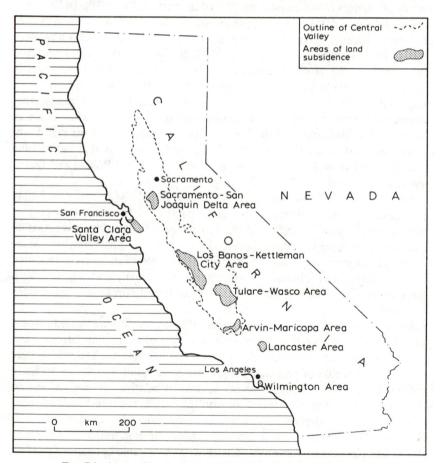

FIG. 7.1. Areas of land subsidence in California (after Poland, 1969)

a weakening of the strength of materials by the creation of voids or by the lowering of fluid pressures can have a similar effect. Clearly, however, there are thresholds that must be exceeded before any subsidence can occur, and these will vary according to local circumstances. In general, the nature of subsidence will be related to the nature of the changes responsible for the disruption of equilibrium and to the response of the materials that are affected.

In examining subsidence, therefore, it is convenient to recognize different types on the basis of the causes of disequilibrium, and to consider human responses in the context of these causes and in case studies. There are several

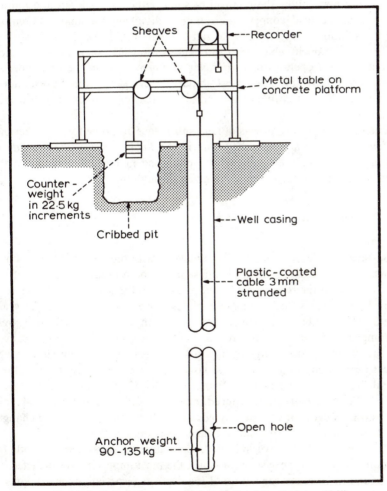

Fig. 7.2. Compaction-recorder installation (after Lofgren, 1961)

aspects of subsidence that are of particular interest: firstly, the geometric characteristics of area and amount; secondly, the rate of subsidence, especially with respect to the rate of stress change; thirdly, the ratio between the amount of subsidence and the amount of stress change; fourthly, the nature of any horizontal movements accompanying subsidence; fifthly, the consequences of subsidence on human activity; finally, the possibility of retarding, reversing, or preventing subsidence.

Undoubtedly the most important cause of subsidence attributable to man is the withdrawal of subterranean fluids, especially oil and water, and, to a much lesser extent, gases. A second cause relates to the compaction of sediments due in some cases to drainage and in others to irrigation. Thirdly, subsidence may follow the removal of solids either by underground mining of, for example, coal, copper, and iron or by dissolving the solids and removing the solutions, as in the case of salt and sulphur. There are a number of other minor ways in which man can cause subsidence, such as by vibration of uncemented, open-textured granular sediments, but in the remainder of this chapter only the three most important categories will be examined. Subsidence due to thawing of ground ice in areas of permafrost is discussed in Chapter 9.

The most important reviews of land subsidence are to be found in *Reviews in Engineering Geology*, 2 (1969), and in the *Publications of the Institute of Scientific Hydrology*, 88 and 89, also published in 1969. References to the specific examples quoted below are given through the text.

7.2. Withdrawal of Fluids

(a) Principles

Surface subsidence is principally associated with the withdrawal of oil and water. The cause of subsidence is the same for both fluids: extraction of the fluids results in the reduction of fluid pressure in the underground reservoir. This leads directly to an increase of 'effective stress' (or 'grain-to-grain stress') in the system, and compaction results. In this context, effective stress is composed of two separate stresses. The first is gravitational stress, caused by the weight of the overlying deposits. The second is a dynamic seepage stress exerted on the grains by the viscous drag of vertically moving interstitial fluid. The effect of these two stresses is additive, and together they increase as fluid pressure is reduced, and they lead to a reduction of void ratios (ratios between volume of voids and volume of solids) and to changes in the mechanical properties of the deposits (Lofgren, 1968).

As several authors have indicated, the nature of subsidence depends on certain variables in the system, of which the most important are the effective stress and its increase, the nature of the deposits (especially their compressibility, lithology, geochemistry, and thickness), the time the deposits have been subjected to increased stress, and whether or not the increased stress is being applied for the first time. A further general consideration is whether the fluids are confined (e.g. artesian conditions) or unconfined: most serious subsidence is associated with confined conditions.

A basic distinction, of considerable practical importance, is between elastic and non-elastic compaction, a distinction which depends largely on the nature of the deposits (Lofgren, 1968; Poland, 1969). Many reservoir sediments respond as *elastic* bodies, in which stress and strain are proportional, independent of time and reversible. Deposits of coarse sand and gravel respond as elastic bodies. That is to say, when fluid pressures fall, compaction is immediate, and if fluid pressures are restored, expansion follows. *Non-elastic* compaction results from permanent rearrangement of the granular structure of the reservoir sediments. Such is the case in fine-grained clay beds, which may occur adjacent to or within the principal fluid-storage sediments. These beds are highly compressible, but the adjustment of pore pressures is slow, time-dependent, and permanent.

To a certain extent the compaction of sediments subjected to fluid withdrawal is predictable, given an understanding of the main variables involved. R. E. Miller (1961) was able, for example, to predict successfully the compaction of a confined aquifer system using the following equation for each lithological unit:

$$\Delta h = \frac{e_1 - e_2}{1 + e_1} h \qquad (7.1)$$

where Δh = compaction, in feet; e_1 = initial void ratio of the sediments; e_2 = void ratio after increased loading (determined for a specific fall of aquifer pressure), and h = thickness of aquifer unit, in feet. A complicating factor, as mentioned above, is the time lag involved. In order to apply this predictive equation, Miller required information on (i) the thickness and nature of the relevant sediments, (ii) the decline in artesian head, which he determined from records of water levels in wells, (iii) the extent that sediments of different lithologies compacted, which he derived from simple, one-dimensional consolidation tests (Terzaghi and Peck, 1948), and (iv) the load imposed by the overlying deposits.

In addition to the similarity of the causes of subsidence due to withdrawal of oil and water, there is another resemblance: both types of subsidence are relatively recent. Subsidence due to oil extraction was first noted in the Goose Creek oilfield of Texas in 1925, and subsidence arising from groundwater exploitation was recorded in the Santa Clara Valley, California in 1933 (Poland and Davis, 1969). It is no accident that the phenomenon is recent, for it arises largely from the novelty of oil extraction and the rapid rate at which water use has grown, especially in industrialized cities (several Japanese cities are seriously affected, for instance) and in areas of extensive agricultural irrigation. Both of these changes, of course, have been made possible by the contemporaneous development of appropriate equipment for exploitation.

But there are differences between the consequences of oil and water extraction. In the first place, groundwater is normally withdrawn from reservoirs that are shallower, have greater porosity and permeability, and are more extensive than those from which oil is extracted. The reduction of fluid pressures is also usually much less in groundwater reservoirs than in oilfields: the former may be no more than 13 atmospheres of pressure, whereas the latter may be as high as 275 atmospheres (Poland and Davis, 1969). As a result of these differences, subsidence in oilfields tends to be greater and more localized than in areas of groundwater exploitation.

(b) Withdrawal of oil: case study of the Los Angeles area

There are two major types of subsidence in the Los Angeles area. There is widespread and slight subsidence due to the reduction of artesian water pressure arising from extensive pumping of shallow aquifers. And there is localized and pronounced subsidence associated with the exploitation of oilfields. The best examples of this second type are the Wilmington and Inglewood oilfields.

Subsidence in the Wilmington oilfield has probably been greater than anywhere else in the world. The area of subsidence is elliptically shaped with its axis running N.W.–S.E., congruent with the axis of the field's elongate structural dome (Fig. 7.3). The maximum vertical subsidence has been 9·3 m in the period 1928–71. Figure 7.4 shows that the rate of subsidence has varied considerably through this period, generally changing in harmony with the rate of extraction: increasing up to 1951, the year of peak production, and subsequently declining, except for slight increases with stress relief following local earthquakes. The volume of oil and water (excluding gas) removed between 1928 and 1962 was approximately 1,397 million barrels. This figure compares with a total volume of surface subsidence during the same period within the two-feet (60·9 cm) subsidence line equivalent to approximately 550 million barrels. In rough terms, therefore, subsidence has been equivalent to some 39 per cent of the volume of fluid removed (Poland and Davis, 1969). It occurred mainly in the reservoir sands.

Subsidence is closely allied to horizontal movements and related problems. In an investigation of horizontal movements in the Wilmington field, Grant (1954) drew an instructive analogy between the 764–917 m thick layer of sediments above the oil-bearing zone and a circular plate with clamped margins that is deformed by a surface centre-point or uniformly distributed load (Fig. 7.5). The deformation of such a layer or plate will produce an inflexion in the subsidence curve separating an inner concave surface from an outer convex one. Horizontal displacement will be most marked along the

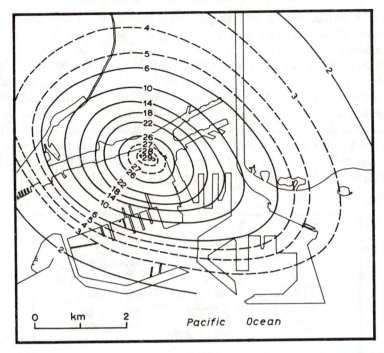

FIG. 7.3. Subsidence between 1928 and August 1971 in the Long Beach area, California (Port of Long Beach, 1971)

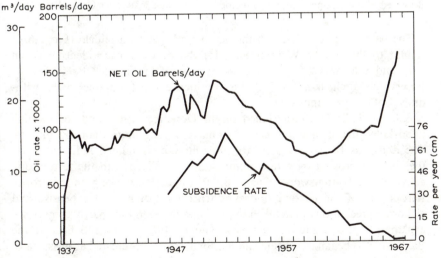

FIG. 7.4. Rate of oil production and subsidence rate in the Wilmington oilfield, California (after Mayuga and Allen, 1969)

line of inflexion, and least at the point of maximum depression and at the margins. Maximum observed horizontal displacement at the surface in the Wilmington field is 2·7 m.

Deformation of the 'plate' also creates a distinctive stress pattern. Firstly, shear stresses are probably greatest along the line of inflexion and in the plane approximately mid-way between the surface and the bottom of the deformed zone. For this reason, rupture of wells by horizontal shear stresses has been observed at depths of 458–88 m. Secondly, there will be surface tension in the deformed sediments outside the inflexion line, and compression inside it. The former has caused tensional rupture of radially directed linear structures such as pipelines, and the latter has caused, for instance, buckling of railroad tracks.

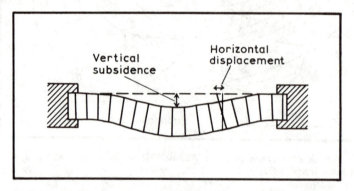

FIG. 7.5. Vertical deflection of a cross-section of a circular plate with clamped margins deformed by a centre-point load or a uniformly distributed load (after Grant, 1954)

The pattern of subsidence in the adjoining Inglewood oilfield (Fig. 7.6) is similar to that in the Wilmington field, and maximum subsidence between 1917 and 1963 was 2·9 m only 0·68 km west of the ill-fated Baldwin Hills Reservoir. Much, if not all, of this subsidence is attributable to the withdrawal of oil, gas, and water.

Baldwin Hills Reservoir failed on 14 December 1963, and the ensuing flood killed five people and caused damage costing more than $15 million. The dam was built on loosely to moderately consolidated sediments, and the site was crossed by a few minor faults associated with the tectonically active Newport–Inglewood zone of uplift. Engineers working on the project were aware of the tectonic and subsidence hazards so that expensive and elaborate precautions were taken to prevent a disaster. One partly conjectural explanation of the failure is as follows (James, 1968). Sediments beneath the Baldwin Hills began to subside and tensional stress was created at the surface, especially near to the perimeter of the subsidence bowl. Eventually, earthquakes occurred, opening up the ground notably along two fault lines beneath

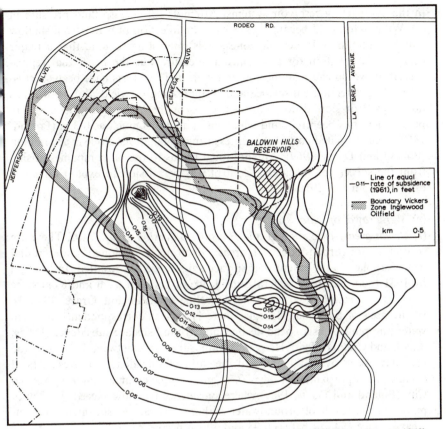

Fig. 7.6. Subsidence rates in the Baldwin Hills area, California (after James, 1968). Isolines in feet

the reservoir. By the spring of 1963 the openings produced were sufficient to permit a slight but perceptible increase of flow into the reservoir's under-drainage system. Water flowing through one crack emerged at the down-stream face of the dam, and erosion proceeded rapidly to the point where the dam was undermined.

Man has responded in three principal ways to subsidence in the oilfields of Los Angeles. Firstly, because parts of the port of Long Beach and adjacent coastal land were formerly above sea level but are now threatened by sea flooding, an extensive programme of coastal protection has been under-taken. Secondly, much effort and expense has been required to repair and maintain oilfield structures, especially oil wells, affected by subsidence and horizontal movements. Thirdly, as damage in the Wilmington field rose to over $100 million, the U.S. Department of Justice attempted to close the oil operations. The oil companies responded by trying to reduce effective stresses in the compacting reservoir systems. Attempts to increase fluid pressure in the Wilmington field began in 1958 with the injection of water from shallow aquifers into the oil-reservoir zones. It is interesting to note that a major economic motivation for this remedial measure was the belief that repressurization would cause an increase in oil yield. The efforts have been successful both in affecting subsidence and in increasing oil yield. Subsidence has now ceased or slowed down in most parts of the Wilmington field, and in places there has been rebound which amounts to as much as 0·3 m (Poland, 1969). The estimated cost of the repressuring installations alone is $30 million (Poland and Davis, 1969). An additional bonus to the oil companies arising from the success of subsidence control was permission to exploit oil sands offshore from Long Beach.

(c) Withdrawal of water

(i) Urban areas: case study of the Houston–Galveston area, Texas. There are several major urban-industrial areas which have exploited underground water supplies to the extent that surface subsidence has resulted. In London by 1931 there had been general subsidence of between 6 cm and 8 cm since 1865 as the result of artesian head decline (Wilson and Grace, 1942–3). In the area of Savanna, Georgia, there is a clear correspondence between water pumping, artesian head decline, and ground-surface subsidence (Davis, Small, and Counts, 1963). Here, maximum subsidence between 1918 and 1955 may have been over 100 mm, and the ratio of subsidence to artesian head decline was approximately 1 : 300. A more spectacular example is Mexico City (Poland and Davis, 1969) where subsidence has now exceeded 7·5 m in places as the result of groundwater withdrawal; rate of subsidence has increased since 1948 to between 25 and 30 cm per year.

Subsidence in urban areas has many varied consequences. In places, it may be so slight that it has little obvious direct effect on urban structures and activities. Such is the case in London, although even here subsidence may contribute in a very small way to the increased likelihood of flooding along the River Thames. Where subsidence is pronounced, flooding may become a serious problem, as in Venice. Also, problems of water transport, drainage, building construction, and 'piped' facilities may all be exacerbated, as in Mexico City.

During and since the Second World War there has been rapid industrial development along the Gulf Coastal Plain of Texas, especially in the Houston –Galveston region where much of the growth has been associated with oil, port, and aerospace industries. Demands for water from industry, the urban population, and rice irrigators have been met largely by the aquifers beneath the area (Winslow and Doyel, 1954) although surface-water supplies became available from Lake Houston after 1954. The conditions for artesian water supply are favourable in the gently dipping, thick, relatively unconsolidated sands and clays of the Gulf Coastal Plain (Fig. 7.7). In the Houston area, most water has been withdrawn from the Lissie and Willie formations (up to 700 m thick) which are composed mainly of sands with some interbedded clays. The 33–100 m thick Alta Loma Sand is the main aquifer in eastern Harris County and Galveston County whereas the principal aquifer for Texas City is the Beaumont Clay (up to 400 m thick), a formation with less sand and more clay than the other aquifers. Heavy withdrawals of water from these aquifers have resulted in a decline of artesian pressure head (Fig. 7.7), and the rate of water-level decline has increased from about 1·21 m per annum in 1954–9 to about 2·13 m per annum in 1959–64 (Gabrysch, 1969).

Subsidence began before the Second World War, but it was only monitored precisely following the establishment of a suitable network of levelling stations in 1943. By 1964, the pattern of subsidence was clear (Fig. 7.7). Maximum subsidence was 1·52 m in the Houston–Galveston region. The rate of subsidence increased from about 6 cm per annum during 1954–9, to about 7·6 cm per annum between 1959 and 1964. The subsidence undoubtedly arises largely from water abstraction, although removal of oil from many fields in the area may have played a small part. The relations between land subsidence and declining water levels are exemplified in Figure 7.7 (Gabrysch, 1969). The relatively greater subsidence south-east of Houston (Fig. 7.7) may have been the result of both oil and water extraction. And subsidence related to oil extraction has undoubtedly occurred in the Goose Greek field north of Galveston.

In principle, the response of the aquifer system to water withdrawal is straightforward—reduction in fluid pressure increases the load on the skeleton of the aquifer and compaction results. It seems probable that most compaction occurs not in the sands, but in the clay aquicludes (confining beds bounding the aquifers) and in other clay beds within the aquifers. As water is withdrawn, hydrostatic pressure in the sands falls, creating a pressure difference between the clays and the sands; water thus moves from the clays into the sands; the overlying sediments cause compaction in proportion to the reduction of pressure, and most of the compaction occurs in the clays because they are less competent than the sands. Evidence to support this argument comes from an inspection of curves showing ground-surface subsidence and artesian pressure decline. In the northern part of the area where water comes mainly from sands, the ratio of surface subsidence to decline of artesian pressure head is 1 : 100; in the south, where the proportion of clays in the water-bearing formations is greater, the ratio is also greater, probably indicating that subsidence may be attributed mainly to the compaction of the clays. The ratios may not be the true ratios because there is a time lag between water withdrawal and ground subsidence due to the slow rate at which water drains from the clays, and the length of the lag is unknown.

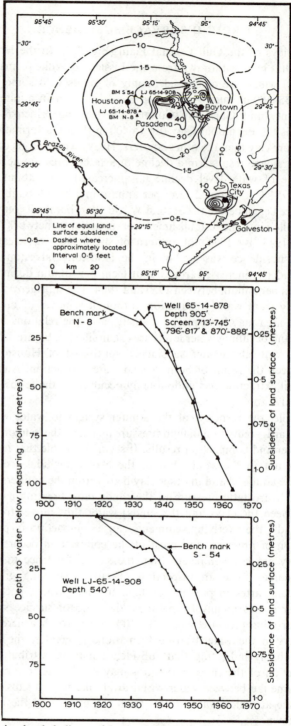

FIG. 7.7. Artesian head decline and land subsidence in the Houston–Galveston region, Texas, 1943–64. The lower two graphs show relations between subsidence of bench marks and water levels in nearby wells (locations on map), (after Gabrysch, 1969)

(*ii*) *Rural areas: case study of the Central Valley, California.* Ground-surface subsidence is less common in rural areas, largely because the rate of water withdrawal, and hence the difference between rate of withdrawal and rate of replenishment, is usually less than in urban-industrial regions. It is found, however, where water has been removed in great quantities to promote surface irrigation in arid and semi-arid lands. The Central Valley of California is such an area.

The pattern of subsidence in the southern Central Valley is shown in Figure 7.8. The main areas are south-west of Los Banos, between Tulare and

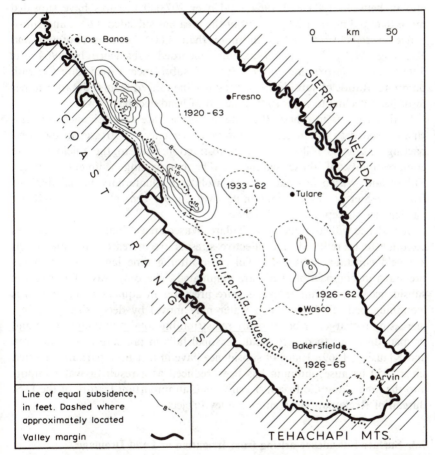

FIG. 7.8. Magnitude and extent of subsidence in the southern Central Valley, California (after Poland, 1969)

Wasco, and south of Bakersfield. Subsidence has affected over 900,000 hectares, and it has been caused largely by groundwater withdrawal from confined aquifers, although an important additional reason in some areas is hydrocompaction (see below). Maximum rates of subsidence range from 14 cm to 45 cm per annum, and maximum recorded subsidence exceeds 8·53 m. The main areas of subsidence have been studied in detail (Bull, 1964; Poland and Davis, 1956 and 1969; Lofgren, 1963; Lofgren and Klausing, 1969).

In the Tulare–Wasco area irrigation acreage has grown rapidly during this century. In 1886, some 16,000 hectares were irrigated; by 1921 the figure had increased to 74,000 hectares; and by 1958, some 50 per cent of the area, or 184,000 hectares, was irrigated. Groundwater levels declined by up to 60·9 m between 1905 and 1964; and over 207,000 hectares have subsided more than 0·3 m, over 53 per cent of the area has subsided 0·6 m, and subsidence exceeds three metres in more than 4,000 hectares (Lofgren and Klausing, 1969; Fig. 7.9). Maximum recorded subsidence by 1964 was 3·9 m. The approximate annual rate of subsidence varied directly with pumping. Ratios of subsidence to head decline ranged from $0·5 \times 10^{-2}$ m per m of head decline to $5·0 \times 10^{-2}$ m per m of head decline.

In this area, as in others, the magnitude and rate of subsidence are principally related to the increase of intergranular effective stresses in the compacting sediments following water abstraction, and to the thickness and compressibility of those sediments. The compacting sediments between Tulare and Wasco are mainly unconsolidated Quaternary fluvial deposits from the Sierra Nevada containing over 25 per cent clayey-silts and are particularly susceptible to subsidence.

Subsidence will continue in the Tulare–Wasco area as long as fluid pressures continue to be reduced, and effective stresses are increased. If water levels are held constant, subsidence will cease after a time lag in which excess pressures in compactible beds are eliminated. The only way of preventing subsidence is to eliminate excess pore pressures in aquicludes or aquitards (confining beds through which water may move) by decreasing pumping, increasing recharge, or both; or by redistributing total pumping in time and in space in order to reduce local, seasonal falls in pressure heads (Lofgren and Klausing, 1969). Rates of subsidence have in fact been partially arrested since 1950 because pumping has been reduced as a result of water import into the area along the Friant–Kern Canal, an important feature of the U.S. Bureau of Reclamation's Central Valley Project.

7.3. Surface Subsidence arising from Irrigation or Land Drainage

Subsidence related to the extraction of fluids is usually fairly extensive and rather slow, and it reflects surface adjustment to changes at considerable depths. In contrast, subsidence caused by the wetting and compaction of certain types of surface sediment (hydrocompaction) or drainage of others results from near-surface compaction, and it is rapid and localized.

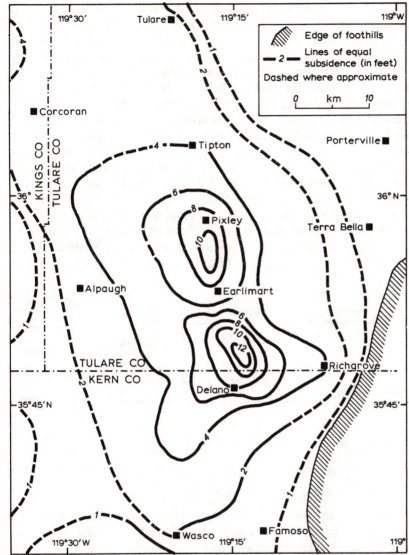

FIG. 7.9. Land subsidence in the Tulare–Wasco area, 1926–62 (after Lofgren and Klausing, 1969)

(a) Hydrocompaction and surface subsidence

In some areas, moisture-deficient, unconsolidated, low-density sediments have sufficient dry strength to support considerable effective stresses without compacting. When such deposits are wetted for the first time, by percolating irrigation water for example, or when the overburden load is significantly increased, the intergranular strength of the deposits is weakened, rapid compaction occurs, and ground-surface subsidence follows (Lofgren, 1969).

The main requirement for this kind of subsidence is therefore low-density, unconsolidated, moisture-deficient, surface sediments. It is satisfied principally in those arid and semi-arid lands where there are either wind-blown loess deposits or certain alluvial sediments which are above the water table, are not normally wetted below the root zone, and have high void ratios. Areas of such sediments where hydrocompaction has been described are mainly in the dry lands of the United States and the U.S.S.R.

In the creation of alluvial fans in dry lands, a layer of sediment may be deposited in a flood and then rapidly dried. The dried layer may have many voids, such as intergranular openings held in place by clay bonds, bubble cavities, and desiccation cracks (Bull, 1964). Nevertheless, it may have sufficient strength to withstand the imposition of another increment of sediment in the next flood, except perhaps for slight compaction of the upper part as the result of wetting by percolating water from the second deposit (Fig. 7.10). When such a deposit is first wetted by percolating irrigation water,

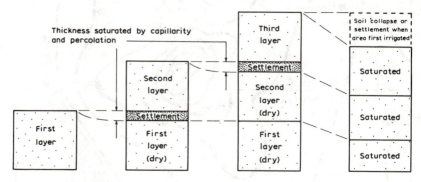

Fig. 7.10. Development of relatively strong, dry, alluvial-fan deposits (after Roberts and Melickian, 1970)

the strength of the dry material is reduced (for example by the weakening of clay bonds), the void ratio is reduced, and the surface subsides. In these circumstances, the amount of subsidence depends mainly on the overburden load, the moisture conditions (such as depth of water penetration) and the amount and type of clay in the deposits. For example, Bull (1964) showed in the south-western Central Valley of California that hydrocompaction increases with an increase in overburden load and that maximum compaction occurs at a clay content of about 12 per cent. Total amount of compaction here may be over 4·5 m. In areas of loess deposits there is considerable lithological and structural variety, but the low-density sediment normally has a high void ratio, with voids often consisting of intergranular openings sustained by clay bonds, clustering of silt aggregates and tubular channels, and it responds to wetting by compacting (Lofgren, 1969).

Bull (1964) described in detail the nature and consequence of hydrocompaction in 21,000 hectares on certain alluvial fans of western Fresno County, California. Subsidence is quite variable, being greater for example where water penetrates more deeply or more continuously, such as along canals and ditches. Often field surfaces become quite irregular, and subsidence cracks may develop, for instance, along unlined canals and ditches or in irrigated gardens.

The consequences of subsidence in Fresno County are numerous. To the farmer, subsidence in fields is a major problem, because canal-irrigation water accumulates in hollows and some plants are flooded, and water supply is reduced to plants on higher ground. It may cost as much as $10 per acre to relevel the land. The farmer also faces the problem of damage to well cases, pipelines, ditches, and canals, with the attendant difficulties of leakage and repair. Re-elevation of these structures, or even pumping may become necessary. One partial solution for the irrigator is to adopt a sprinkler system, which avoids the necessity for careful relevelling and distributes water uniformly (but which is nevertheless itself susceptible to subsidence). To the engineer, subsidence brings local problems of damage to buildings, roads, pipelines, power-transmission lines and canals. For example, subsidence along a 2·5 km length of cement-lined canal caused damage that to repair required as much as $5,000 a year for sand and gravel alone.

The most effective method of preventing hydrocompaction and its associated problems is to compact the vulnerable sediments by wetting before development.

(b) Subsidence due to drainage of organic deposits: Case study of the Florida Everglades

Many organic soils subside when they are drained, by shrinkage from drying, compaction with tillage, loss of groundwater buoyancy, wind erosion, burning, and biochemical oxidation (Stephens and Speir, 1969). In general, subsidence rates are correlated with groundwater depths; the lower the water table, the greater the subsidence. The problem has been encountered in many areas, including the Sacramento–San Joaquin delta in California, the Florida Everglades, the English Fens, certain Dutch polders, and near Minsk in the Soviet Union.

Stephens (1958; Stephens and Speir, 1969) has recorded the details of land subsidence in the Florida Everglades, the largest single body of organic soils in the world, covering over 8,000 km². Three methods of recording subsidence were used: periodic resurveying of profile lines, controlled water-table experiments on plots, and comparison of land elevations from contour maps made at different times. The progress of subsidence along one profile

line across peaty muck soil is shown in Figure 7.11. Very rapid subsidence due
to shrinkage and loss of buoyancy occurred in the five years following
drainage; as the surface fell to a level where gravity drainage was not so
effective, subsidence rate declined; when pumping in 1927 further lowered
the water table, better drainage and initial tillage compaction promoted more
rapid subsidence; and a steady rate of subsidence has followed, due mainly

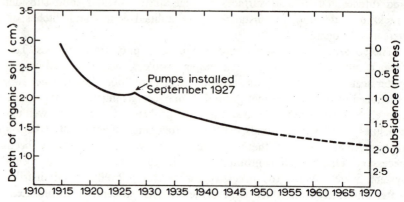

Fig. 7.11. A periodically re-surveyed profile near South Bay, Florida, indicates the normal
pattern of subsidence on drained organic soils (after Stephens and Speir, 1969)

to slow oxidation from biochemical action. As subsidence continues, and it is
estimated that over 88 per cent by volume of cross-sectional area from pre-
drainage days could be lost by A.D. 2000, water-control problems increase.
Mole drainage may become impractical; farm ditches may have to be cut in
bedrock; and eventually the soil may be economically unusable. As a result
of these problems, and in order to conserve the agricultural productivity of
the Everglades, a far-reaching programme of flood-control and water conservation has been prepared.

7.4. Subsidence due to Extraction of Solids

Extraction of solids from beneath the surface does not introduce new
principles into this discussion, but it is an important activity in many areas,
particularly in the older coalmining fields of Europe. Here subsidence is a
problem that follows removal of coal without stowing or packing of waste in
the underground spaces created. The principal components of mining subsidence in areas where stowing has not been practiced are shown in Figure 7.12,
a general diagram based on research in Germany's Ruhr coalfield (Wohlrab,
1969). The extent of the subsidence zone and the nature of subsidence will
vary from area to area, depending mainly on the form of mining, the thickness of the seam(s), and the depth and nature of the overlying strata. The consequences of solid removal are predictable (H.M.S.O., 1949; National Coal

Board, 1966). For example, a Royal Commission reported in 1929 that 'in the absence of special precautions, the amount of subsidence to be expected is about two-thirds of the thickness of coal removed and that the area affected may extend to a distance beyond the vertical line equal to about half of the depth of the seam being worked' (H.M.S.O., 1929, p. 5). As Figure 7.12

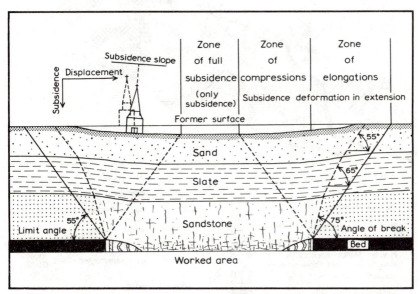

FIG. 7.12. Components of mining subsidence (after Wohlrab, 1969). The 'angle of break' depends on the geological characteristics of the rock formations. The 'limit angle' is defined by the line joining the limit of collapse to the limit of surface subsidence

shows, the 'angle of break' varies according to geological characteristics of the rocks and the 'limit angle' will clearly be defined by local conditions.

Within the zone of subsidence, features described in previous sections will occur—horizontal and vertical displacement, compressional and tensional phenomena, and deformation of the surface by bending or fracture. In Britain, one common problem is the inundation of subsidence areas and disruption of surface hydrology. A clear example of such disruption has occurred near a colliery on the floodplain of the River Stour, near Canterbury, England (Coleman, 1955). Figure 7.13 shows the growth of the subsidence lake on the floodplain from 1933, following mining of a 1·2–1·5 m thick seam which had been packed after extraction with rubble to reduce subsidence. Subsidence was normally no more than 60 cm. The 'limit angle' was here about 21 degrees. Another example of subsidence from mineral extraction that leads to flooded depressions occurs in the Cheshire salt field where brine pumping has led to the creation of 'flashes'. In this area subsidence at various times has damaged farmlands, broken piped services and

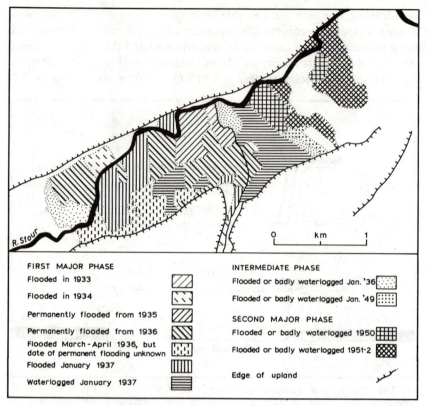

FIRST MAJOR PHASE

Flooded in 1933

Flooded in 1934

Permanently flooded from 1935

Permanently flooded from 1936

Flooded March-April 1936, but date of permanent flooding unknown

Flooded January 1937

Waterlogged January 1937

INTERMEDIATE PHASE

Flooded or badly waterlogged Jan. '36

Flooded or badly waterlogged Jan. '49

SECOND MAJOR PHASE

Flooded or badly waterlogged 1950

Flooded or badly waterlogged 1951-2

Edge of upland

FIG. 7.13. Growth of a subsidence lake along the River Stour, near Canterbury, England (after Coleman, 1955)

retarded their provision, damaged canal and rail communications, and injured buildings (Wallwork, 1960a, b).

7.5. Conclusion

Ground-surface subsidence is a particular type of slope instability in which equilibrium is disturbed by a relative increase in downward directed stresses, a decrease in the strength of crustal materials, or a combination of both. In terms of environmental management, the most serious types of subsidence are those caused by man, for they are often both relatively rapid and recent, and they normally have serious direct and indirect consequences on human activity. Subsidence due to the creation of subterranean voids, for example by mining of solids, is generally localized; that arising from withdrawal of fluids is commonly a regional phenomenon. Management of local and regional subsidence has tended in the past to focus on the repair of its damaging effects and on a continuous adjustment of activities to changed ground conditions by individuals or groups. But recent developments of techniques designed to retard, prevent, or even reverse subsidence while resource exploitation continues point the way towards rational control of the phenomenon.

8

COASTS

Cuchlaine A. M. King

8.1. Introduction

Coastlines are of great importance to the countries which possess them. It is here that many industrial enterprises may occur, especially those linked to ocean or river transport, and it is at the coast that some of the most rapidly operating geomorphological processes are to be found. Knowledge of these is relevant to many aspects of the human occupancy of coastlines and estuaries. Not only are coasts susceptible to erosion and flooding but they have value as sites of inorganic (e.g. sand and gravel) and organic (e.g. seaweed and shellfish) resources. Some coastal areas have been reclaimed from the sea, while others are being considered for reclamation. Many coastlines are used for recreation (e.g. beaches and marinas, sailing, surfing, underwater sport, and fishing). Different uses often conflict with each other or with controls which are necessary either to combat erosion or to encourage reclamation.

The coastal defences around the shores of England and Wales are the responsibility of the River Authorities. Their engineers are often assisted by the Hydraulics Research Station scientists when specific problems require laboratory testing or other advice. The Hydraulics Research Station has extensive experimental facilities notably at its Wallingford research station. An example of its contribution is the model experiments carried out to assess the optimum pattern of groynes for reducing the serious erosion along the coast of Suffolk at Dunwich. It is also concerned with fundamental research as well as applied work, much of the latter being concerned with harbour and estuary improvements, such as the work on the Mersey Estuary, for which it builds large models.

Coastal work in France is carried out in part under the auspices of the Comité d'Océanographie et d'Études des Côtes, while the laboratory at Delft has facilities for coastal work in the Netherlands, where the low-lying nature of the hinterland makes effective coastal defence work essential.

In the U.S.A. the Coastal Engineering Research Center (CERC) has authority to undertake any measures considered necessary for the protection of the public coast, and it carries out fundamental research, including field studies and experimental work. CERC assists the civil works programme of the Corps of Engineers, in devising measures to prevent erosion, in improvement of harbours and channels for navigation, and in protection against flooding and wave damage by hurricanes and other storms. A U.S. federal interest in shore protection first started in 1922 when the state of New Jersey set up an advisory board, which included officers of the Corps of Engineers, to study its erosion problems. In 1929 the Chief of Engineers organized a body known as the Board of Sand Movement and Beach Erosion. In 1930 Congress initiated the Beach Erosion Board (BEB). The BEB was superseded by CERC in 1963, and CERC was given a broader mandate, including all the coastal engineering and civil activities of the Corps as well as shore protection. The services of CERC are available to all Federal agencies and also under certain conditions to State agencies and private concerns. The Sea Grant College also provides funds for coastal research in Universities, while other projects are supported by the Geographical Branch of the Office of Naval Research. Many other bodies are also concerned with coastal work, including Scripps Institute in California, and the Virginia Marine Science Institute.

A number of books have been published recently on coastal geomorphology, and in the field of applied geomorphology the 1966 edition of the CERC Handbook is particularly useful. (A new metric edition is in preparation at present.) The published proceedings of the conferences on coastal engineering that take place about biennially also provide a great deal of very valuable material. They cover the 1950s and 1960s, the twelfth conference having been held in Washington (1970). Separate volumes of proceedings are devoted to pure research and applied research respectively. The BEB and CERC also publish much useful information. Between 1940, when it started publishing, and 1963 when it became the CERC, the BEB published 135 Technical Memoranda, five miscellaneous papers, four technical reports and seventeen volumes of Bulletins. CERC has continued active publication (Allen and Spooner, 1968). In France the COEC publish an annual volume on coastal research, and much is published by the Institute of Oceanology in the U.S.S.R. Of the textbooks available in English, that by Bird (1968) treats the subject of coasts at a more elementary level than that by King (1972) or Davies (1972). Russell (1967) concentrates on low coasts while Zenkovich (1967) is useful for examples of coastal studies in the U.S.S.R. Several significant papers, formerly published in journals, are reprinted in Steers (1971a, b).

8.2. Processes in the Coastal System

(a) Introduction

Coastal processes are characterized by their extreme variability and generally high level of activity. The forms (Fig. 8.1) and materials within the coastal zone are thus subject to considerable, and often rapid, change both in time and in space. Where mobile beach material is plentiful the changes can be rapid, but on hard-rock cliffed coasts changes may be imperceptible on a human time scale. In evaluating the process–response systems along a coastline it is important to identify distinct portions or stretches of coast within which specific responses by the coastline to the physical processes can be identified. For convenience these portions may be called *land units*, in anticipation of the discussion on land-systems analysis in Chapter 13. These units can also be related to their use by man (Sect. 8.5).

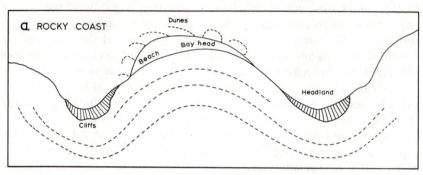

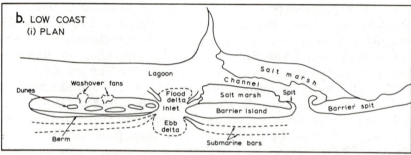

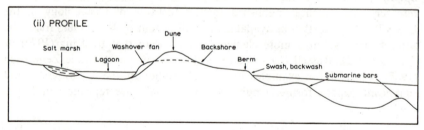

FIG. 8.1. Coastal forms on (a) a rocky coast (b) a low coast (i) in plan (ii) in profile

The processes which affect a coast include waves, tides, and winds. These operate as process variables on beach materials to influence the response variables which are the forms of both the beach and the coast in profile and in plan.

(b) Waves

Waves are the most important process element on most, though not all, coasts. Wind-generated waves can be subdivided into two major types. These are the waves actually being generated at the time, known as *sea*, and the longer, lower waves that have travelled far from the generating area and have been modified to form *swell*.

The wave characteristics at any one place along the coast depend on its aspect with respect to generating winds and swell, and on the offshore and coastal plan. Waves are generated by the wind through transference of energy from air to water, whereby waves grow in size with increasing wind strength, wind duration, and fetch of open water (Russell and Macmillan, 1952; Kinsman, 1965; Ocean Wave Spectra, 1963; Deacon, 1949). The process is

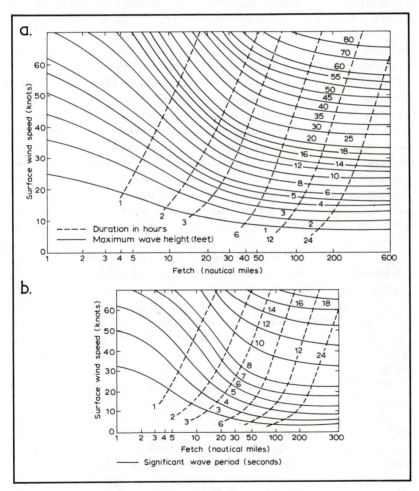

FIG. 8.2. Graphs relating wave height (above) and wave period (below) to wind speed, wind duration, and fetch in oceanic waters (after Darbyshire and Draper, 1963)

not yet fully understood, although empirical relationships have been established and curves have been presented that allow the significant wave height and period to be estimated for any specific combination of wind variables (Coastal Engineering Research Center, 1966; King, 1972; Fig. 8.2). The significant height, which is defined as the mean of the highest one-third of the waves, hides the variability inherent in nearly all wave trains, as these include a spectrum of waves of many heights and periods, moving in a wide range of directions (Fig. 8.3). The three-dimensional wave spectrum may be analysed by Fourier techniques, whereby the energy assigned to the various periods present can be obtained (Deacon, 1949). The significant wave height and period are statistical values obtained from the complex spectrum. The relationship between the wave height and wave period gives the wave steepness, which is an important relationship in assessing the behaviour of waves on the beach. The direction of wave approach is also of great significance, in that it controls in part at least the longshore transport of beach material.

The world approach by Davies (1964) to coastal classification in terms of wave types (Fig. 8.4) is of value in this respect. He divided the world coastlines into four main groups. The storm-wave environment occurs in high latitudes in the zones where strong winds frequently blow, creating high, steep storm waves that reach the coasts, often accompanied by strong onshore winds, before they are converted into swell. The storm-wave coasts often have long stretches of rocky cliffs and wave-cut platforms. These rocky coasts are of importance from the ecological and human point of view because they provide suitable habitats for the seaweeds that are being increasingly exploited, for example on the rocky coasts of western Scotland and Ireland, and for important shore species, such as lobsters and crabs.

The lower latitudes are dominated by the swells, created in the stormy high latitudes but converted to swell by the time they reach the coasts of the lower latitudes. The west-coast swells, which are higher and hence have greater energy, are differentiated from the east-coast swells. These long swells have a mean period of 14 seconds, with a length of 305 m, and are therefore considerably refracted before they reach the coast. The fourth wave type is the low-energy one, which is found either where the fetch restricts the growth of waves or where sea ice, or some other offshore feature, prevents effective waves from reaching the coast. Thus most polar coasts, as well as those of enclosed seas, are low-energy in type.

Under deep water conditions it is possible to predict wave characteristics from wind data. The modifications that waves undergo as they approach shallow water are important, however, in evaluating their effect on a beach. In deep water waves move at a velocity dependent on their length:

$$C = LT \tag{8.1}$$

where C = wave form velocity, L = wave length, T = wave period, and the

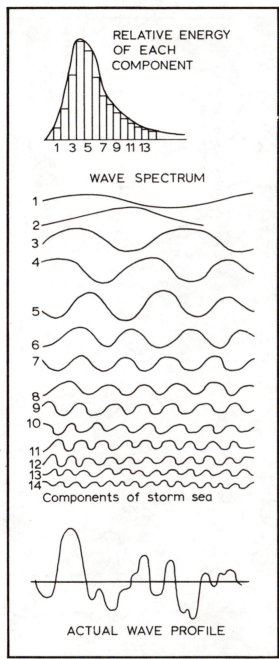

FIG. 8.3. An actual wave profile shown at the bottom is divided into its 14 components, to form a wave spectrum. The relative energy in each of these components is shown in the upper diagram (after Deacon, 1949)

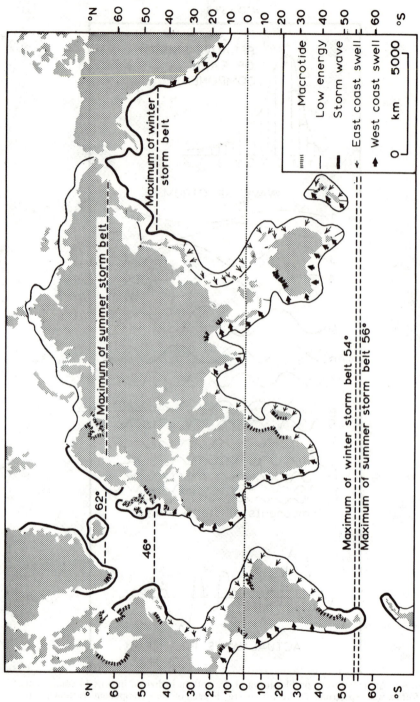

Fig. 8.4. Classification of world coastlines according to the distribution of wave types (after Davies, 1964)

relationship holds that

$$L \text{ (feet)} = 5 \cdot 12 \; T^2 \text{ (seconds)}$$
$$\text{or } L \text{ (metres)} = 1 \cdot 56 \; T^2 \text{ (seconds).} \qquad (8.2)$$

Waves do not influence the seabed at depths greater than their length. As they approach water shallower than this their length and velocity are reduced according to:

$$C^2 = \frac{gL}{2\pi} \tan h \frac{2\pi d}{L} \qquad (8.3)$$

where

C and L are as above;
$\quad g$ = acceleration due to gravity;
$\quad d$ = depth of water;
$\tan h$ = the hyperbolic tangent.

The bottom relief is, therefore, reflected in the bending of the wave crests as they enter shallow water by refraction (Davis and Fox, 1972; King, 1953). This process is particularly conspicuous in waves that are long, and therefore feel the bottom in relatively deep water. These long, low waves can travel immense distances over the ocean without losing their identity. They are the long swells that play a constructive role on beaches (Fig. 8.5). Experiments have shown that the mass movement in these waves in shallow water produces a forward thrust towards the land along the bottom that increases in

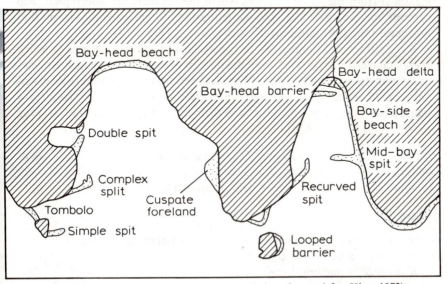

FIG. 8.5. Constructional forms on a highly indented coast (after King, 1972)

intensity as the breakpoint of the waves is reached. Because of the forward thrust the flat waves move sand landwards. They continue to be constructive inside the breakpoint, up to the limit of the swash.

The long constructive swells contrast in their action on the beach with the storm waves, which are destructive. Storm waves have a high steepness value (height-to-length ratio), and are often accompanied by a strong onshore wind, which increases their destructive capacity by creating a seaward current along the bottom that reinforces the seaward flow associated with mass movement under these conditions.

The direction of wave approach is of great importance in many aspects of coastal processes and their control. Short waves can approach at a considerable angle to the shore, and cause consistent downwave currents, especially on a straight or gently curving shore. This can result in serious erosion in areas where there is no adequate supply of beach material reaching the coast from updrift, for example along the Holderness coast, or at Selsey Bill in Sussex (England).

Long waves, on the other hand, suffer considerable refraction before they reach the coast, and currents in general will flow from zones of convergence of energy to zones of low energy. Waves are higher in the zones of convergence and lower in zones of divergence of energy, where longshore currents converge they often flow offshore through the surf zone as rip currents. These rip currents, which flow seawards through the breaker zone, can be a serious danger to bathers as well as playing an important part in the cellular nature of longshore current and sediment movement. The rip currents can be identified from the shore as zones where the breakers are interrupted. Rip currents are also related to a cellular pattern of inshore water movement that influences longshore current velocities (Cook, 1970).

Wave refraction diagrams can be constructed to assess the degree to which energy converges or diverges from the pattern of orthogonals, which are lines at right angles to the wave crests. The wave refraction diagrams are an essential step in assessing the alongshore wave power, which also requires a knowledge of the wave dimensions and their angle of approach. The following relationships have been established:

$$I_1 = K(EC_n)_b \sin a_b \cos a_b \qquad (8.4)$$

and:

$$I_1 = K''(EC_n)_b \cos a_b . (V_t)/U_m \qquad (8.5)$$

where

$I_1 =$ is the immersed load being transported along the coast;
$K = 0 \cdot 17$, a numerical constant;
$K'' = 0 \cdot 28$, a numerical constant;
$E =$ wave energy;

a_b = breaker approach angle;

C_n = wave form velocity;

V_t = longshore current velocity;

U_m = maximum horizontal velocity of the water particles in the wave orbit. The relationships fit empirical data derived from measurements of material movement by tracers (Komar and Inman, 1970; Komar, 1971). Various other theoretical relationships relating to the longshore current velocity and sediment transport alongshore have been discussed by Galvin (1967).

Longshore currents, as well as being instrumental in determining zones of erosion and accretion, also play an important part, together with tidal scour, in the dispersal of sewage effluent discharge into the sea. Beaches down-drift of the sewage discharge points may well suffer pollution for this reason.

Waves are a constantly varying force acting on the beach material. They vary through space and time, both because of changes in coastal aspect and offshore relief, and because of changes in wind direction and force over a very wide area of sea. This variability means that the beach is rarely in equilibrium with the waves, despite the fact that it may take only a few hours (relaxation time) for equilibrium to be reached under a new set of forces. This almost continuous state of disequilibrium is even more likely to occur where the tidal range is considerable, as the waves are always acting at a different level on the beach. It is, therefore, necessary to consider the range of variability within which the processes operate, and the resultant range of beach form. The time scale must also be borne in mind. The day-to-day variability will usually be less than that between seasons, and still less than that between extreme storms and normal calm weather.

These variations are important from many practical points of view. For example, during periods of storm when destructive waves attack the shore, the beach is lowered, and under these conditions the backshore zone often suffers erosion. This may take the form of damage to coastal defences under severe erosional conditions. Under natural conditions the cliff-foot support may be removed, resulting in slumping or erosion by hydraulic action, in unconsolidated and hard rocks respectively. The foreshore foundation may also be eroded and lowered if the beach is completely removed. In this case even if the same quantity of beach material returns after the storm the profile will be permanently lowered and the coast made more vulnerable. In some areas where seasonal changes are marked, tidal or river inlets through barriers may be completely closed during the period of maximum accretion, especially if this coincides with low river flow. This may result in transport problems. The coastal barriers of parts of west Africa exemplify this situation (Galliers, 1969). Thus in some areas inlet controls have to be built, which may have adverse repercussions along the adjacent shores.

(c) *Tides*

The variability of the waves is great compared to the regularity of the tidal processes. The tide may, nevertheless, play an important part in beach processes. The tidal range ensures that wave action is spread over an area of foreshore commensurate with its value, and tidal beach profiles may be differentiated from tideless ones for this reason. More important in many areas, however, is the effect of tidal streams. They are particularly important in areas where there is a residual movement resulting from differences between rectilinear ebb and flood velocity vectors, and where there is abundant loose sediment for the tidal streams to transport. The significance of the tidal processes will be exemplified in Section 8.7.

(d) *Wind*

Wind must not be ignored, as it can create very important currents in shallow water. When the wind is blowing strongly onshore these currents greatly enhance the destructive effect of steep waves. The wind also plays an essential part in the creation and maintenance of coastal sand dunes, which are important in the protection of many low coasts.

8.3. Materials in the Coastal System

Beaches and coastal forms depend on the interaction between processes and materials. This interaction can be considered from either direction: (a) influence of the material on the processes may be studied, or (b) the character of the material can be analysed to give information concerning the process. Some examples will make these two approaches clearer.

Storm waves will act differently according to the material on which they are operating. Steep storm waves breaking on solid rock may under certain conditions set up shock pressures that lead to the disintegration of the rock, followed by attrition of the blocks so formed. Similar waves breaking on shingle will throw some of it to the backshore zone to create a storm-beach ridge, which may extend far above the high-tide level. For example, Chesil Beach (in southern England) reaches 13·1 m above high tide level at its south-eastern end. Storm waves may also drag much shingle down the foreshore to create a step at their breakpoint. On sand beaches storm waves are entirely destructive, creating a steep beach scarp at the limit of their action and carrying much sand offshore to their break-point where they create a submarine bar, while some sand is also carried to deeper water offshore. An exception occurs where waves can wash over a low sandy barrier island, carrying sand into the lagoon behind. This process usually only takes place when the water level is raised abnormally by surge activity (Dolan, 1973; Godfrey and Godfrey, 1973).

Because of the different response of sand and shingle to waves, it is important to appreciate that dominant waves are those producing the most permanent forms. The nature of these forms will differ, however, with the type of beach material. Storm waves will generate major backshore ridges if the beach is composed of shingle; while on sandy beaches storm waves can create a submarine bar at their break-point. This latter is especially the case under tideless conditions, as in the Mediterranean and Baltic seas, and in the Gulf of Mexico.

The type of material also influences the effect of tidal streams and wind action. The wind can only directly affect sand and finer material, as it cannot move pebbles. Wind is most important in relation to the building of coastal dunes where there is an abundance of sand available on the foreshore, a constructive wave regime and a predominantly onshore wind direction. These conditions apply, for example, on the coasts of California, Oregon, and Washington (Cooper, 1958, 1967). De Jong (1960) and van Straaten (1965) have similarly considered the coastal dunes of the Netherlands. One important aspect of coastal control is the stabilization of natural dunes and their artificial creation, although, as mentioned later, this is not always desirable. Woodhouse and Hanes (1967) discuss dune stabilization, with reference to the outer banks of North Carolina.

Tidal streams are often fast enough to move fairly coarse sediment (King, 1964), but the forms normally associated with tidal streams in the offshore zone or in tidal estuaries usually consist of sand and only occur where there is sufficient sand available. In quieter areas, where the tide ebbs and floods over large expanses of flat ground bringing muddy water periodically over the area, tidal mud flats and salt marshes can be established, formed of fine sediment accreting vertically. Under some circumstances mud can be deposited on the open foreshore, to the detriment of recreational activities on holiday beaches. An example is cited in the case study in Section 8.7. Moderate- to fine-sized sand is the optimum for holiday beaches, because finer sand produces flatter beaches and hence a greater expanse of foreshore. Steep shingle beaches are the least desirable.

The different coastal processes operate in such a way that they leave their mark on the sediment-size distribution and character. Grain size is normally measured on a ϕ (phi) scale which has been so designed that the normal range of beach sediment sizes have positive values. The ϕ scale is a negative log. scale (to the base 2) but it can be directly related to grain sizes measured on a metric scale. Thus sand lies in the range $-1\,\phi$ to $4\,\phi$ (i.e. 2 mm to 0·0625 mm) silt sizes are $+4\,\phi$ to $+8\,\phi$ (i.e. 0·0625 mm to 0·0039 mm), and clay has phi values greater than $+8\,\phi$ (i.e. less than 0·0039 mm). It is possible to recognize various sedimentary processes from the nature of the size distribution curve plotted on logarithmic probability paper. The curve ideally should consist of three major log. normal sections (Fig. 8.6). The 'coarse' end represents the traction load, moved mainly by rolling along the bottom; the 'fine' end represents the suspension load and is usually limited in beach foreshore sediments, but may be important in offshore, lagoon, and dune deposits. The main, central part of the curve represents the saltation load, which may be divided into two parts, the swash and backwash respectively, each of which is deposited under slightly different hydrodynamic conditions. The great fluctuation in velocity in this zone allows finer particles to become

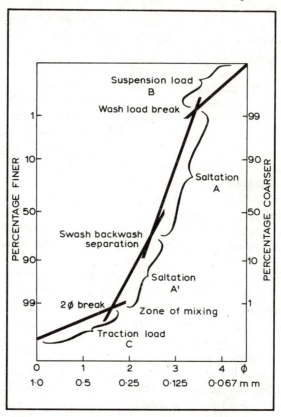

Fig. 8.6. Sediment-size distribution in terms of processes (after Visher, 1969)

trapped amongst the generally coarser ones, so that this zone tends to show poor sorting. In general the zones of maximum energy expenditure are those where the material is coarsest, and where fine material is in short supply. This accounts for the negative skewness (King, 1972) on the ϕ scale due to a tail of coarse grains. Fine material cannot settle in these highly active zones except where it becomes trapped amongst coarser grains. Thus the breaker, surf, and swash zones are characterized by large mean size, poor sorting, and negative skewness on the ϕ scale. On the other hand, dune sand and some unidirectional current sediments are characterized by positive skewness, indicating a tail of fine sediment. For these purposes skewness may be defined as:

$$Sk = \frac{\phi_{16}+\phi_{84}-2\phi_{50}}{2(\phi_{84}-\phi_{16})} + \frac{\phi_5+\phi_{95}-2\phi_{50}}{2(\phi_{95}-\phi_5)} \qquad (8.6)$$

where ϕ_{84} is the ϕ value coinciding with the 84 per cent point on the size distribution curve and similarly for the other ϕ notations (Folk and Ward, 1957).

Sand is by far the most common material to be found on beaches, particularly in low latitudes. Much sand is formed of quartz material, because this mineral is particularly resistant to weathering. In some areas, however, sand consists of suitably-sized basalt fragments, such as the beaches of Iceland and Hawaii. Other sands are organic, such as the foraminiferal sands of western Ireland and the coral sands of the tropical seas. Some beaches contain heavy mineral sands, including gold. The sorting action of the waves concentrates the heavy minerals in places forming rich reserves. These have been located on currently existing beaches and in both raised beaches (as in western New Zealand) and submerged beaches (as off Alaska), where they have been exploited.

Other more mundane, but more essential, resources in the form of sand and gravel for construction purposes are available on beaches and have been exploited. It is now realized, however, that the beach sediments are one of the most vital, but decreasing, elements of coastal protection and conservation schemes. Instead of being removed, these materials are now frequently put on the beach (Sect. 8.6c). In seeking sand and gravel resources, contractors are now looking to the offshore zone to supply their growing needs. Material has been obtained from offshore for some decades, but now a systematic survey of resources has been undertaken in many areas, such as the eastern United States coast (Duane, 1969), and the fluvio-glacial deposits off north Lincolnshire have also been assessed for sand and gravel supply (Robinson, 1968). These deposits contain both sand and coarser material.

The shingle that forms such a conspicuous feature around the coast of Britain, including Chesil Beach and Dungeness, is more plentiful around the high latitude coasts, probably because much of it was derived from reworked glacial drift deposits. Few rivers now bring shingle-sized material to the coast, but fluvio-glacial sediments and other drift deposits are widespread in the shallow marine zone in many parts of the coastal zone of north-west Europe. Eroding coasts in drift also supply mixed size sediments, for example Holderness and north Yorkshire in eastern England and the cliffs of the northern part of Long Island (U.S.A.).

8.4. Beach Form

The forms of coastal features are related both to process and material. These relationships are determined by the whole environment, which controls the nature of the processes. It is possible to analyse the relationships both by studying the effect of the processes and materials on beach form, or by using the form to infer the processes.

Fine sediments, for example, can only be deposited in suitable environments, the resulting morphology depending on the environment. Mud can only accumulate in sheltered areas, such as salt marshes, lagoons, runnels landward of high ridges, the lower foreshore where shelter is provided by offshore banks, and the quieter shelf areas (such as the northern Celtic Sea off southern Ireland). Each environment differs from the other and each can be differentiated in terms of the local controls including climate, tidal range, type of sediment, type of vegetation, etc.

Sand can be deposited on a wide variety of forms in many different environments, including several types of coastal dune; beach features such as berms, ridges, cusps; and many types of bars and barriers, spits, and tombolos, as well as tidal features, offshore banks, sand waves, and sand ribbons (Figs. 8.1 and 8.5; Table 8.1). Sand is the most widespread and versatile beach material, partly because it includes the sizes most easily picked up by moving air or water.

Shingle is less common than sand, but it does give rise to recognizable beach forms. The most important are the storm shingle ridges that are conspicuous in the backshore region. Such ridges can cover extensive areas of the backshore, for example at Dungeness and Orfordness, in England. A careful study of such shingle structures, which are often clearly revealed on air photos, provides valuable evidence for deciphering the development of the coastal zone through time, thus adding another dimension to the study.

By studying the form of coastal features it is often possible to derive useful evidence concerning the processes that formed them. A good example of the relationship between process and form is the way in which a beach becomes aligned to the direction of approach of long swell waves that are refracted by the bottom morphology according to their length and direction of approach. This is a two-way feedback relationship, because the extent of wave refraction depends on the wave direction of approach and length as well as the bottom relief, so that both process and form are involved in explaining the final beach morphology, both in plan and profile. Davies (1959, 1960) described examples from the south coast of Australia, where long swells have moulded the beach outline until it is in equilibrium with their action and lies parallel to the offshore contours. Similar examples occur on the west coast of Donegal, Ireland (King, 1965). Once equilibrium has been established the longshore drift will be reduced to a minimum as the waves approach parallel to the shore at all points on the beach.

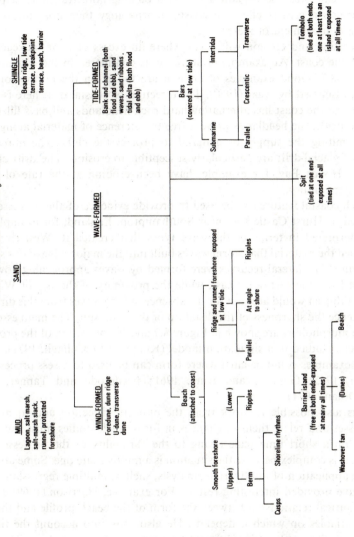

TABLE 8.1
Material and related forms

MUD
Lagoon, salt marsh, salt-marsh slack, runnel, protected foreshore

SAND

- **WIND-FORMED**
  Foredune, dune ridge U-dune, transverse dune

- **WAVE-FORMED**
  - **Beach** (attached to coast)
    - Smooth foreshore
      - Berm (Upper)
      - Ripples (Lower)
    - Cusps
    - Shoreline rhythms
    - Ridge and runnel foreshore exposed at low tide
      - Parallel
      - At angle to shore
      - Ripples
    - Barrier island (free at both ends-exposed at nearly all times)
      - Washover fan
      - Beach
      - (Dunes)
  - **Spit** (tied at one end-exposed at all times)
    - **Bars** (covered at low tide)
      - Submarine
        - Parallel
      - Intertidal
        - Crescentic
        - Transverse
    - **Tombolo** (tied at both ends, one at least to an island - exposed at all times)

- **TIDE-FORMED**
  Bank and channel (both flood and ebb), sand waves, sand ribbons, tidal delta (both flood and ebb)

SHINGLE
Beach ridge, low tide terrace, breakpoint terrace, beach, barrier

A very important relationship between form and processes is that between wave direction, wave length, and coastal outline, as these variables determine the amount and direction of longshore transport of material, and through this process the nature of many coastal forms. Thus zones of accretion and erosion can often be explained in terms of longshore transport, a process that exerts a greater effect on coastal morphology than does the on- and off-shore movement of material.

Accretion and erosion will occur where an excess of material reaches or leaves the coast. An example of accretion is considered in the case study in Section 8.7. Good examples of erosion are found on the north Yorkshire coast, discussed by Agar (1960). The direction of coastal drift here is to the south and the coast has alternating hard rock headlands and bays filled with glacial drift. The headlands prevent free transference of material alongshore, thus limiting the supply of material to protect the cliffs. The unresistant cliffs of glacial drift are particularly susceptible to erosion. The drift cliffs in Robin Hoods Bay, for example, have been eroding at the rate of about 28 m per century.

Each coastal feature can be used to provide evidence of the processes that formed it. Hurst Castle spit, near Southampton, England, for example, can be interpreted in terms of the wave types that created it. Westerly waves supplied the material that storm waves built into the major ridge of this shingle structure. The lateral recurves were formed by waves approaching down the Solent from the east-north-east, while the proximity of the Isle of Wight to the south-east would account for the absence of long waves from this direction and hence the sharpness of the distal tip of the main spit. The main essentials of the morphology are shown in Figure 8.7 and the operation of the processes has been studied in a simulation model (King and McCullagh, 1971). Many other examples could be cited where form can be used to assess process (e.g. Schou, 1945; Guilcher and King, 1961; Niederoda and Tanner, 1970; and Dolan, 1971).

It is also possible to work from the opposite direction and to measure processes and relate them to changes in form. Such studies must usually be made on a short time scale owing to the variability of the processes. The analysis is complex because the situation is a multivariate one. Some attempts at the application of multivariate analysis, such as multiple regression analysis, have provided interesting results. For example, Harrison (1969) derived an empirical relationship between the form of the beach profile and the process variables on which it depends. He also took into account the time to reach equilibrium by using different time intervals in the analysis. This is an empirical approach, but in the present state of knowledge concerning the complex and active beach environment this has some advantage over an oversimplified theoretical approach.

Harrison's analysis showed the importance of the water level in the beach

in determining the amount of sand moved and the beach profile adjustment.
It also showed the significance of wave height and steepness in affecting beach
profiles. Spectral analysis of the data by King and Mather (1972) showed the
importance of the tidal cycle on the beach characteristics through the tidal

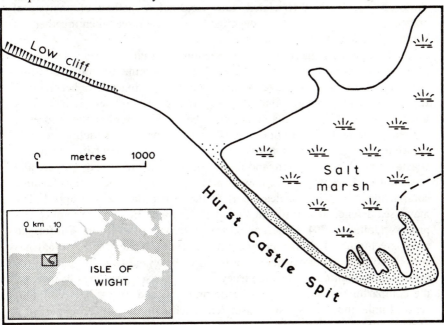

FIG. 8.7. The form of Hurst Castle Spit

influence on the wave dimensions and the water level in the beach. One of
Harrison's predictor equations indicates the type of result such studies can
provide:

$$\Delta Q_f = 5{\cdot}803 - 113{\cdot}151(\bar{H}_b/g\bar{T}_b{}^2)^{\frac{1}{2}}{}_{0{\cdot}0} - 52{\cdot}111(h/b)_{3{\cdot}0}$$

$$-7{\cdot}269(\alpha_b)_{6{\cdot}0} + 2{\cdot}724(\bar{D}/\bar{z})_{6{\cdot}0} - 35{\cdot}067(\bar{m})_{12{\cdot}0} \qquad (8.7)$$

where

ΔQ_f = quantity of sand on the foreshore;
$\bar{H}_b$ = breaker height;
$\bar{T}_b$ = wave period;
h = hydraulic head;
b = breaker run-up;
α_b = acute angle between the breaker crest and the shore;
$\bar{z}$ = mean trough-to-bottom distance in front of the breaking wave;
$\bar{D}$ = grain diameter;
$\bar{m}$ = beach slope.

The subscripts outside the brackets refer to the time lag before low water.

8.5. A Land-unit Approach to the Analysis of Coastal Areas

The previous sections have stressed the processes, materials, and forms individually with reference to one of the other two major elements. These three are so closely interrelated, however, that a method of analysis of coastal problems based on an integrated view of the three taken together is a useful approach in coastal zone management.

In adopting the suggestion made in Section 8.2a that coasts may be studied in terms of a classification into discrete land units some difficulties do arise. Although these units may have specific responses (e.g. in terms of beach form) to the physical processes that they experience, nevertheless it has to be appreciated that beach materials may be highly mobile and processes extremely variable. In addition, along a coastline the interaction between adjacent units is usually dynamic. This property is of great importance in coastal control, preservation, and improvement. The building of groynes and breakwaters prevents the free flow of material along the shore, thus forming compartments in which at least part of the beach material is trapped. The many sea-defence structures on the south coast of England exemplify this process (Jolliffe, 1964). The compartments along part of the Californian coast are discussed by Bowen and Inman (1966), while Cherry (1965) has considered coastal compartments in terms of heavy minerals on part of the north Californian coast. The tendency in some coastal measures to increase the compartmentalism may well be detrimental, and the subdivision into natural units must always be considered. Each natural unit forms part of a larger dynamic system, and control measures must be based on an understanding of the whole system.

Land units (as in the *land-systems* approach described in Chapter 13) may be distinguished in two directions: normal to the coast and parallel to it. In the first category are the backshore, foreshore, nearshore, and offshore zones. Further subdivision is possible, for example the swash, surf, and breaker zones forming important elements of the nearshore and foreshore zones (Fig. 8.1).

The units parallel to the coast are best delimited by significant changes in morphology. Thus the rocky headland must be distinguished from the bay-head beach (Figs. 8.1 and 8.5), these two often forming the units of a small cell or system, in that movement of material is restricted to the bay. A large system is characteristic of a more open, smooth coast, and may consist, for example, of a headland, a barrier spit, barrier islands, tidal inlets, lagoons, offshore submarine bars, and the offshore shelf zone, all of which interact in a complex, often very extensive system, stretching for up to 100 or more kilometres along the coast. The offshore barrier itself consists of separate units: the berm, backshore zone, dunes, and wash-over area may all be present (Fig. 8.1). Fire Island on the south side of Long Island, New

York, provides a good example of a barrier island of this type; it is 51 km long and rarely more than 0·8 km wide. Other similar barriers along the Outer Banks of North Carolina at Cape Hatteras and Cape Look Out are described by Dolan (1973) and Godfrey and Godfrey (1973). The former is a man-altered barrier, while the latter is natural. They are compared in Section 8.6, to illustrate the deleterious interference of man along the coast.

In examples of this type the beach affects the dunes and these affect the wash-over area, while parallel to the shore a tidal inlet through the barrier island influences both ends of the barrier islands that flank it. A headland influences the pattern of wave energy and through this the movement of material into the neighbouring bay and the form of the headland and bay-head beaches. These have been shown to develop a form that represents the state of equilibrium between the processes operating and the material on which they operate. Thus there is mutual adjustment between process, material, and form, and an interaction of energy exchanges between natural land units.

The logarithmic spiral beach form illustrates this type of interaction clearly, and produces a recognizable morphological result. Yasso (1965) has shown that a headland beach develops a logarithmic spiral form with gradually increasing curvature towards the headland. This curve can be defined as:

$$r = e^{\theta \cot a} \tag{8.8}$$

where r = vector length at angle θ from the log. spiral centre, and a = angle between a radius vector and the tangent to the curve at that point, and is a constant for each log spiral. The expression also can be stated in linear form by taking logarithms of both sides of the equation (Yasso, 1964, 1965; King, 1971). In this form it can be solved by the method of least squares to assess the optimum position of the log. spiral centre and the value of a. The close fit of the shoreline to this form in suitable areas shows that such a geometrical form is an equilibrium one in which a balance between the wave energy distribution, the size of material, and the beach profile and plan is achieved. Under these conditions energy increases in the down-drift direction from the headland.

Half Moon Bay in California (Fig. 8.8) illustrates the pattern clearly and also shows how the size of material is adjusted to the energy distribution, as are the beach form and gradient. The material increases in size from a value of 2·56 ϕ in the lee of the headland to 0·63 ϕ on the open coast down-drift of the headland as the energy increases. The increase in sand size is also related to an increase in beach gradient in the same direction. Thus plan form, profile, and material character can be related to wave energy distribution. This is the most important process variable in this area, where waves come predominantly from the north-west. The close relationship between material, process, and morphology can best be demonstrated where, as in Half Moon Bay, a state of dynamic equilibrium can be established in a relatively small closed system, consisting of the bay between controlling headlands.

The same relationship can be demonstrated in a more open coast, where spits can also be shown to conform to the log. spiral form. This form fits well the coastal outline of Sandy Hook spit bar and also the spit at Gibraltar Point in Lincolnshire, England. Both respond to wave refraction around the distal tip of the spits, to form an equilibrium curve in which form is adjusted to process.

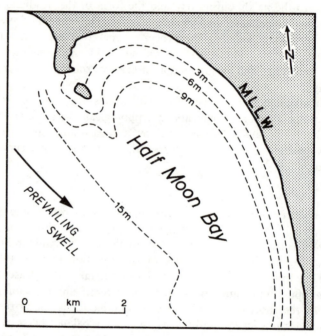

FIG. 8.8. The form of Half Moon Bay

Silvester (1960) has used this form, which he called a half-heart, to assess the direction of longshore drift from the morphology of the coast as shown on maps and charts. He assumed that the coastal outline becomes adjusted by the movement of material alongshore to give the log. spiral form, in which longshore movement is towards the open part of the curve. These forms will be best developed on sand beaches which become adjusted in outline to the long, refracted swells that create the log. spiral form. Such studies of the beach can provide valuable information concerning processes and also data that can be usefully applied in coastal protection work (Sect. 8.6). A practical example is given by Yasso (1964) who shows that along the spit at Cape Cod a log. spiral form develops between groynes in response to the pattern of wave attack.

Coastal land units can be mapped to delimit homogeneous morphological units, in such a way that the character of their material and the processes that formed them can be readily identified and tabulated. In nearly all instances the three elements of form, material, and process are closely interrelated, and information concerning one provides data by which to assess the others. Table 8.2 is an attempt to identify the units that together make up the system forming the zone of accretion in south Lincolnshire (Fig. 8.9). In a dynamic,

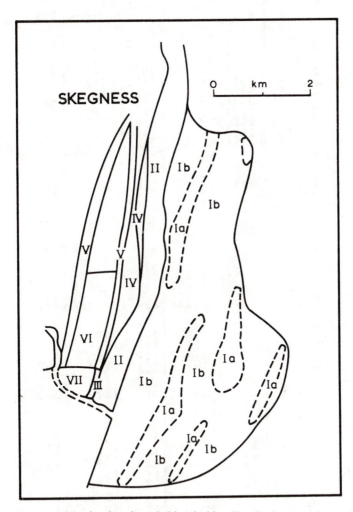

FIG. 8.9. The coastal land units of south Lincolnshire. For the key to unit numbers see Table 8.2

TABLE 8.2

Land units for the zone of accretion, south Lincolnshire coast, England

Unit	Genetic description	Morphology	Material	Process	Diagnostic property	Variability	Some actual rates of change
I a	Offshore tidal stream bank	Elongated parallel to tidal streams	Sand—moderate, very well sorted, low skewness, but generally +ve	Rectilinear tidal streams where sediment abundant	Covered except at low spring tide, elongated with asymmetrical sand waves, elongation parallel to tidal streams	Slow horizontal movement	Probably about 50 m/yr (data are not precise)
I b	Offshore tidal stream channel, ebb and flood	Channel that shallows in direction of dominant tidal stream	Sand and gravel—coarse, rather poorly sorted, slight positive skewness	Strong tidal streams, up to 200 cm/sec. Either flood- or ebb-dominated	Flood channel shallows in direction of flood stream. Ebb channel shallows in reverse direction	Slow horizontal movement	
II a	Foreshore wave-built ridges	Elongated almost parallel to coast, slight divergence to south	Sand—moderate size, generally well sorted, negative skewness	Constructive waves building equilibrium gradient on flat overall slope	Covered at high tide, exposed at low tide on foreshore, steep landward slope and equilibrium seaward slope of firm smooth sand	Regular landward movement, width of sweep zone dependent on ridge height. Superimposed sweep zones show accretion and vertical upgrowth of foreshore	Maximum 104 m/yr, normal 38 m/year maximum accretion 89 m³/m/yr
II b	Foreshore runnels, wave-and tide-formed	Elongated landward of ridges draining south	Sand, mud, and vegetation—fine size, very poorly sorted, positive skewness	Sheltered from wave action, tide flow concentrated	Rippled or muddy sand, covered at high tide, exposed at low tide on foreshore		
II c	Smooth foreshore fine sediment	Low gradient, lower foreshore where bank shelter prevents ridge growth	Sand and silt—fine size, poorly sorted, positive skewness	Wave action very slight, tidal silt sedimentation	Lower foreshore, muddy or rippled surface	Narrow sweep zone shows slow accretion and little movement	Approx. 8 cm/year up-growth

III	Wave-built spit	Elongated and distally recurved, low, broad proximal end	Sand, shingle, and vegetation —mixed material	Wave action— longshore drift, constructive waves	Continues line of coast where this turns abruptly west	Trends of length, area, volume, and horizontal movement show development	Volume increase 600 m³/year Length increase about 9 m/year Horizontal increase 6m/year at 3·3 m
IV a	Foredunes, wind-formed on foreshore ridges	Curved ridges, convex seaward	Sand and vegetation— similar to fore-shore ridges	Wind and vegetation	Exposed above high spring tide, with vegetation	Successive profiles show rapid upgrowth in suitable areas	30 cm/year up-growth approx. where growth is rapid
IV b	Salt marsh slacks, tidal deposition in foreshore runnels	Elongated strips between foredunes, low gradient	Silt and vegetation— similar to fore-shore runnels	Tidal sedimentation and some wind action, vegetation	Flooded at high spring tide only, with salt-marsh vegetation	Successive surveys show slow, regular accretion	
V	Dune ridges, wind-formed from fore-dunes	Elongated or arcuate ridges, single or complex	Sand, vegetation— similar to fore-shore	Wind and vegetation, soil formation	Mature vegetation, behind foredunes and well stabilized	Stable, show little change	
VI	Mature salt marsh, tidal deposition in dune ridge shelter	Flat area with low concave interfluves between creeks with levees	Silt, sand, some shingle and vegetation —mixed sediments, some black mud	Tidal sedimentation occasional wave wash-over, old beach ridges, vegetation	Covered only by high spring tide from well-developed creeks, pans, and mature salt-marsh vegetation	Very slow growth, apart from local areas and storm wash-over	
VII	Immature salt marsh in spit shelter	Flat area with marginal creeks	Silt, sand, and vegetation— similar to fore-shore runnels	Tidal sedimentation wave wash-over at low proximal end of spit; wind and vegetation	Covered by all high tides, with creeks developing at margins; immature salt-marsh vegetation	Successive surveys show regular and fairly rapid upgrowth and vegetation spread and development, maximum at creek margins	Average 2·3 cm/year upgrowth

as opposed to the classificatory sense, this is not an entirely closed system as material enters the system from outside in greater volume than it leaves it. In terms of interaction between the units, however, the zone of accretion can be considered as a system of land units, all of which are interrelated, but which can be differentiated in terms of their morphology, material, and processes. In the first column (Table 8.2) the genetic description of the unit is indicated, and the second gives the essential elements of the morphology of the units. The third column indicates the dominant material and those characteristics of it that relate to the morphology and process. The next column lists the processes. Thus fine sediments are associated with low-energy processes that operate in sheltered environments to create low-gradient morphology. The coarser sediments are associated with higher energy processes in the more exposed areas, or those in which energy is concentrated, the spit and ridge crests and seaward slopes and the tidal channels. In the next column some of the diagnostic properties are suggested in terms of surface character and relation to water level. Some of the units can best be subdivided into distinct subdivisions, for example the foreshore may consists of three distinct units: (i) the smooth lower foreshore, (ii) the ridges, and (iii) the runnels on the middle and upper foreshore, in places extending onto the lower foreshore. These last two form the foundation for the backshore features, the foredunes and marsh slacks respectively, while an earlier spit could well be the foundation of the main dune ridges. Thus the units are all interrelated to form a whole system. The last two columns in the table suggest the variability in the units and how this can be best illustrated, as well as giving some actual quantitative data on the rates of change in this particular area.

8.6. Human Interference in the Coastal System

(a) Introduction

Man can interfere with the natural dynamic coastal system either intentionally or unintentionally. The former implies an awareness of how the system operates, the latter does not. Intentional interference usually aims to prevent or reduce erosion, to conserve the beach or reclaim land from the sea. Unintentional interference is usually the result of a human use of the coastline, such as for recreational activities. Measures designed to control port approach channels through estuaries or inlets often have injurious results. A good example of the latter is the erosion that followed the opening of an inlet for navigation at Thyboroen in Denmark (Bruun, 1954). The opening was cut through an originally straight barrier along the coast of west Jutland in 1825. Since then the coast has suffered continuous erosion on both sides of the inlet. Between 1921 and 1950 the northern barrier retreated 4·77 m/year at the maximum and on the southern barrier the maximum was 2·74 m/year, despite attempts to stabilize the area by building groynes. The beach has become steeper during the recession.

Estuaries are often a zone of conflict in land use and have thus been studied from the management point of view by politicians, economists, planners, and ecologists. The ecological implications are often the last to be considered (Cooper, in Hite and Stepp, 1971). Under natural conditions the temperate zone estuaries are among the most productive areas, with their marshes and shallow waters in which shellfish and other fish, often in their young phases, abound. The ecology is naturally complex and specialized in the different habitats available; it is, however, a delicate and easily disturbed environment, and dredging, filling, or similar interference prevents the re-establishment of the natural ecosystem. Thus it is essential that the planning of environmental management systems for these areas should put ecological considerations first. Margins of error should be on the conservative side to prevent irreparable damage in an environment for which many basic data remain unknown and for which there are many competing land uses.

(b) Coast protection

The best protection for any coast is a wide, high beach. On low and eroding coasts, however, it is sometimes necessary to build solid sea-defence structures, including sea walls, breakwaters, jetties, and groynes. Sea walls are designed to protect the backshore zone from direct erosion, but because of their impermeable character, which increases the backwash, they tend to produce a destructive wave effect that builds up by positive feedback. Groynes are built normal to the beach and are designed to trap beach material moving alongshore. In that they are designed to raise the level of the foreshore they are beneficial to their immediate hinterland, but, in that they reduce the amount of material passing down-drift, they are detrimental to areas down-drift. Their effects show clearly the necessity of taking into account the whole coastal system, rather than just one unit. The up-drift units must also be considered because groynes will fail in their purpose if there is no material moving into the area from up-drift or offshore.

Breakwaters can be built either normal or parallel to the coast. Those built normal to the coast act as large groynes and trap longshore transported material. Their aim is usually to prevent silting of tidal inlets or estuaries, or to provide shelter for shipping on an open coast. The type normal to the shore have the same effect as groynes, often with serious consequences on the down-drift beaches, as exemplified by the erosion east of Santa Barbara breakwater in California. The breakwater was built in 1928 and erosion spread as much as 16 km down-drift after only a few years, as the breakwater trapped the longshore drift of 214,360 m^3/year. Breakwaters built parallel to the shore provide shelter and have a different effect in that sand will accumulate in their shelter as waves refract and diffract around them. Silvester (1960) has suggested that concrete defences could be so placed as to take advantage of the natural log. spiral form of equilibrium beaches (Sec. 8.5). The defences should be placed at suitable intervals along the shore, as indicated in Figure 8.10, to allow the log. spiral form to develop between them. In this way a dynamic equilibrium could be maintained, with minimum land loss and a relatively stable coastal configuration.

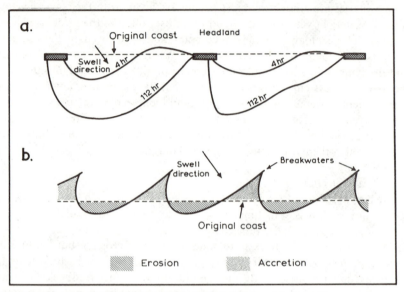

FIG. 8.10. Stages in the development of a stable model log spiral coastline
a. Model experiment after 4 hr and 112 hr
b. Theoretical stabilization of coastline with one predominant swell direction (Source: Silvester, 1960)

(c) Beach improvement

Methods of beach improvement which avoid possible detrimental effects elsewhere are being increasingly used. These methods are based on the knowledge that a wide, high beach is the best form of coastal defence and improvement. Sand bypassing and beach replenishment are two of the most widely used methods. In bypassing, sand is moved artificially from up-drift beaches to down-drift ones. For example, sand is pumped by suction from one side of South Lake Worth inlet in Florida to the other. Sand bypassing was started when groynes failed to prevent erosion induced by breakwater construction. About 48,000 m³/year were moved, where the estimated long-shore drift is about 225,000 m³/year. After five years stability was reached over a mile down-drift, although erosion started when pumping was stopped. Now pumping passes about one-third of the littoral drift alongshore.

Beach replenishment by dumping material of a suitable size on the fore-shore is becoming a useful means of coastal improvement. A good example of successful beach filling is that carried out at Virginia Beach, Virginia, United States. The natural beach sand has a median diameter of 0·3 to 0·4 mm. A total volume of over 1 million m^3 of sand was used to restore the beach between 1952 and 1953, and smaller amounts have maintained stability thereafter. In assessing suitable beach fill it is necessary to consider the size distribution of the natural sand and artificial fill, and to select a material that is at least as coarse as the natural sand (Krumbein and James, 1965).

Dune stabilization can be a useful means of improving the backshore zone, although sometimes this can lead to reduction of the width of the foreshore, as recorded by Dolan (1973) along the Outer Banks near Cape Hatteras, North Carolina, United States. Artificial dunes concentrated hurricane wave energy on the foreshore with deleterious effects on the beach. The lower natural dunes allowed the wave energy to be spread across the whole barrier island, which gained material by this process of overwashing. The beach width is maintained because the backwash is reduced (Fig. 8.11). Thus in some circumstances a normally valuable backshore improvement method can be detrimental. This illustrates the complexity of coastal processes and the need to consider each coastal system as unique.

Coastal hazards, such as flooding (see also Chap. 5) and the damage caused by hurricanes, and the human reaction to them along the shores of Megalopolis, the heavily populated north-eastern coast of the United States, have been studied by Burton et al. (1969). Barrier islands dominate this stretch of coast, which they have divided into four types according to their human adaptation: village, urban, summer, and empty shores. The environment is a fragile one, as the natural processes by which barrier islands migrate inland by wash-over flooding in storms cannot readily be combined with human occupation. Attempts at artificial control, as already mentioned, are not conducive to natural stability, which demands flexibility and not the rigidity imposed by man. The old village shores have become adjusted to flood hazards, but some of them, such as Chincoteague in Virginia are built less than 3 m above mean sea level, and do suffer severe flooding under storm conditions, such as the flood of 6–7 March 1962. The flooding comes mainly from the bay side, as the dunes on the barrier protect the village from direct marine action. Local people are well aware of the danger during a storm, for their houses are built on piles, but even so evacuation follows storm warnings. The urban and summer shores are often protected by sea walls, as for example the summer shore at Sandy Hook, New Jersey, an area densely developed for recreational purposes. The artificial protection is necessary because the beach is only 60 m wide and dunes and their vegetation have been destroyed over much of the area, which is liable to heavy storm damage. Major floods reach 3 m elevation and 86 per cent of the area is below this level. The 1962 storm made eleven breaches in the Sandy Hook spit. There is a high degree of storm hazard awareness in the zone and many private adjust-

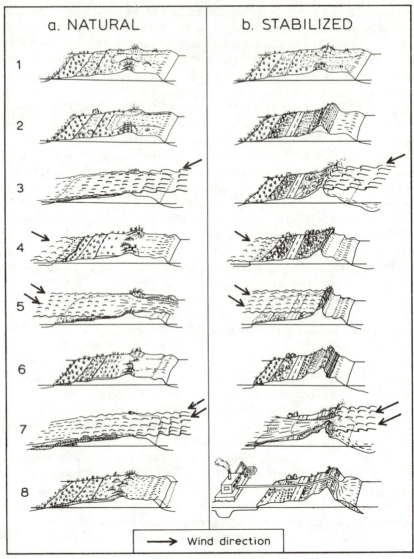

FIG. 8.11. Natural and man-controlled barrier island shorelines (Source: Godfrey and Godfrey, 1973)

To begin with both the natural and the artificial barrier systems are alike (stage 1), thereafter they may differ as follows:

Stage	Natural	Controlled
2	No change	Stabilization of an artificial dune to protect a road
3	Storm has overwashed the natural barrier	Storm waves have eroded the artificial dune, foreshore and nearshore zones, with some positive feedback

ments are made, while the major public adjustment is the 3 m high sea wall and a number of jetties to control sand movement. Sensible planning is essential if such coasts are to be used properly. In this area further intensification of land use is liable to take place, and this will require the maintenance and strengthening of coastal defences that are in many ways undesirable, since they prevent the natural processes operating that alone can maintain a dynamic stability in this delicate natural system. The ideal is to create zones in which the natural processes can operate unhindered, and this can be achieved by designation of National Seashore zones, in which building and human interference are prohibited. The contrast between the barrier islands at Cape Look Out and Cape Hatteras, natural and man-controlled areas respectively, clearly reveals the dangers of interfering with natural processes (Godfrey and Godfrey, 1973; and Dolan, 1973). Figure 8.11 illustrates the response of both natural and man-controlled barrier islands to the effects of storms of varying intensity and direction.

8.7. Case Study: The Lincolnshire Coast, England

The coast of Lincolnshire suffers both from erosion and accretion. The central part of the coast between Mablethorpe and Skegness has been subject to erosion since the protective offshore barrier, behind which the outmarsh built up, was destroyed. Its remnants moved onshore in the stormy thirteenth century to form the dunes, which are the natural protection of the reclaimed marshland behind them. Where the dunes do not form an adequate protection seawalls and banks have long been an essential part of the sea defences, particularly as the sea level has been rising slowly so that now the reclaimed land is well below the high-tide level. The sea walls, however, have a deleterious effect during storm conditions. The storm surge of 1953 illustrated the danger clearly. Sea level was raised by about 2·5 m above the normal high tide and the water overtopped the walls and washed out their foundations

4	Overwash terrace is raised and re-colonized	Badly eroded stabilized dunes and a rebuilding berm
5	Waves from the sound overwash the natural barrier	Sound storm tide erodes the land side of the dune, damaging the property and vegetation supposedly protected by it
6	Natural barrier undamaged by the storm	Further artificial rebuilding of the dune
7	Violent storm causes overwashing but no permanent damage to the natural barrier	Same violent storm breaches artificial dune with disastrous results to human installations
8	New salt marsh growing on overwash sediment and stability of the natural barrier is increased	Artificial beach nourishment at great cost, attempts to repair the damage on the narrowed seaward slope and erosion of the marsh further narrows the barrier from the land side

from behind, resulting in their collapse and destruction by the storm waves. Beach material was washed through the breaches with the flood water, and carried seawards by the destructive wave action, thus exposing the clay foundation of the beach to erosion. Since the storm stronger and higher walls have so far proved adequate protection, although they have yet to be tested under major surge conditions. Groynes built across the foreshore have raised the beach level about one metre, creating an apparently stable beach profile, in that it now has a narrow and static sweep zone.

The whole Lincolnshire coast, during the storm surge of 1953, illustrated very clearly the beneficial effect of a wide high beach in coast defence, as the major breaches all took place where the beach was low and narrow (Fig. 8.12). Where it was wide and high, in the zone of accretion, only minor dune cliffing occurred. The value of the dunes, in providing a flexible coastal defence, has been recognized in the work done to promote and stabilize them by planting marram grass, and placing brushwood fences to trap sand and prevent human interference.

In the zone of accretion south of Skegness the beach poses different problems that affect its function as a recreational asset (Fig. 8.13). On this stretch the coastal morphology becomes more complex and more coastal units occur. They are listed and described in Table 8.2. The essential reason for the abrupt change from erosion north of Skegness to accretion south of the town is the pattern of offshore tidal banks that start at this point on the coast. The flood channel between the Skegness Middle Bank and the shore allows material to move onto the lower foreshore just south of Skegness where it is built into ridges by wave action (King, 1964). Runnels develop landward of the ridges and the ridges move up the beach owing to their southerly divergence from the shore. The ridges provide shelter in the runnels in which mud can be deposited.

The belts of mud represent a problem on this recreational shore. The ridges, which are built by long constructive waves coming from the direction of maximum fetch to the north, represent the action of waves in creating an equilibrium gradient on a beach that has an overall gradient lower than the equilibrium one. As the ridges become stabilized on the upper foreshore by the development of new ones to seaward, they collect wind-blown sand and become foredunes, while the runnels build up with tidal silt, aided by vegetation, to become salt-marsh strips. In this form the foredunes and marsh strips become an asset on the coast, so that attempts to overcome the problem of the mud on the foreshore are based on speeding up the natural processes of foredune formation and marsh development.

As long as man works with and not against the natural processes, the results are likely to be beneficial. In order to accomplish this a sound knowledge of how natural processes operate is essential, and one important element of this is an appreciation of the variability within which natural

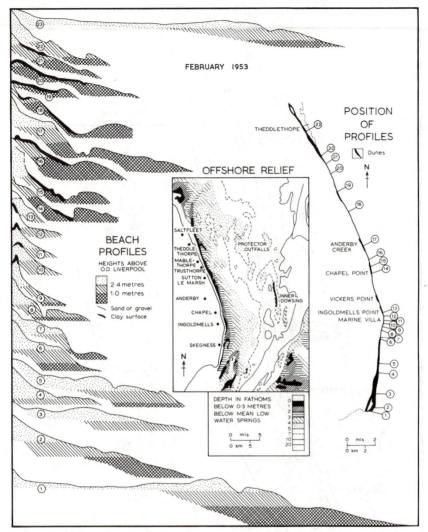

FIG. 8.12. Profiles of the Lincolnshire beaches surveyed after the storm surge of 1953 (after King, 1972)

processes operate, and their rate of growth and change. In order to provide such information, which is particularly relevant in the active coastal environment, a measure of the dynamic range within which the different coastal units vary is valuable. The final columns in Table 8.2 attempt to provide such information. In the foreshore units such information can be usefully indicated by the sweep-zone method, which provides information on both the short-term variability and longer-term development of the foreshore. The sweep zone is defined as the zone across which any one beach profile varies

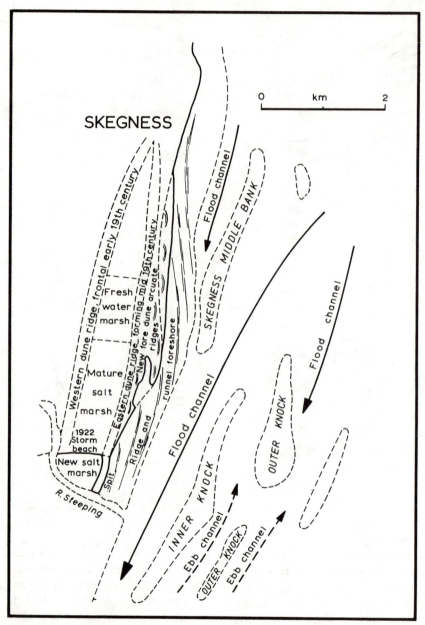

FIG. 8.13. The Lincolnshire coast to show the morphological character of the offshore, foreshore, and backshore zones (compare with Fig. 8.9)

through time. A wide sweep zone indicates a variable beach, and a narrow one a stable beach. Sweep zones on ridge and runnel beaches are wide because the upper limit connects the ridge crests, while the lower one joins the runnels. Thus the higher the ridges, the wider the sweep zone (King, 1973).

In developing dunes and salt marshes, time-series data, shown by successive profiles or maps, can provide growth rates. In the offshore zone the horizontal movement of the banks and channels through time is more significant than vertical growth, so the dimensions require change. On the spit that prolongs the main outer dune ridge trends of growth in position, volume, area, and length all provide information concerning the dynamic development of the units of the system that make up the zone of accretion south of Skegness.

All the units interact and are mutually interdependent. Thus as the offshore banks migrate the zone of accretion on the foreshore moves. The ridges control the runnels, and form the foundations for foredunes, while the runnels develop into the marsh slacks. The mature dunes protect the mature marsh and are developed from foredunes. The eastern dune ridge is prolonged to form the spit, in the shelter of which the immature marsh is developing. Thus although separate units can be readily delimited, they must be treated as one whole dynamic system. This one example provides a pattern that could be applied to a wide variety of coastal types. It is generally true that the data necessary for a sound understanding and control of any stretch of coast, can be presented by means of its component units so long as the characteristics of the individual units are defined in terms of their materials, morphology, and the processes that create them. In addition information is required about the interaction which occurs between the component units.

8.8. Conclusion

A useful way in which to consider coasts on a broad basis in the world context is in terms of their energy pattern. Coasts can be divided into high-energy, medium-energy, and low-energy coasts (Price, 1955; Tanner, 1960). The high-energy coasts include the storm-wave coasts of Davies (1964) and those on which the swells are long and relatively high, for example the coasts of south-east Australia and south-west Australia and Africa. These are coasts normally dominated by vigorous wave action. The medium-energy coasts occur in the more sheltered areas, such as some of the east-coast swell zones, and the larger enclosed seas, such as the Mediterranean Sea. The low-energy coasts occur in very sheltered areas with very small fetches, such as the Irish Sea and parts of the southern North Sea, as well as ice-protected polar seas. In parts of these low-energy coasts reclamation of salt marsh can take place, where sheltered conditions allow the deposition of silt. Good examples of the latter include the extensive reclamations in the Wash and around the coasts of Holland and Denmark (Blumenthal, 1967).

The coast is a geomorphologically active and variable zone in which process elements vary over a wide range of intensity, both in time and space. The processes interact with the coastal materials, whether they be solid rock or drift, or mobile beach sediments, to create a wide range of morphological units. The character of the unit varies according to the processes and materials that form it. In order to control and improve this fragile and vulnerable, but important, environment, man must learn to work with and not against the powerful marine processes. Thus an understanding of the operation of the processes in different environmental settings is essential.

9

FREEZE, THAW, AND PERIGLACIAL ENVIRONMENTS

9.1. Ground Ice and Man

The disruption of roads and water-supply systems under conditions of freezing and thawing is a common enough problem in many temperate lands. In higher latitudes, and particularly in the Arctic, problems relating to the freezing and thawing of water are much more common and more serious. They have persistently restricted development in these areas, and man has responded to the problems by devising numerous technical solutions. This chapter is concerned with these problems and their solutions. Perhaps two quotations will summarize its purpose.

Muller, in one of the few reviews in English of high-latitude engineering problems to consider the extensive Russian literature, observed:

Costly experience of Russian engineers has shown it to be a losing battle to fight the forces of frozen ground simply by using stronger materials or by resorting to more rigid designs. On the other hand, this same experience has demonstrated that satisfactory results can be achieved if the dynamic stresses of frozen ground are carefully analysed and are allowed for in the design in such a manner that they appreciably minimize or completely neutralize and eliminate the destructive effect of frost action. Mastery of this working principle, however, can be achieved only if the natural phenomena of frozen ground are thoroughly understood and their forces are correctly evaluated (Muller, 1947.)

Price, in a geographical appraisal, alluded to problems that commonly arise from viewing alien environments from a 'mid-latitude' standpoint:

An understanding of these processes [of the periglacial environment] is vital to a rational development of the north since the entire history of man's exploitation of marginal environments—tropical rainforest, deserts, tundra—has been to apply middle latitude technology and approaches to land utilization (Price, 1972, p. 1).

Man's permanent occupation of high latitudes is not new, of course, as the history of various hunting and trapping communities testifies; what is relatively new is the penetration of technically advanced groups into these areas in order to exploit their animal, vegetable, and mineral resources, and, perhaps as important, to secure the defence of remote frontiers. The groups mainly involved are those from the northern hemisphere circumpolar nations—the U.S.S.R., the U.S.A. and Canada, and the Scandinavian countries. The problems of freezing and thawing in high-latitude development have been longest recognized and most extensively overcome in the

Soviet Union, where experience stretches at least from the construction of the Trans-Siberian Railway at the turn of the last century to the building of modern towns such as Norilsk in areas of permanently frozen ground. Since 1938 a thorough survey has been mandatory before any structure can be erected on permanently frozen ground in the U.S.S.R. Soviet literature on the problems of high-latitude development is extensive (e.g. Shvetsov, 1959).

North American experience, rudely initiated by the Klondike Gold Rush in 1896 and slowly accumulated in the early decades of this century (e.g. Taber, 1929), was rapidly extended during the Second World War, with the construction of airfields and of transportation routes such as the Alcan Highway. Since the war, with changing political and military alliances and the discovery of great mineral wealth in the Northlands, the impetus for defence and development has increased, and studies of environmental problems have burgeoned (e.g. Legget, 1966). In particular, the discovery of oil at Prudhoe Bay, Alaska, in 1968 has promoted enormous interest in these problems (Mackay, 1972).

Academic and applied studies of frozen-ground phenomena have flourished for over four decades and they have recently become more closely allied. There are several sound and recent reviews of geomorphological problems in high latitudes, notably those by Embleton and King (1968) and by Washburn (1972). A useful set of essays is that edited by Price and Sugden (1972). Studies relating frozen ground and development problems have been surveyed by, for example, Muller (1947), Black (1950), Krynine and Judd (1957), Linell (1960), Corte (1969), Swinzow (1969), R. Brown (1970), and Price (1972); and several conference proceedings provide valuable information (e.g. National Academy of Science, 1966; Brown, 1969; Péwé, 1969; Legget and MacFarlane, 1972).

Some research results are published in generally available journals such as *Arctic, Arctic and Alpine Research, Quaternary Research, Geografiska Annaler* (*A*), and the *Biuletyn Peryglacjalny* (Łódź, Poland), but much published information is probably not readily available to students. Much of the most important work, for example, comes from government and other agencies, some of which only produce reports for limited circulation—for instance, the Cold Regions Research and Engineering Laboratory (CRREL) of the U.S. Army (now incorporated in the U.S. Army Terrestrial Sciences Center), the U.S. Highway Research Board, the National Research Council of Canada's Division of Building Research, and other institutions such as the V.A. Obruchev Permafrost Research Institute (Moscow), and the Norwegian Geotechnical Institute. There are several books on the development of the Northlands, a few of the more recent being those by Naysmith (1971), MacDonald (1966), and Hansen (1967).

Faced with such a large library of research, this short review will inevitably be selective. It will concentrate on the patterns of freezing and thawing, the principle processes of frost action and solifluction, the consequences of these processes, and means of overcoming them.

9.2. Definitions and Distributions

(*a*) *Definition of terms*

The terminology of freezing and thawing in the ground is often both controversial and etymologically monstrous. Some of the major terms are defined below in their most generally accepted form, and the features are illustrated schematically in Figure 9.1.

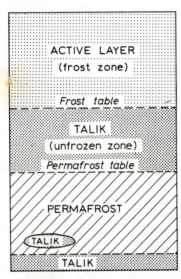

FIG. 9.1. Terminology of some features associated with permafrost

Permafrost is perennially frozen ground, that is, ground frozen continuously for two or more years. It is defined on the basis of temperature (i.e. ground below 0 °C) rather than on the presence or absence of ice. *Dry permafrost* occurs where there is no water present, as in certain bedrock areas of the Brooks Range, Alaska. Normally, however, ice is present. The upper limit of perennially frozen ground is a surface of some importance because it is relatively impermeable; it is called the *permafrost table*. Permafrost is often divided into at least two categories. *Continuous permafrost* normally has a mean annual temperature at 10–15 m depth of less than −5 °C, and the permafrost table is usually no more than 0·61 m below the surface (although it may be up to 1·8 m deep in granular material). *Discontinuous permafrost* is thinner, is broken by thawed areas, has a lower permafrost table, and its mean annual temperature at 10–15 m depth is between −5 °C and −1·5 °C.

The *active layer* comprises the ground above the permafrost table. At least part, if not all of this zone is subjected to intermittent freezing and thawing, where freezing can occur from the top down and the bottom up, often on a seasonal basis: this highest zone, the zone of greatest temperature fluctuation, is called the *frost zone*, and its uneven lower surface is called the *frost table*. If the frost zone and the active zone do not coincide, there may be a residual zone of thawed ground between the frost table and the permafrost table: this ground is called *talik*, a term also applied to any thawed area within and beneath the permafrost. Talik may occasionally act as a viscous liquid and flow. The lower limit of the active layer may vary from year to year.

In areas beyond the limits of permafrost, where freezing occurs from the top down, and thaw occurs from the top and bottom of frozen ground (Corte, 1969), freezing and thawing may be *seasonal* or *sporadic*. A phrase commonly allied to permafrost is *periglacial environment*. This may be taken to mean, broadly, the environment where frost processes predominate (or, more strictly, where permafrost occurs).

(b) Distribution of frozen ground

Approximately a fifth of the world's land surface is underlain by frozen ground of one kind or another. Descriptions of the distributions of frozen ground and related features are being refined continuously. Figure 9.2 is based on several recent studies. Because there is more land in high latitudes of the northern hemisphere than the southern hemisphere, the permafrost area in the north is greater than in the south, being 22·4 million km² compared with 13·1 million km². Over 80 per cent of Alaska, 50 per cent of Canada, and 47 per cent of the Soviet Union are underlain by permafrost. In general, permafrost extends further south on the eastern and more continental land areas. Beyond the permafrost limits, most of the land north of 30 °N is affected by seasonal or sporadic freezing and thawing.

The vertical distribution of frozen ground and thawed ground also varies spatially. In Figure 9.3, longitudinal cross-sections through Eurasia and North America show variations in the thicknesses of permafrost and the active layer. Permafrost reaches its maximum known thickness, 1,500 m, in Siberia, but it is not normally thicker than 600 m. In Eurasia the average thickness is 305–460 m, and in North America it is 245–365 m (Black, 1954). The permafrost layer thins from the zone of maximum thickness northwards beneath the Arctic Ocean and southwards into warmer latitudes. Discontinuous permafrost is normally less than 60 m thick. The active layer may vary in thickness from a few centimetres over continuous permafrost to 4 m and more on discontinuous permafrost.

Thickness of different zones associated with frozen ground, often an important consideration in engineering studies, can be determined in a variety of

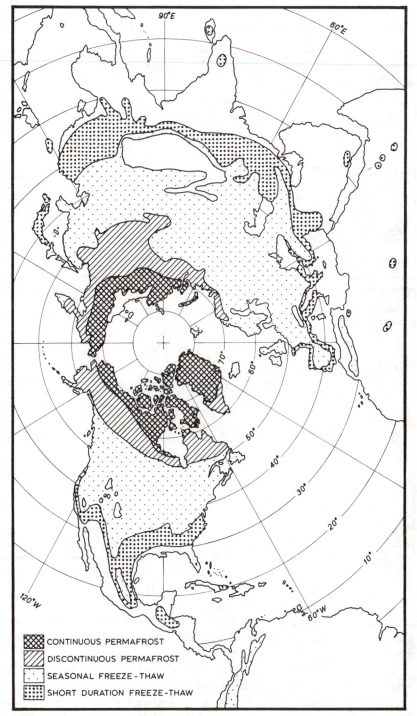

Fig. 9.2. Distribution of permafrost and related phenomena in the northern hemisphere (based on Corte, 1969, Mackay, 1972, Brown, 1967, and others)

ways. Drilling is perhaps the most widely used method, but various geophysical methods (Barnes, 1966), such as electrical resistivity and seismic techniques, have been employed.

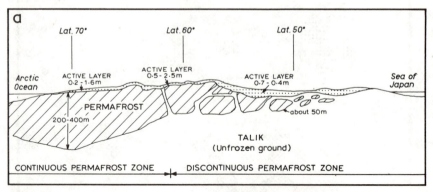

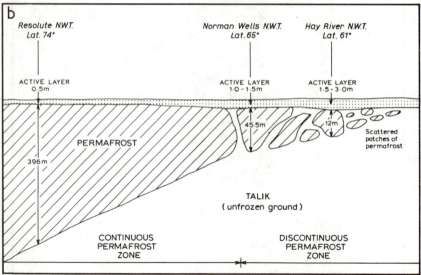

FIG. 9.3. Vertical distribution of permafrost and active zones in longitudinal cross-sections through (a) Eurasia and (b) North America (after Brown, 1970, and Muller, 1947)

(c) *Thermal profiles of frozen ground*

Numerous devices are available for monitoring thermal conditions above and within frozen soil (e.g. Hansen, 1966), including simple glass thermometers, thermocouple resistance thermometers, and thermistors.

Figure 9.4a shows a typical temperature profile through the active and permafrost layers. This diagram shows several important features. Firstly, temperatures fluctuate most widely at the surface and in the active zone, largely in response to daily and seasonal changes in atmospheric temperature.

Secondly, fluctuations decrease in depth until the level of *zero annual amplitude* is reached. Thirdly, below the level of zero annual amplitude, temperature rises with depth; where temperature rises above 0 °C, permafrost ends.

Daily and seasonal fluctuations of air and ground temperature are exemplified in Figure 9.4b, and c. The daily march of soil temperature in Wright

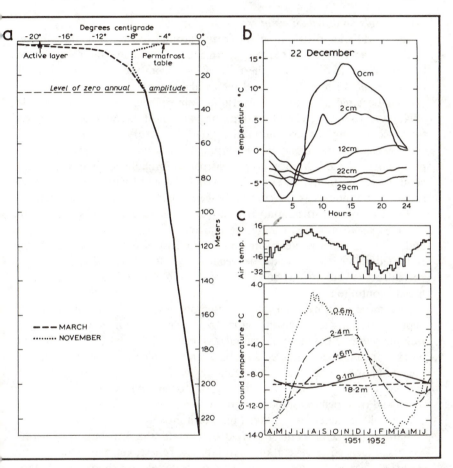

FIG. 9.4. a. Temperature of permafrost in a shaft at Yakutsk, Siberia (from Muller, 1947)
b. Daily variations of air and soil temperature in Wright Valley, Antarctica (after Ugolini, 1966). Cm values refer to depths below the surface
c. Seasonal fluctuations of air and ground temperature at Barrow, Alaska (after Price, 1972)

Valley, Antarctica shows that daily fluctuations tend to decrease with depth and that maximum soil temperatures occur later at lower depths (Ugolini, 1966). Seasonal fluctuations (Fig. 9.4c) show similar trends. During the spring thaw, temperatures rise relatively uniformly throughout the soil; but in the autumn freeze the change is not uniform owing in large measure to the fact that latent heat of fusion is released when water freezes and soil temperature is thus maintained around freezing-point for some time. This so-called *zero curtain* is well shown by the 0·6 m depth curve for October and November in Figure 9.4c.

(d) Controls on distributions of frozen ground

The world-wide distributions of frozen ground described in the previous sections in general reflect a kind of thermal equilibrium arising from relations between climatic conditions, the flow of heat from the earth's interior, the properties of earth materials, and the availability of water.

Climatic conditions are of overriding importance. According to Muller (1947) the following climatic conditions are most favourable for permafrost—cold, long winters with little snow; short, dry, and relatively cool summers; and low precipitation during all seasons. Seen in continental perspective, the occurrence of permafrost is broadly related to climatic conditions. In Canada, for example, Brown (1965 and 1966) showed that there is a broad but not very close correlation between mean annual air temperature and permafrost distribution. Brown (1965) indicated that south of the $-1\cdot1$ °C mean annual isotherm permafrost is rare; between the $-1\cdot1$ °C and $-3\cdot9$ °C mean annual isotherms permafrost near the ground surface is restricted mostly to peatlands; between $-3\cdot9$°C and $-6\cdot7$ °C discontinuous permafrost is widespread; and north of the $-6\cdot7$ °C isotherm permafrost is mostly continuous.

At a more practical level, several attempts have been made to predict the depth of freezing and thawing by means of climatic indices. One index, the *freezing index*, is based on the cumulative totals of degree-days. The freezing index is the number of degree-days between the highest and lowest points on the cumulative curve constructed from degree-days data plotted against time for one freezing season. There is a good correlation between the freezing index and frost penetration of the ground when dry unit weight and moisture content of the materials are taken into account.

Within the overall control exerted by climate, the distribution of permafrost is influenced by several other variables. Of these, vegetation and ground cover, surface water, terrain, and surface materials are fundamental. Most permafrost occurs either beneath the northern *boreal forest* or, north of the tree line, beneath *tundra* vegetation dominated by low-growing sedges, grasses, mosses, and lichens. The boundary between continuous and discontinuous permafrost in some areas approximately coincides with the southern limit of the tundra. The importance of vegetation is mainly that its thermal properties determine the movement of heat into and out of the ground (Brown, 1966), although the precise effect of vegetation on permafrost characteristics is often difficult to disentangle from related environmental

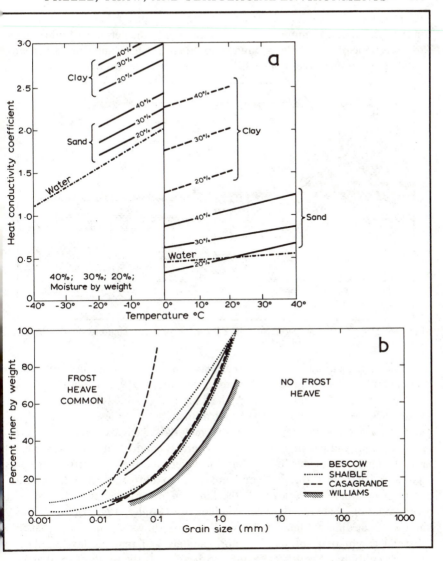

FIG. 9.5. a. Relations between the thermal conductivity of sand, clay, and water at different temperatures (after Evdokimov–Rokotovsky; from Muller, 1947) b. Limits of frost-heave susceptibility with respect to grain size (after Beskow, Schaible, Casagrande, and Williams; from Corte, 1969, and Embleton and King, 1968)

variables. But it is clear, for example, that depth of thaw increases if vegetation cover is removed, and in the zone of sporadic permafrost ice may occur preferentially beneath a blanket of insulating peats (see also Fig. 9.6, below).

Surface-water bodies also influence the occurrence of ground ice (Ferrians, Kachadoorian, and Greene, 1969). Beneath lakes and rivers that do not fully freeze in winter, permafrost is thinner or may be absent, especially in the zone of discontinuous permafrost. Similarly, the permafrost table tends to reflect the shape of the ground, rising beneath hills and falling beneath valleys, and responding to insolation variations due to ground aspect.

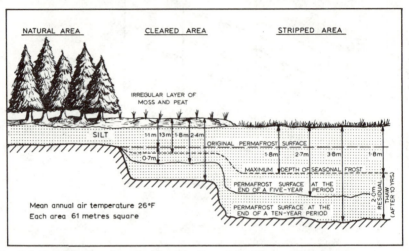

FIG. 9.6. Degradation of permafrost in silt under different land-use conditions near Fairbanks, Alaska (after Linell, 1960)

Finally, the surface materials themselves and their water content relate significantly to frozen ground. According to Corte (1969) some of the important properties of the materials are their thermal conductivity, specific heat, volumetric heat capacity, and thermal diffusivity, and these, in turn, relate to fundamental characteristics of the materials such as texture, mineral composition, and packing of particles. Thermal conductivity, for example, is about four times greater for ice than for water, and therefore frozen ground has a much higher thermal conductivity than unfrozen ground. Equally, thermal conductivity varies with the kinds of material. Figure 9·5a shows some relations between the heat conductivity (Muller, 1947) of sand, clay, and water at different temperatures. This figure shows, for instance, that the thermal conductivity of sand is approximately half that of clay. Similarly, the susceptibility of soils to freezing and heaving as the result of ice segregation varies according to the composition of the material. Many engineers use the percentage of grains smaller than 0·02 mm as a rule-of-thumb criterion. For example, fairly uniform sandy soils must contain at least 10 per cent of grains smaller than 0·02 mm for ice layers to form, and in soils with less than 1 per cent of such particles, no ice layers are likely to develop (e.g. Terzaghi and Peck, 1948). Figure 9.5b shows the limits of frost-heaving susceptibility according to grain size, taken from the work of various authors.

Thus it can be argued that the thermal pattern of frozen ground today reflects a quasi-equilibrium state that is related primarily to the refrigerating effects of climate and the loss of heat from the earth's interior, and is locally conditioned by the nature of vegetation, the availability of water, and the properties of sediments. It is only in a shallow surface zone that temperatures fluctuate markedly, daily or seasonally.

While the thermal equilibrium may appear to arise from present conditions, it is clear that most permafrost has developed over millennia, and some would argue that much continuous and sporadic permafrost is in fact a relic from colder climates. Certainly the well-known evidence of deep-frozen extinct mammals points to the antiquity of permafrost. Equally, permafrost is said to be actively forming in some areas. It is contemporary changes to frozen ground that are of the greatest interest to the resource manager.

9.3. Contemporary Aggradation, Degradation, and the Disturbance of Frozen Ground

(a) Disruption of equilibrium

The thermal equilibrium of frozen ground may have taken millennia to become established, and doubtless its broad pattern is still being modified by secular changes in climate and other controlling variables. But from the practical point of view, it is of the greatest significance that the thermal equilibrium can be rapidly altered, and permafrost may be reduced (degraded) or increased (aggraded) as the result of modifications to existing conditions by man.

Man modifies existing conditions in a great variety of ways (e.g. Haugen and Brown, 1971); many of his deliberate changes have unforeseen and undesirable consequences. The main changes are as follows. (i) Removal of vegetation, perhaps deliberately for timber or to create cleared land for agriculture, or perhaps accidentally, as in the case of a forest fire, or the stripping of vegetation by tracked vehicles. (ii) Modification of drainage conditions by, for instance, draining bogs, diverting rivers, or creating reservoirs. (iii) Construction of buildings, roads, pipelines, and associated ground preparation without appropriate precautions. In all these cases, the effect is usually permafrost degradation, and the change may be irreversible. The tendency, of course, is towards the establishment of a new thermal equilibrium in the ground. With small structures, the relaxation time may be rather short, and the new equilibrium might be established in a year or two; with large structures or extensive land-use modifications it may take decades to re-establish the equilibrium.

The nature and rate of change will vary according to initial ground and climatic conditions and the nature of the modification. Susceptibility to disruption is usually greatest in the extreme north where the active zone is shallow, ice content of the near-surface soil is great, and the insulating organic layer is thin; in general, susceptibility declines southwards.

Three examples will illustrate more precisely the influence of human activities on permafrost destruction: permafrost change with land use in Alaska; the influence of two building types on the permafrost table; and the predicted effects of an oil pipeline on permafrost.

Figure 9.6 shows the degradation of permafrost under different land uses on silt over a ten-year period as observed by the U.S. Corps of Engineers near Fairbanks, Alaska (Linell, 1960). Beneath the natural area of trees, brush, moss, and grass the permafrost table was maintained at its original level of 1·06 m. Where trees and brush were removed, the table was lowered to 1·8 m in five years and to 2·4 m in ten years. And where all vegetation was stripped

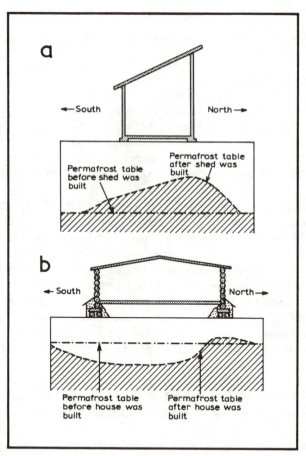

FIG. 9.7. Alterations to permafrost caused by (a) an unheated hut and (b) a heated cabin (after Tsytovich and Sumgin; from Muller, 1947)

and its insulating effect was therefore absent, the table was lowered to 2·7 m in five years and to 3·8 m in ten years. In addition, the maximum depths of seasonal freezing were also lowered.

Two kinds of change to permafrost caused by small buildings constructed without significant precautions are shown in Figure 9.7. If the building is uninhabited and unheated, the permafrost table may rise in the shadow, as it were, of the surface insulation provided by the building. If the building is heated, the permafrost table may fall. In both cases, the depth of permafrost has changed and so has the thickness of the active zone.

Figure 9.8 shows the *predicted* growth of the thawing zone around an oil pipeline, 1·2 m in diameter, with its axis buried 2·4 m below the surface,

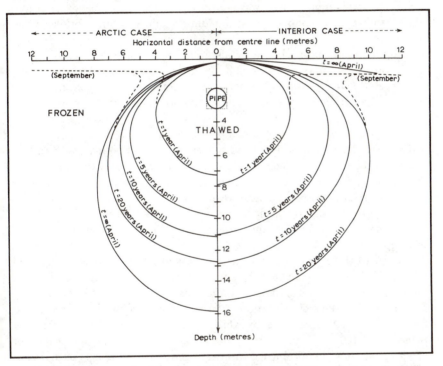

FIG. 9.8. Predicted growth of the thawed cylinder around a 1·21 m diameter oil pipeline with its axis 2·42 m beneath the surface and maintained at a temperature of 80 °C, near the Arctic coast of Alaska (left) and (right) near the southern limits of permafrost (after Lachenbruch, 1970)

passing through 'medium silt' permafrost and carrying oil that maintains its temperature at 80 °C, for Arctic coast conditions (left) and (right) near to the southern limits of permafrost (Lachenbruch, 1970). These calculations were made in anticipation of an underground oil pipeline being constructed south from Prudhoe Bay, Alaska, and their implications are important. In the first place the thawed cylinder would continue to grow for several decades. Secondly, in certain circumstances the thawed zones might seriously affect the stability of the structure. For example, if excess water generated by the growing thawed cylinder cannot be removed as quickly as it is formed, part of the cylinder may behave as a fluid and the pipe might settle or be broken, and the rate of thawing be increased.

The consequences of changing thermal equilibria are extensive. Water-logged channels may be created, erosion may extend along tracks, foundations may be disrupted, and the nature of frost action and solifluction may be altered. The properties of the ground surface, such as bearing strength, may be changed. Almost all types of human activity are likely to be affected, including water-supply and waste-disposal systems; drainage; runways, roads, railways, telephone lines; buildings; mines, dams, bridges, and reservoirs.

Although man causes change by disrupting the established frozen-ground conditions, it is important to emphasize that such changes generally only influence his activities through processes that are in any case operating in the periglacial environment. These processes are briefly introduced below.

(b) Ground ice and geomorphological processes

The most destructive processes of concern to man in permafrost areas are related to the formation and thawing of ground ice, either within the perma-frost zone or in the active zone above it. The two principal groups of processes are *frost action* (processes arising from ice formation in soils) and *solifluction* (processes of soil flow in the active zone). Frost action includes *frost shattering*, *frost heaving*, and *frost cracking*; the first of these is examined in Chapter 11.

Critical to these processes are the nature and mode of origin of ground ice. There have been many attempts to classify ground ice (e.g. Corte, 1969; Shumsky and Vtyurin, 1966), and the classification adopted here is by Mackay (1972). This classification is based on the origin of water prior to freezing and on the processes by which water is transferred to the freezing plane. In all, ten ice forms are recognized (Fig. 9.9). From the practical point of view, the most important ice forms are those arising from thermal contraction of frozen ground, and those associated with segregated and intrusive ice.

Briefly, the characteristics of the main ice forms are as follows. *Open-cavity ice* is formed by sublimation of ice crystals directly from atmospheric water vapour. *Single-vein ice* forms when an open crack penetrating perma-frost becomes filled with water and freezes. *Ice wedges*, like vein ice, arise from thermally induced cracking of frozen ground. *Tension-crack ice* grows in cracks produced by the mechanical rupture of the ground (usually by growth of segregated or intrusive ice). *Closed-cavity ice* is relatively unimportant, and forms by vapour diffusion into enclosed cavities in permafrost. *Segregated ice* includes *epigenetic ice* that grows as lenses in material predating the lenses and may grow into massive bodies. The second form of segregated ice is *aggradational ice* which develops when the permafrost table gradually rises and may incorporate epigenetic ice lenses. *Intrusive ice* is formed by intrusion of water under pressure, and its freezing, causing uplift of the ground above it. One type, *sill ice* is formed when water is intruded between horizontal sheets, as along bedding planes. *Pingo ice* occurs where intrusive ice domes the overlying surface. Finally, *pore ice* is that which holds soil grains together. The amount of ice in frozen ground may be more than available pore space (super-saturation), about the same as pore space (saturation),

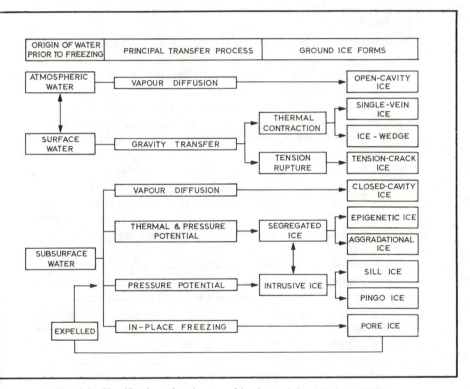

FIG. 9.9. Classification of underground ice forms (after Mackay, 1972)

less than pore space (under-saturation), or the sediment may be cemented by ice that does not form veins or granules. Ground may be hard frozen (grains are embedded in an ice cement), plastic frozen (with some water remaining unfrozen), or granular frozen (in which grains are in contact and excess ice is absent).

9.4. Frost Action and Solifluction: Processes and Control

(a) Frost heave

Frost heave associated with segregated ice in super-saturated material involves two separate processes. The first causes vertical displacement of the ground surface and arises from growth of ground ice. The surface disruption may be in the order of many centimetres, and the consequences for buildings, lines of communication, and other engineering structures may be serious. This process produces several distinctive landforms. The second causes the vertical sorting of particles, in which large stones in mixed sediments migrate upwards to the surface and poles, pilings, fences, etc. are forced out of the ground. The most common natural features reflecting this process are the stone pavement, and stripes, nets, circles, and other patterns of surface stones.

(i) *Vertical displacement of the ground surface* on a small scale due to the growth of ice crystals arises to a slight extent from the 9 per cent increase in volume of water when it is converted to ice, but largely by the force of crystallization of ice. The continuation of the latter requires the transfer of water to the ice crystal or growing ice lens, and there has been much discussion of the process since it was first recognized by Taber (1929 and 1930). Differential swelling of the ground may arise for three main reasons: unequal distribution of load, areal differences in the texture of the ground and ground cover, and differences in the amount of water available for ice growth (Taber, 1929; Muller, 1947). Two of the factors that make soils particularly susceptible to frost heave have already been mentioned: particle size (Fig. 9.5b) and water availability. Also important are pore space, capillary properties of soils, rate and frequency of freezing and thawing, and depth of frost penetration (Taber, 1929).

Ground hummocks (known by various names such as *palsas* and *mima mounds*) are rather small features often attributed to ground heaving. They are usually in the order of 0·5–2 m high (and perhaps as high as 10 m) and up to about 10 m in diameter, and are commonly associated with poorly drained areas and peaty material (Lundquist, 1969). The origin of these features in periglacial and some mid-latitude regions has long been a matter of controversy. Their dimensions and composition are varied, and undoubtedly no single explanation is satisfactory in all areas. Some argue that hummocks were elevated by frost heaving following segregation of ice lenses (e.g. Taber, 1952); others have suggested that expanding patches of frozen ground squeezed earth between them into mounds (e.g. Sharp, 1942); there are those who consider that degradation of permafrost (by removal of vegetation for instance) might cause ice wedges to thaw, causing subsidence and material to slump into the hollows, leaving mounds between them (e.g. Péwé, 1954).

A large feature, associated with intrusive ice, is the ice-cored mound called a *pingo*, which may be up to about 100 m high and 600 m wide. They occur in zones of continuous and discontinuous permafrost. Two main types have been distinguished—the *closed-system type* (Mackenzie type) and the open-system type (*East Greenland type*). The *closed-system type* (Mackay, 1966) usually develops as follows. If a lake that is not normally fully frozen and is underlain by unfrozen saturated sands becomes fully frozen (as the result, for example, of draining or sedimentation), permafrost will encroach upon the zone beneath the lake and create a closed system of unfrozen material. Growth of permafrost around this core increases water pressure in it, and the surface layer is heaved to relieve this pressure. Ultimately, the closed system is entirely frozen and the excess water forms an ice core beneath the dome. The *open-system pingo* usually occurs on sloping ground and usually in discontinuous permafrost where hydraulic pressure in taliks causes water to approach the surface and freeze. A continual supply of water builds up the ice mass and this domes the surface (Muller, 1963). Occasionally tension cracks develop on the crests of pingos. And the phenomenon of 'drunken forest' is often associated with them—raising of the ground causes the original trees to be inclined away from the vertical.

(*ii*) *Vertical upwards migration of large objects* in the soil or other unconsolidated material has been investigated under laboratory conditions, and Washburn (1969 in Price, 1972) recognized two types of mechanism. The *frost-pull mechanism* depends on the ground expanding during freezing and carrying the large objects (e.g. stones) upwards; on thawing, the fine material moves down and beneath the large objects so that they do not return to their original position. The *frost-push mechanism* depends on the fact that the thermal conductivity of stones is greater than that of the soil. As a result, stones heat and cool more quickly than the soil and ice would form first beneath them, forcing them upwards. During thawing, fine material would move beneath the stones and prevent them returning to their original positions.

(*iii*) *The control of frost heave* depends initially on a knowledge of climatic, material, and water conditions, and particularly on the nature of freezing and thawing. As discussed previously, it is useful to predict the depth of freeze–thaw in terms of freezing indices, for example, so that appropriate design measures can be adopted.

Frost heaving can be reduced in three main ways (Corte, 1969): by chemical treatment of the soil, by filling of soil voids, and by structural design precautions. Many different techniques have been tried, and the examples below illustrate some of the more successful methods in these three categories. The injection of calcium chloride (CaCl) has been shown to reduce effectively freezing temperatures of water in the soil and to reduce the loss of strength of materials arising from freeze–thaw cycles. A neutral solution of waste sulphite liquor from the paper industry in Canada has been used to reduce soil capillarity (and thus reduce water flow to growing ice lenses) and to slow the rate of ice growth. Voids in the soil can be filled with a variety of often rather expensive cements, such as polymers, resins, portland cement, and other fine material. Chemicals have also been used to aggregate or disperse fines, and efforts have been made to wash out fine material from the soil. Design methods include increasing the thickness of pavement and sub-grade materials to prevent frost penetration, and insulation of the surface with natural materials such as straw, moss, or peat, or artificial materials such as cellular glass blocks. In some circumstances it may not be necessary fully to control frost heave; simpler measures may be satisfactory. For example, it may be adequate to create a pavement surface for a road that has sufficient bearing capacity during the thawing season when the soil is at its weakest.

(b) Frost cracking, ice wedges, and patterned ground

Permafrost areas are very frequently characterized by polygonal ground patterns formed by cracks that are the location of ice wedges. The importance of these features is that they provide evidence of ground-surface conditions (such as the type of material, and nature and amount of ground ice), and, if the ice wedges are caused to melt for any reason they can become the loci of subsidence (Mackay, 1970) and seriously disrupt superincumbent structures. The cracks may be up to ten metres deep, and the diameters of the polygons may exceed 100 m. The patterns of polygons are usually dominated by orthogonal junctions between cracks, but non-orthogonal patterns are commonly found.

Lachenbruch (1966) has reviewed the mechanics of frost-wedge formation. Figure 9.10 illustrates four stages in the evolution of an ice wedge according to the thermal contraction theory. During the winter, because the thermal coefficient of contraction of frozen ground is rather high, tensile stress in the material is likely to exceed the tensile strength of the material and cracking occurs through both active and permafrost layers (a). By the following autumn (b), water has penetrated the cracks and a vein of ice has been formed. Compression caused by re-expansion of the melting permafrost in summer deforms the near-surface strata. In the following winter, the crack is reopened, first in the top of the ice wedge beneath the active layer. The next spring thaw fills the new crack with water, and so on. Analysing the stress in permafrost material, Lachenbruch suggested that it depends mainly on the amount and rate of cooling, and on the temperature at the time of cracking.

The centres of polygons between ice wedges may be higher or lower than their edges. Where growth of the ice wedges causes upheaval of material adjacent to them, as for instance in areas where wedges are actively growing, the edges of the polygons will be higher than their centres. But where ice wedges are thawing and may be acting as loci of drainage concentration, the centres of the polygons may be higher. This contrast therefore reflects the condition of the ground, for actively growing ice wedges are largely restricted to zones of continuous permafrost, and they are less active or melting in the zones of discontinuous permafrost (e.g. Péwé, 1966). There are numerous other patterned-ground phenomena in periglacial areas (see for example, Washburn, 1956, 1969) and most but not all of them are associated with frost heaving and frost cracking.

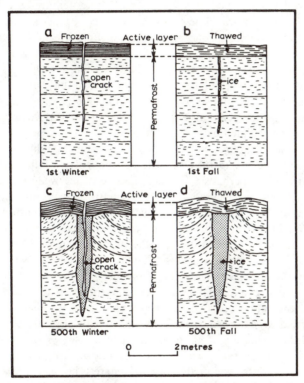

FIG. 9.10. Schematic representation of ice-wedge evolution according to the contraction-crack theory (after Lachenbruch, 1966)

(c) Thawing, settling, and solifluction

The properties of thawed ground are usually markedly different from those of frozen ground. For instance, the mechanical properties of frozen ground tend to approach those of ice, having higher shear strength and thermal conductivity than similar but unfrozen material. When sediments in the active zone thaw, therefore, they behave differently from the frozen ground. Frequently, and especially beneath a superincumbent load such as that of a building, they *settle* or subside, and behave like a viscous fluid. Certain materials are particularly susceptible to settling: for example, gravel embedded in large quantities of ice is inherently unstable when thawed, and silty frozen soil with a high proportion of ice generally becomes waterlogged and spongy when thawed (Krynine and Judd, 1957). If the ground surface is inclined, the thawed material may flow, even on very low slopes. This process is called *solifluction*.

Features arising from the degradation of ground ice in permafrost areas are collectively called *thermokarst landforms*. They include certain kinds of pits, dry valleys, and lakes. All these subsidence features are formed by thawing of super-saturated icy soils at the top of permafrost (Mackay, 1970), and they are most pronounced where melting affects abundant ground ice in the upper permafrost. Clearly the cause of thawing is important. Climatic changes may cause regional thaw. Kachurin (1962), for example, has pointed out that thermokarst is most extensive in the southern permafrost areas of the Soviet Union, where regional climatic change may be responsible. Seasonal thaw can also lead to thermokarst development. Local thawing is often initiated by man through, for example, vegetation clearance, compaction of peat, fire, bulldozer tracks, ploughing, ditch-digging, and construction (Mackay, 1970). Often the thawing pattern reflects the location of near-surface ice, perhaps following the line of ice wedges or coinciding with thawed pingos. Degradation of permafrost may proceed laterally, usually as a result of horizontal erosion by water. The extension of certain lakes (*thaw* lakes) is attributed to this cause. Vertical degradation is more common and may be illustrated by the development in Siberia of *alases*, large, flat-floored, vegetated basins (Czudek and Demek, 1970). These are initiated by the melting of ice wedges that become the loci of lakes. As the lakes deepen and coalesce, they may become too deep to freeze entirely in winter and the rate of permafrost degradation beneath them is increased. Eventually the large lakes may be filled or drained, vegetation may cover the floors, and the formation of alases is complete.

Solifluction is important because it can severely disrupt surfaced structures; features produced by solifluction are of interest because they can reveal something of the nature of surface conditions (especially stability). Solifluction comprises two main processes, namely the flow of water-soaked debris, and the creep of surface material by freeze–thaw action (Embleton and King, 1968). The first involves the thawing of ground in the active zone or the addition of water to the thawed surface by snow melt. The presence of permafrost is not essential, but if it exists its relative impermeability helps to maintain water saturation in the near surface material. As the ground thaws or water is added to it the weight of material increases, and its shear strength is reduced because water reduces both internal friction and cohesion. The nature of flow that results depends mainly on the characteristics of the material (such as its texture and viscosity), water availability, depth of thaw, slope of the ground, and vegetation cover. Solifluction is especially facilitated by a high silt content and an abundant supply of water. Vegetation insulates the ground beneath and thus probably reduces the depth of the thawed zone and restricts the rate of *surface* flow. The second process, creep induced by freeze–thaw, is caused by the heaving of particles on a slope upwards normal to the ground surface and the lowering of them vertically upon thawing under the influence of gravity. This process, like frost heaving, depends on the frost-susceptibility of the materials, the availability of water, vegetation cover, and the nature of freeze–thaw activity.

Rates of solifluction have been measured in several areas, mainly by recording the translocation and vertical deformation of linear objects, such as pipes and cables, inserted through the mobile layer. Most movement usually occurs during the spring thaw, and field measurements show that rates may range from less than a centimetre to over 30 cm a year (e.g. Embleton and King, 1968; Corte, 1969). Solifluction produces several distinctive landforms. The most common are lobes and terraces of material, of which some are terminated by stone banks and some are covered with vegetation. These features normally occur on slopes over five degrees, their treads may be several tens of metres wide, and their risers may be a few metres high.

Various other forms of mass movement occur in periglacial areas such as mudflows, landslides, and debris avalanches. These phenomena have been considered in Chapter 6.

9.5. Solving Permafrost Problems

(a) Terrain evaluation

Evaluation of permafrost conditions and the terrain associated with them normally involves reconnaissance surveys followed by detailed site investigations. Because most permafrost terrain is remote and inaccessible, reconnaissance surveys are usually undertaken in the first instance by an analysis of available remote sensing imagery, notably panchromatic aerial photographs (e.g. Frost, 1950 and Frost et al., 1966). The purpose of most such surveys is to determine the nature of terrain, material, and ground-ice conditions and sources of construction materials, to plan access routes and provisional layouts for structures, and to select sites for detailed studies on the ground. Naturally these purposes will vary according to the specific project. For example, Mollard and Pihlainen (1966) identified five objectives for an airphoto interpretation exercise related to the building of an arctic road: to classify terrain, to locate the most likely route, to map and evaluate terrain conditions, to map granular materials, and to comment on terrain conditions and their engineering implications at previously selected sites. Often air-photo surveys will be carried out first on smaller-scale photographs

TABLE 9.1

Common topographic features that indicate ground conditions in arctic and near-arctic regions

Feature and description	*Associated ground conditions*
Polygonal ground (ice wedges)—Usually indicates the presence of a network of ice wedges. Wedge networks are also common in wet tundra where no surface expression occurs. (Subject to extreme differential settlement when surface is disturbed.)	Typically indicates relatively fine-grained unconsolidated sediments with permafrost table near the ground surface; also known from coarser sediments and gravels where wedge ice is less extensive.
Stone nets, garlands, and stripes—Frost heaving in granular soils produces netlike concentrations of the coarser rocks present. If the area is gently sloped, the net is distorted into garlands by down-slope movement. If the slope is steep, the coarse rocks lie in stripes that point downhill.	Strong frost action in moderate well-drained granular sediments that vary from silty fine gravel to boulders. Superficial material commonly susceptible to flowage.
Solifluction sheets and lobes—Sheets or lobe-shaped masses of unconsolidated sediment that range from less than a foot to hundreds of feet in width that may cover entire valley walls; found on slopes that vary from steep to less than 3 °.	An unstable mantle of poorly drained, often saturated sediment that is moving down-slope largely by seasonal frost heaving. On steeper slopes they often indicate bedrock near the surface, and on gentle slopes, a shallow permafrost table.
Thaw lakes and thaw pits—Surface depressions form when local melting of permafrost decreases the volume of ice-rich sediments. Water accumulates in the depressions and may accelerate melting of the permafrost. Often form impassable bogs.	Poorly drained, fine-grained unconsolidated sediments (fine sand to clay) with permafrost table near the surface.
Beaded drainage—Short, often straight minor streams that join pools or small lakes. Streams follow the tops of melted ice wedges, and pools develop where melting of permafrost has been more extensive.	A permafrost area with silt-rich sediments or peat overlying buried ice wedges.
Pingos—Small ice-cored circular or elliptical hills that occur in tundra and forested parts of the continuous and discontinuous permafrost areas. They often lie at the juncture of south- and south-east-facing slopes and valley floors, and in former lakebeds.	Silty sediments derived from the slope or valley, and ground water with some hydraulic head that is confined between the seasonal frost and permafrost table or is flowing in a thawed zone within the permafrost. Those in former lakebeds indicate saturated fine-grained sediments.

Source: Ferrians, Kachadoorian, and Greene (1969).

(e.g. 1 : 40,000–1 : 100,000) and then pursued in greater detail on larger-scale imagery (e.g. 1 : 20,000).

It is in such reconnaissance surveys that techniques of land-system mapping (Chap. 13) and site classification are especially valuable. The methods described in Chapter 13 have been more or less explicitly employed in many environmental surveys of permafrost areas (e.g. Hughes, 1972; Mollard, 1972). In preparing reconnaissance maps an understanding of landforms and drainage patterns, together with a knowledge of ecological conditions, is indispensible. In particular, as the discussion in previous sections has shown repeatedly, features such as pingos, ice wedges, high-centred and low-centred polygons, thermokarst phenomena, and solifluction lobes and terraces all provide valuable information on fundamentally important considerations of ground-ice and permafrost conditions, and the nature of superficial materials. Table 9.1 summarizes ground conditions commonly associated with different periglacial landforms. The distribution of water bodies is also a valuable guide to the same variables and to such things as the possibilities of water supply. At the same time, an understanding of peri-glacial ecology also helps terrain evaluation. Plant communities are often sensitive indicators of soil-moisture conditions and communities differ in the insulation they provide to the ground; easily-recognized peat bogs (known as *muskeg* in Canada) pose special engineering problems; and active surface disturbance is often revealed by, for instance, 'drunken forests' and the curvature of tree trunks.

The field studies that invariably follow reconnaissance surveys are nor-mally directed towards validating the air-photograph predictions and analysing in detail the conditions at specific sites. More information is generally required on the precise nature and distribution of permafrost, talik, the active zone, vegetation, surface materials, drainage, water supply, atmospheric and ground temperature and moisture conditions, and the per-formance of pre-existing buildings (Johnston, 1966a). The techniques used for these field investigations are numerous, and commonly include geo-physical prospecting by drilling and other methods, test-pit studies, geo-logical and geomorphological mapping, and meteorological observations.

(b) Engineering responses to permafrost problems

There are four main engineering responses to the presence of frozen ground. It can be neglected, eliminated, preserved, or structures can be designed to take expected movements into account (Brown, 1970). Permafrost can often be safely ignored where there is good drainage and ground materials are not susceptible to serious frost action and solifluction. Where permafrost is thin, sporadic, or discontinuous, and where thawed ground has a satisfactory bearing strength, the possibilities might be examined of removing the frozen ground by, for example, stripping the insulating cover of surface vegetation, by excavation, by treatment of the ground, or by thawing of the ice with steam. Material not susceptible to frost action may be placed on the surface. This approach is known as the *active method*. In regions of continuous permafrost elimination of frozen ground is likely to be impracticable, and the *passive method* may be adopted, whereby efforts are made to preserve the permafrost by, for instance, insulating the surface with vegetation mats or gravel blankets, ventilating the undersides of heat-generating structures, and preserving vegetation.

The techniques of design and construction in permafrost areas are numerous, and they have been extensively considered in other reviews (e.g. Brown, 1970; Crory, 1966; Krynine and Judd, 1957; Muller, 1947; Price, 1972 and various publications of the Highway Research Board). Only a few important considerations will be mentioned here.

In the first place, it is axiomatic that successful responses will be based on an understanding of permafrost, and a local knowledge of permafrost conditions. Secondly, whether an active or a passive approach is adopted, efforts will normally be made to disturb as little as possible or to control the thermal regime established before construction begins. Thirdly, although there is inevitably likely to be some settling or heaving after a development has been completed, the engineering aim is usually to minimize the effects of such changes either by directly modifying the process or by designing structures capable of withstanding them. Fourthly, lateral instability caused by solifluction is most commonly countered by avoiding sites where such flow is operating or by firmly fixing structures in permafrost beneath the active zone.

In selecting the locations for buildings, certain conditions are preferable: a thin active layer, bedrock near to the surface, thawed material with adequate bearing properties, good drainage, lack of frost-susceptible materials, and a stable site away from possible areas of water seepage and icing. Because most building construction problems are associated with the active zone, foundations are frequently fixed in the underlying permafrost. A most important consideration here is the *adfreezing strength* of frozen ground. This is the resistance to the force required to pull apart the frozen ground from objects, such as foundations, to which it is frozen. In general, this strength is greater in sands than in clays.

Pilings are commonly used as foundations for buildings and the importance of adfreezing strength is usually recognized in their emplacement. In general, the aim is to freeze the pilings into the permafrost layer to a depth normally at least twice the thickness of the active zone, and to avoid adfreezing in the active zone by lubricating, insulating or 'collaring' the piles in that zone. In this way, it is hoped that the downward-acting force in the permafrost zone (together with the load of the building and the weight of the pile) will be sufficient to overcome the upward-heaving force exerted on the protected piling in the active zone, thus preventing movement of the pile.

Other techniques frequently used in building construction include the creation of a trench around the building (possibly filled with material not susceptible to frost action) in order to eliminate possible horizontal stresses; the raising of the floor above the ground by 0·5–1 m to allow air circulation; the insulation of the floor to restrict temperature rise beneath the floor;

and the provision of skirting insulation between ground and floor with air vents that can be closed in summer, and opened in winter to allow free air movement (e.g. Swinzow, 1969). Jacks may be used occasionally to adjust the building level if movement occurs. Gravel, wood, or concrete pads are often placed on natural vegetation to provide an insulating foundation for small and often temporary buildings.

The provision of utilities to communities also poses special problems in permafrost areas, problems that are mostly related to seasonal freezing and the presence of active and permafrost zones. A suitable water supply may be difficult to find—it usually comes from surface sources or from groundwater within, above, or beneath the permafrost. The disposal and decomposition of sewage may be a problem, leading to pollution and health hazards. Freezing and thawing of such services creates additional problems. One novel solution to some of these problems is for all linear utilities—water, sewerage, electricity, etc.—to be placed in a single insulated and heated conduit (known in North America as a *utilidor*) that may be placed above or in the ground.

Problems encountered in building roads, runways, and similar features are allied in many ways to those accompanying building construction (e.g. Linell, 1960). A passive approach demands the creation of adequate natural or artificial insulating layers, and an active approach often requires replacement of surface material with material that is not frost susceptible. The aim is invariably to provide a stable track across what is frequently an inherently unstable surface. In addition, problems may arise because the route crosses varied permafrost terrain in which a variety of responses are necessary. Drainage is frequently difficult, especially where the route crosses drainage lines, and it is desirable to remove as much water as possible from accumulating beneath the road, usually by providing ditches some distance away from it or preventing infiltration. Ditches may become eroded by flowing water. Road or railroad cuts may penetrate permafrost, lower the permafrost table, and cause landslides or solifluction in the newly-created active zone— hence they are to be avoided as far as possible. In cuts and elsewhere, water may seep to the surface and freeze, producing treacherous *icings*. One way of avoiding this hazard is to transfer the icing away from the road into specially prepared ditches (Muller, 1947).

9.6. Case Study: Inuvik, a New Canadian Arctic Town

Aklavik was created in 1912 on the Mackenzie River delta by the Hudson's Bay Company. By the 1950s, its permanent population had grown to 400 and this was increased in the summer season to some 1,500 by an influx of Eskimos and Indians. The town had become an administrative, educational, and strategic focus in the North West Territories. In the 1950s it was ripe for further expansion (Robertson, 1955). But the site of Aklavik was in many ways unsatisfactory; it was built on fine-grained, ice-rich deltaic deposits, and frost heaving and settling posed serious building hazards; it was liable to flooding; ground drainage was poor; the river bank was being eroded; and local gravel resources for further development were limited; there was little room to expand. Thus in 1952 the Advisory Committee on Northern Development suggested that a new site should be sought (Merrill, Pihlainen and Legget, 1960; Pritchard, 1962; Robertson, 1955).

A search was made in the neighbourhood of Aklavik for a site which had economic and social potentialities, ground conditions suitable for permanent development, access to a navigable river channel, proximity to a suitable airfield location, and an adequate water supply. Adequate provision for sewage disposal, and local sand and gravel supplies for building were additional considerations. An aerial-photograph reconnaissance survey identified some nine or so possible sites in the area of the Mackenzie delta, and the subsequent helicopter reconnaissance and field survey by a team of engineers, geographers, and geologists led to the selection of an area on the Mackenzie River about 48 km from Aklavik (Merrill, Pihlainen, and Legget, 1960). Building began here in 1955 and today there is a population of about 1,500. The building of Inuvik provides an excellent example of how adequate understanding of permafrost and knowledge of local conditions, together with detailed planning, can successfully overcome problems of building in a permafrost environment.

The site of Inuvik has several important characteristics that satisfy the initial planning requirements (Johnston, 1966b; Pihlainen, 1962). Being on a terrace adjacent to the Mackenzie River it is accessible to navigable water and yet not liable to flooding. Water supply can be obtained from a lake above the site and from the river. An airfield location is available a few kilometres away. The gently undulating topography is dominated by spruce and birch, and the hummocky ground is everywhere covered with moss from 10 to 15 cm thick. Sixty or more centimetres of peat underlie the moss, and beneath this are gravels, fine sands, and silts. A typical section of deposits is shown in Figure 9.11. Extensive gravels for construction are locally available. The whole area is underlain by permafrost up to 90 m thick. Above it, the active layer varies from 30 to 150 cm in thickness, being thinnest in peat, and thickest in gravel with a moss cover. As data collected in 1957 show (Fig. 9.12) depth of thaw also varies according to the degree of vegetation disturbance (Pihlainen, 1962). The nature of ground ice varies over the site (Fig. 9.11), ranging from small ice lenses and coatings to massive ice segregations. The peats and the fine-grained soils usually have the higher ice contents. Clearly, the major problem in developing Inuvik is the presence of permafrost and an active layer.

The need to build Inuvik with as little disturbance as possible to ground conditions was recognized at the outset by several planning decisions. Firstly, the natural moss covering was to be left intact in order to retain its insulating benefits. Secondly, all permanent structures were to be placed on piles securely embedded in permafrost. Thirdly, road cuts and ditches were prohibited in order to prevent permafrost degradation, and culverts were to be installed in gravel fill to accommodate surface runoff. Finally, all traffic routes and some temporary structures and storage areas were to be built up with gravel pads placed on top of the natural vegetation.

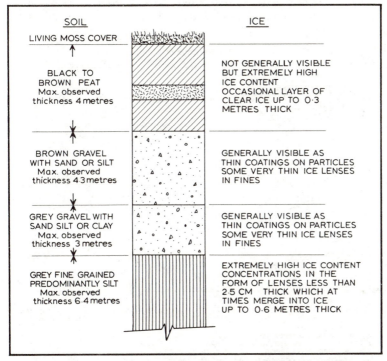

FIG. 9.11. Typical sequence of sub-surface materials at Inuvik (after Pihlainen, 1960)

The use of piles posed several problems (Johnston, 1966b). Each site had to be carefully prepared with a minimum of ground disturbance—the moss cover was maintained, vehicle movement was restricted, the piling equipment was placed on insulating gravel pads, and where vegetation had to be cleared, it was done by hand. Over 20,000 piles were used; most of them were of local spruce, but some larger wooden piles, and a few concrete and steel piles, were imported. As the upper part of timber piles would be in the active zone, they were treated with a preservative. Steam thawing was used to emplace most piles, except if the piles were to be closely spaced for carrying heavy loads, when drilling was used to prevent too extensive thawing of permafrost.

A major problem was to decide how deep the piles should be buried. At Inuvik, as a general rule, piles were frozen into permafrost to a depth twice the maximum thickness of the active zone—that is to say, they were normally buried 4·5 m below the surface. Refreezing time for piles in permafrost is most important and depends on such variables as time of emplacement, ground temperature, and soil moisture. It was found that piles driven in the early spring could be loaded within two or three months, whereas those driven in autumn required above six months refreezing time. The fact that only a very small number of piles moved after they had been emplaced, and then only by a very small amount, is a tribute to the efficiency of the pile-construction techniques.

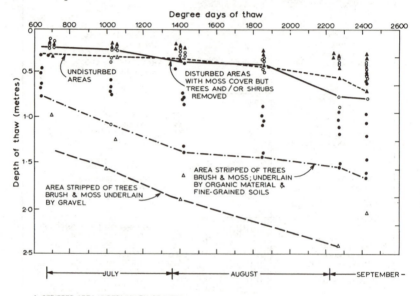

FIG. 9.12. Extreme limits of depth of thaw at Inuvik in areas (a) stripped of vegetation and underlain by gravel, (b) undisturbed and underlain by gravel, (c) undisturbed and underlain by fine-grained soils and (d) stripped and underlain by fine-grained soils (after Pihlainen, 1960)

Other efforts were made to insulate the ground from the effects of new structures. For example, all major buildings had air spaces beneath them of at least a metre. All heating, water-supply, and sewage-disposal lines were placed in an elevated and insulated utilidor (Fig. 9.13). The heating supply also warmed the utilidor and thus prevented freezing of the other services. But the utilidor was expensive, costing $735 per m; utilidettes, leading from the utilidor to individual houses, cost $440 per m (Yates and Stanley, 1966).

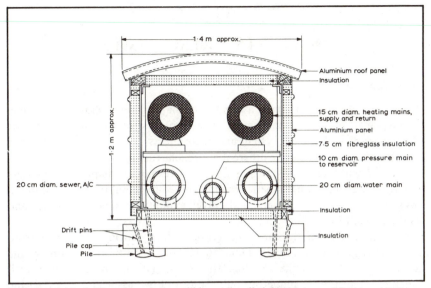

FIG. 9.13. Section through the utilidor at Inuvik (after Yates and Stanley, 1966)

Water supply comes from the Mackenzie River when it only requires chlorination—it is pumped through the system to the lake behind the town. The lake also has its own catchment. Sewage is taken from the town to an artificial lagoon, where it is stored, treated, allowed to digest, and discharged into the river when there is no danger of contaminating water supplies.

The decision to relocate Aklavik at Inuvik was dramatically justified shortly after the new town was completed: in June 1961 an ice-jam on the Mackenzie River caused serious flooding of the old town.

9.7. Conclusion

Rational management has been a serious consideration in many recent developments of periglacial environments. Geomorphologists, often trained in geology, geography, or engineering, have participated in numerous environmental surveys and development projects. Geomorphological interest has often focused on two major themes: the processes of frost action and solifluction and the relations between these processes, ground disturbance, and surface stability; and the landforms resulting from these and other processes. The study of processes and the variables that control them may be useful in preventing accidental and often problematic environmental disturbance, and may also help towards the formulation of successful development plans and engineering designs. The study of landforms is also valuable because the forms frequently reflect surface and sub-surface processes and conditions, and they can therefore be used in reconnaissance and subsequent surveys to predict those conditions.

10
MATERIAL RESOURCES

10.1. Introduction

Chapters 1–9 have considered a number of major geomorphological systems important in resource management. The emphasis throughout has been on the processes within each system and their effects upon man's environment. In this and Chapter 12 quite different approaches are adopted. This chapter is concerned with earth materials as a natural resource; Chapter 12 considers landform scenery as an aesthetic resource linked to geomorphology. Material resources occur mainly as sand and gravel, exploitable mineral deposits, and as the agricultural soils of an area.

General studies of the sand and gravel industry include that by Barratt *et al.* (1970), Beaver (1968), Dunstan (1966), Institution of Mining and Metallurgy (1965), Lenhart (1962), and Young (1968). In addition there is much of interest in the numerous publications of two associations of the sand and gravel industry: the Sand and Gravel Association Ltd. (London) and the National Sand and Gravel Association (Washington). Numerous detailed resource appraisals have been prepared by various geological surveys and planning authorities, of which those of the Illinois State Geological Survey are excellent examples. Several pertinent observations are also to be found in Flawn (1970).

Mineral resources at or near the earth's surface in relation to geomorphology are reviewed by McKinstry (1948) and Thornbury (1954), while the work done in the U.S.S.R. is typified by Piotrovski *et al.* (1972).

Texts concerned with the relationship between soils and geomorphology include the early work by Jenny (1941) which is still much used, as well as the more recent books by Donahue *et al.* (1971), Fitzpatrick (1971), and Foth and Turk (1972). Bridges (1967) provides an interesting review of the relationship between geomorphology and soils in Britain. The most comprehensive field manual is that produced by the Soil Survey Staff (1951). The importance of soil site is exemplified in *Soil Surveys and Land Use Planning* (Bartelli *et al.*, 1966).

10.2. Sand and Gravel

(a) Industrial requirements and problems

In terms of tonnage, sand and gravel are now the two most important materials extracted from the earth. Both are defined, geologically, on the basis of particle size. Sand is particulate material with diameters between 0·06 and 2 mm, and gravel particles have diameters between 2 and 64 mm. In practice, naturally occurring deposits of sand and gravel normally include particles within a wide range of sizes.

Sand and gravel are required mainly for use in the construction and building industry. Aggregates for concrete are in the greatest demand. Other requirements include gravel for road and railroad beds, macadam preparations and ballast, and sand for plastering, filters, and other uses.

One of the most important requirements in the sand and gravel industry, and one which the geomorphologist can help to meet, is for precise knowledge on the nature and distribution of available reserves. There are several reasons why this requirement is so important. In the first place industrial demands for sand and gravel have been rising rapidly, as Figure 10.1 shows for England

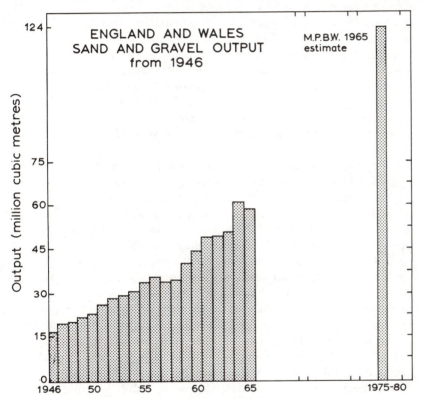

FIG. 10.1. Sand and gravel output in England and Wales, 1946–65, and predicted production based on the Ministry of Building and Public Works forecast, 1965 (after Dunstan, 1966)

and Wales. Closely related to this rise is a trend towards more specific and rigorous requirements for particular purposes. For instance, the size and composition of material for a job may now be precisely specified, as may its resistance to abrasion and weathering processes. In short, the need is for greater quantities of higher-quality products. At the same time increased demand has led in many areas to depletion of the best and most suitably located reserves.

Sand and gravel are a bulky commodity of low intrinsic value which is extremely expensive to transport, so that proximity of supplies to markets is of fundamental importance (e.g. Wooldridge and Beaver, 1950). It has been estimated, for example, that the cost price of gravel at a point some 20–27 km from a pit may be double the cost price at the pit (e.g. Dunstan, 1966; Flawn, 1970).

It follows that sand and gravel reserves within and around major urban areas, where needs are greatest, are the most valuable. And yet it is precisely in these areas that competition for land resources is keenest and the conflict of interests is most pronounced. Urban development may itself preclude the use of sand and gravel. The building of Heathrow Airport (London), for example, resulted in the sterilization of about half of the 9,311 hectares of potentially workable sand and gravel in west Middlesex. In addition, many river-terrace sand and gravel deposits on urban fringes coincidentally provide optimum locations for those agricultural activities, such as market gardening, which also benefit from proximity to urban markets. Thus, in order to help resolve such conflicts, the nature and extent of sand and gravel reserves must be known.

The extraction of sand and gravel near urban areas is seen by some to be objectionable on several environmental grounds. The plant is not only unsightly, but screening, washing, and crushing equipment is also often noisy. Heavy local traffic may be generated, and many hectares of land may be consumed. In the United Kingdom more than 1,200 hectares are used each year, with over half of them becoming water-filled pits. But dry or wet pits are not without their uses. Sites for the disposal of waste are scarce in most urban areas, and old sand and gravel workings may be suitable receptacles, especially for inert debris such as rubble and pulverized fuel ash. Once the pits are filled, they can be used in many ways, such as for forestry, recreation, building, or agriculture. Wet pits can often be transformed into recreation areas that provide facilities for water sports, as the Lee Valley Regional Park Scheme (London) demonstrates (Anon., 1969). Finally, the sand and gravel industry not only demands water, but it also may pollute rivers, and affect groundwater conditions (for example, by locally lowering the water-table).

These various characteristics of the industry pose many planning problems to which varied solutions have been devised. In the United States, the designation of 'Sand and Gravel Zones', the issue of conditional use permits, and other planning devices have been used for some years (e.g. Young, 1968). In the United Kingdom, the working of minerals was first brought under general planning control in the Town and Country Planning Act (1947), and the general aims of mineral planning were set out in *The Control of Mineral Working* (Ministry of Housing and Local Government, 1960). Responsibility for resolving problems of conflicting land use and ensuring future sand and gravel supplies rests largely with the planning authorities, although the task of exploitation remains in the hands of private operators.

As reserves have declined and as industrial requirements have become greater and more sophisticated, the sand and gravel industry has responded in a variety of ways. One response has been the increasing use of crushed rock as a substitute for sand and gravel in road-bed construction and similar activities. More importantly, mechanization and the development of such equipment as excavators have permitted a great increase in the rate of production from individual quarries. 'Beneficiation' of sand and gravel deposits by improved methods of crushing, screening, grading, and the elimination of unwanted material has helped to meet the specific requirements of industry and permitted the exploitation of 'poorer' deposits. Unwanted elements in many natural deposits include clay and silt (which are not only useless but may harm concrete and asphalt mixes), porous particles (which are susceptible to weathering processes, Chap. 11), and other unsound particles. Where deleterious material has a specific gravity that differs significantly from that of the required sand and gravel, the Heavy Media Separation process can be used (Lenhart, 1962). Thus, as the technology of the industry has advanced, so poorer reserves have become usable (albeit, more expensively), dependence on specific deposits has relaxed, and the range of deposits suitable for a specific purpose has been extended.

But the fundamental requirement of knowing precisely the nature and distribution of potential reserves remains. And it is here that the geomorphologist has a contribution to make.

(b) Sand, gravel, and the geomorphologist

Sand and gravel supplies normally come from two major sources: stratified sedimentary rock formations and superficial sedimentary deposits. Only the latter are of major concern to the geomorphologist. Suitable sand and gravel resources often occur in four major geomorphological contexts: (i) the channels, floodplains, and terraces of rivers; (ii) fluvio-glacial environments, where a wide variety of material may be laid down on, in, under or in front of ice sheets or glaciers, or carried in meltwater derived from them; (iii) marine environments, especially in the littoral zone; and (iv) on slopes, where suitable materials may be found in screes, etc. Thus in England and Wales approximately 10 per cent comes from Triassic deposits in the Midlands; about 60 per cent of sand and gravel supplies is recovered from river deposits; a further 25 per cent originates from fluvio-glacial deposits, and some 5 per cent is drawn from coastal localities (Dunstan, 1966).

Each of the principal sources of superficial sediments has distinctive characteristics. In the context of demands placed on such sediments by industry, several of these characteristics should be emphasized. Firstly, the better washed and sorted the deposit, by and large, the less processing will be required. Thus river deposits, in which natural sorting and washing have often been extensive, are usually preferable to glacial tills and other similar deposits. Particle-size distribution is a further important feature. For example, the presence of fine material (silt and clay) within and between beds of sand and gravel is undesirable, and large boulders may make working more difficult. In addition, *both* sand and gravel are normally required in the construction industry and a mixed deposit is needed, the optimum proportion of gravel being normally between 40 and 60 per cent. Deviations from this proportion in a deposit may mean that some material is wasted. Closely allied to this point is the requirement that the quantity of material available for exploitation at one location should be adequate for sustained production for a period sufficient to justify the large capital investment needed in a modern operation. Also of importance are the shape, composition, and physical properties of the particles. For example, coarse, 'sharp' (angular) sand—such as that often found in fluvio-glacial deposits—may be suitable for concrete, whereas 'soft' (rounded) sand may be preferable for mortar, and fine sand is required for plastering. Finally, the thickness of useless overburden on the deposit and the depth of the water table affect the suitability of a reserve. In the case of the former, too much overburden may limit the economic viability of a deposit; and in the case of the latter, special extraction techniques may be necessary where the water table is high.

The appraisal of superficial sediments with a view to their possible exploitation normally involves the following phases: preliminary survey of possible source areas from available geological and topographic maps and from air photographs (Dawe, 1965; Rawiel, 1965); field survey of deposits, including mapping of their areal extent and determination of their depth and volume; sampling of deposits from available exposures, boreholes, and trial pits, etc.; and laboratory analysis of samples to determine such things as grain-size characteristics, particle shape, mineralogy, weathering characteristics, abradibility, and shrinkage values.

The techniques required in each of these phases are those widely used by the geomorphologist. In the preliminary survey, land-systems mapping (see Chap. 13) and geomorphological mapping (see Chap. 14) are extremely useful; the field survey can be adequately based on morphological or geomorphological mapping techniques; the distribution and depth of a deposit, and the position of the water table can at times be determined at the reconnaissance stage or in more detailed surveys with electrical-resistivity or seismic equipment, especially if supporting borehole information is available (McLellan, 1967; Vann, 1965), or by drilling methods; and appropriate techniques of sediment analysis are normally included in the training of a geomorphologist.

There are many examples of surveys of sand and gravel resources in which the techniques and expertise of the geomorphologist have been used successfully (e.g. Ministry of Town and Country Planning, 1948). It is also true to say that many research theses by geomorphologists in recent years, such as those on glaciated landscapes in Britain, contain much data pertinent to the interests of the sand and gravel industry, but unfortunately the authors have too often failed to pursue the applied potential of their studies. Two brief examples from the United Kingdom illustrate the contribution of geomorphologists to the assessment of sand and gravel resources.

S. W. Wooldridge was the principal geomorphologist on the Ministry of Town and Country Planning's Advisory Committee on Sand and Gravel (Ministry of Town and Country Planning, 1948), and studies by Wooldridge and others (e.g. Wooldridge and Linton, 1955) on the geomorphological history of south-east England were of great value in assessing the resources of the London region and elsewhere. For many years the main sources of sand and gravel had been the river-terrace deposits, especially west of London, associated with the most recent major phase in the evolution of the River Thames and its tributaries. Studies of geomorphological evolution had shown, however, that the Thames formerly followed a more northerly course and that along this line and elsewhere at elevations above the main terraces there were extensive fluvio-glacial and plateau deposits. Thus the report identified these deposits as the principal gravel reserves of the Metropolis and it pointed especially to the Vale of St. Albans between Watford and Ware (Fig. 10.2). It was also recognized that these reserves were poorer in quality than those of the traditional sources west of London. The main reason for this is that older deposits of the northern area are relatively poorly sorted and poorly bedded fluvio-glacial deposits, often containing silt and clay, and locally overlain by till, whereas the younger terrace deposits are often derived from the older material and, in the process, have become progressively washed and sorted. The older deposits clearly require greater beneficiation than the terrace sediments, but as the traditional sources have become exhausted or sterilized, so the exploitation of the less satisfactory reserves has increased.

A second example illustrates more precisely the ways in which the geomorphologist can explore the nature of sand and gravel resources. In a study of west-central Scotland, McLellan (1967) showed that the vast majority of sand and gravel reserves were of fluvio-glacial origin and that an understanding of the geomorphological characteristics of such deposits facilitated their study from the point of view of the extractive industry. McLellan identified two main groups of deposits: (i) those laid down in front of the ice by running water and forming relatively smooth 'valley trains' of well-

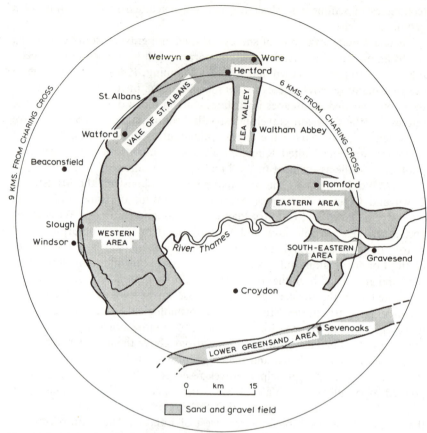

FIG. 10.2. The major sand- and gravel-producing districts in the London area (based on the Ministry of Town and Country Planning, 1948)

sorted and well-stratified sand and gravel with some undesirable local pockets of finer material; (ii) those laid down in an ice-contact position by water flowing on, in, or under an ice mass, and producing such features as kames, kame terraces, and eskers, in which the material is stratified, but not well sorted, generally mixed, and of a wider size range than deposits in the first group. (For details of fluvio-glacial geomorphology, see Price, 1973.) In addition to these deposits, there are morainic deposits (which often underlie fluvio-glacial sediments or occur on higher ground), recent alluvia, and some marine sediments. The nature of the fluvio-glacial material was found to vary according to its origin and location within the geomorphological environment prevailing when it was deposited: for example, the thickness of deposits was greatest in valley floors and on lower valley slopes, and deposition was particularly extensive where the ice margin had been relatively stable for protracted periods.

A preliminary reconnaissance survey based on air-photograph interpretation and fieldwork provided a general map of sand and gravel areas in west-central Scotland (Fig. 10.3a). More detailed fieldwork in central Lanarkshire, taken together with borehole and other information, permitted the distribution and volume of reserves to be described in this area (Fig. 10.3b). Analysis of sediment samples revealed, in a preliminary way, several important aspects of sand and gravel quality. For example, the proportion of gravel in fluvio-glacial deposits was shown to vary considerably, but most of it did not require crushing to reduce it to suitable size. Again, it was suggested that much gravel had relatively poor shrinkage qualities and that a high proportion of stones from the Highlands improved the shrinkage qualities.

10.3. Mineral Resources

The location of minerals is often closely related to the geomorphology of an area (Table 10.1). Perhaps the clearest example of this lies in the relationship between alluvial placer deposits and features of fluvial activity, such as river terraces. However, as is particularly shown by the work carried out in the U.S.S.R. (Piotrovski, *et al.*, 1972), geomorphological investigations have also played a leading part in the search for other types of minerals. The relationship of mineral deposits to relief is summarized in Tables 10.1 and 2. Groundwater of various origins may be classified in one or other of the groups defined, depending upon local conditions.

TABLE 10.1

The relationship between major categories of minerals and relief

I. *Minerals directly related to relief*
(i) Placer deposits (e.g. gold, diamonds, tin (cassiterite), and other heavy minerals).
(ii) Weathering products (e.g. enriched copper, limonite, manganese, bauxite, cobalt, and kaolin).
(iii) Basin deposits (e.g. coal, iron and manganese ore, peats).

II. *Minerals indirectly related to relief*
Deposits associated with structures that are revealed in the relief, such as down-faulted blocks or areas of upwarping (e.g. gas, oil, salt, coal, mineralized magmatic bodies, and buried placer deposits).

III. *Minerals not related to relief*
(Geomorphological analysis of an area is only incidental.)

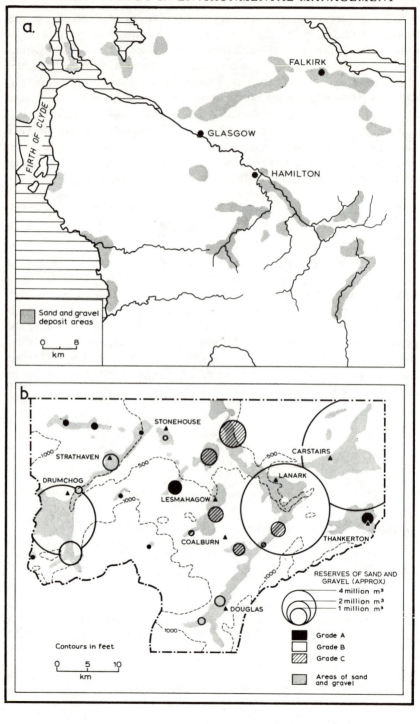

In searching for the association between minerals and relief it is important to establish as much as possible about the geomorphological history of an area. This is because some associations may arise not so much from the contemporary processes operating in the area but more from the recognition of former relief conditions. For example some ore deposits are associated with weathering crusts, having been derived from the chemical breakdown of the original parent material. These crusts tend to be related to planation surfaces of low relative relief. The residual iron ores of the Mayari and Moa Districts (Oriente Province), Cuba, were probably formed as part of a weathering residue of serpentine rock on an ancient land surface (Leith and Mead, 1911).

Where river dissection has followed planation the weathered crust will have supplied to the rivers minerals which may have become concentrated as placer deposits within river terraces. These may subsequently be buried (e.g. by hillslope colluvial material) or may themselves become the source for a colluvial feature. This hypothetical case illustrates some of the complexities that may occur in a landscape, and that need to be unravelled by geomorphological analysis during a search for likely sites of mineral deposits. Such an analysis could do much to eliminate from a field investigation those sites which are unlikely to yield minerals.

A recent example of studies linking geomorphology and the supergene alteration of minerals is the work of Clark, et al. (1967) in the southern Atacama Desert, Chile. In this area, enriched copper deposits have been known for many years, and they have been essential for the economic progress of small-scale mining operations. Such enriched deposits, characterized by chalcocite and other copper-sulphide minerals, are associated with chemical changes in the vicinity of present or former water tables. They are normally found below a zone of oxidation or leaching, and above relatively unaltered hypogene mineral assemblages. As the shape and position of the water tables at present and in the past are closely related to the form of the ground, an attempt was made to establish the relations between enrichment zones and topography, so that the latter could be used to predict where new enriched copper deposits might be found.

The identification of enriched, leached, oxidized, and hypogene mineral zones showed that there had been at least three major periods of supergene mineral alteration, and that the resulting zones are roughly horizontal and sheet-like in form. This evidence suggested that their development was

---

FIG. 10.3. a. Potential sand- and gravel-bearing areas in west-central Scotland (after McLellan, 1967)
b. The distribution of, and estimated volume of, sand and gravel reserves in central Lanarkshire. The quality grades are based on the presence of certain defects: till and clay inclusion; high percentage of 'fines' (sand/silt/clay over 75 per cent); presence of large quantities of coal or lignite; poor shrinkage qualities. A deposit with none of these defects is classified as grade A; a deposit with one defect, grade B, etc. (after McLellan, 1967)

TABLE 10.2:   *Possible associations between geomorphological features and mineral deposits*

RELEVANT GEOMORPHOLOGICAL FEATURES	Placers (continental) (e.g. gold, diamonds)	Placers (littoral, marine and lacustrine)	Peat	Clays and sandy loams	Gravel sand	Brown coal	Black coal	Bauxite and refractory clays	Residual deposits of weathering (nickel, cobalt, iron, magnesite)	Infiltrated weathering deposits (sulphur, uranium, gypsum)	Deposits of translocated weathered crust (e.g. cobalt, iron)	Salt (e.g. potassium, sodium, gypsum)	Sedimentary, ferromanganese	Magmatic ore (metals, diamonds etc.)	Pegmatitic (e.g. muscovites, optical raw materials)	Carbonate (e.g. rare minerals)	Hydrothermal	Metamorphic (e.g. iron, manganese)	Groundwater	Fissure water	Oil and natural gas
Tectonic-structural	*	+	+	+	+	*	*	*	*	*	*	+	+	*	+	+	*	+	+	*	*
Relief expression of deep intrusions	+							*	*		+			*	*	+	++	**		++	
Structures in sedimentary and metamorphic rocks	+	+			+		*	+	+	+	+	+	+	+	+	+	+	+	+	*	*
Valley pattern reflecting a) folding	+	+			+		*	+	+	+	+	+	+	+	+	+	+	+	+	+	*
b) dome structures	+	+			+		+	+	+	+	+	+	+	+	+	+	+	+	+	+	+
c) faulted blocks	*	+			+		*	*	*	*	+	+	*	*	+	*	*	*	*	*	*
Old valley system, changes induced by: a) tectonics	*	+	+	+	+	+	+				+			*				*	*	*	*
b) exogenous factors (e.g. glaciation, sedimentation)	*	+	+	+	+		+							*				*	*	*	*
Longitudinal valley profiles	*									+											*
Slope steepness	*		*	+		+	+	+				+	+	+	+	+	+	+	+	+	+
River terraces (and their sediments)	*		*	*	*			*													*
Floodplains (and their sediments)	*	*	*	*	*				+												+
Riverbeds (and their sediments)	*	*			*																
Planation surfaces	*	*	+	*	+	+		*	*	*	*	+	+					*			
Eluvium and weathered crusts	*	*		*	+	+		*	*	*	*	+	+		+		+	+	+	+	*
Present-day karst	*	+	+	+				*	*	*	+	*		+					*	*	+
Fossil karst	*		+	+				*	*	*	*	*	*						*	*	*
Lake basins, lacustrine and marshy accumulations a) humid regions	+	+	+	+	+	+			+·		+	*	+						*	*	*
b) arid and semi-arid regions		+	+	+	+		+				+	*							*	*	+
Marine and lacustrine littoral forms (primarily of wave and accumulation types)		*	+	+	*	+					+	+						*			
Glacial and fluvio-glacial forms of mountain glaciation	+			+	*			*										*			+
Glacial and fluvio-glacial forms of continental glaciation	+		*	*	*			*										*	+		
Cryogenic forms	+		*	*	+			*										*	*	*	
Relief buried by poorly consolidated and young volcanic rocks	*	*			*			*	*	*							*	*			

+ Association possible - feature mapping useful

* Association probable - feature mapping indispensable

Highly modified from Demek (1972)

controlled by horizontal or gently sloping surfaces, such as the major pedi-plains recognized in the region. But the relationship between enriched ores and pediplains was not simple. For instance, the lowest pediplain in places clearly truncates one of the enriched zones, exposing rich ores at the surface. It seemed probable that supergene enrichment was related to dissection phases between periods of uninterrupted pediplanation, but that the less dissected areas of the planation surfaces permitted both the development and preservation of enriched ores. Canyon development followed the formation of the pediplains and major enriched zones, causing water tables to be lowered and oxidation of the upper parts of the enriched zones. Later gravel deposi-tion in the canyons raised water tables adjacent to valley floors and led to the formation of 'sooty' chalcocite in a zone above the present water table. Subsequent incision of these gravels was accompanied by renewed oxidation.

This study led to the suggestion of guidelines for determining the location of new enriched deposits. For example, copper deposits within canyons could largely be ignored because they are likely to have had their enriched zones eroded or oxidized. And undissected remnants of the lowest pediplain are the most promising localities for the discovery of enriched ores relatively near to the surface.

Another example occurs in the Urals where a series of Tertiary and Quaternary terrace and flood-plain diamond placers has been formed follow-ing the weathering and erosion of fossil marine placers (Piotrovski et al., 1972). The latter occur in Palaeozoic littoral sandstone and coarse gritstone beds, and the primary source of the diamonds has never been found. In South Africa, on the other hand, diamond placers occur within rivers draining from the kimberlite sources in the Kimberley area. The kimberlites rise as gentle domes above the general surface and can themselves be recog-nized by direct reference to the relief.

These examples show that a knowledge of the geomorphological history in each case can guide very materially the continued search for further diamond-bearing placers in these areas.

In mineral prospecting it is also important to know something of the reaction of the minerals themselves to geomorphological processes. For example, those with a high specific gravity (e.g. gold) tend to be concentrated in specific locations and are better sorted than those with a lower specific gravity (e.g. zircon and diamonds). Heavy minerals may also accumulate in placers nearer their sources than minerals of lower specific gravity. The resistance of minerals to weathering and abrasion is also important. Diamonds are very resistant, can be carried far and even recycled, as in the example quoted above from the Urals region. Cassiterite (SnO_2) on the other hand quickly disintegrates during transport, and sites of alluvial and colluvial tin in south-west Uganda, for example, are close to the original source on the valley-sides. In the valley shown in Figure 10.4 the distance from source to

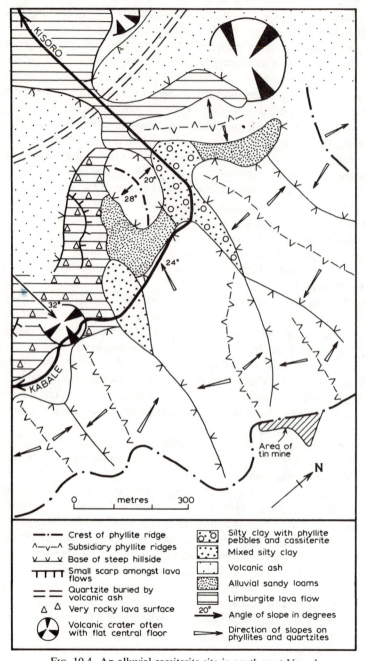

FIG. 10.4. An alluvial cassiterite site in south-west Uganda

payable alluvial deposit is less than 500 m. Beyond a distance of 800 m no deposits have been worked, and no one has yet thought it worthwhile to look for such deposits beneath the cover of lava and volcanic material which buries the lower part of the valley containing these cassiterite-bearing deposits.

On a larger scale whole ore bodies may be directly responsible for particular relief features because their resistance to denudation differs from that of the rocks around them. For example, the lead–zinc lode at Broken Hill, Australia, is marked by a conspicuous ridge. Conversely, calcite veins at Oatman, Arizona coincide with surface depressions (Thornbury, 1954). If the geomorphological expression of a specific ore body can be identified for any one area then the recognition of similar ore bodies may become a comparatively simple task. In such cases aerial photographs become an invaluable aid in the recognition of sites worthy of detailed field examination.

A similar approach has been used in the U.S.S.R. where recent folding and upwarping has been detected by geomorphological means. These have included the recognition and mapping of upwarped planation surfaces and depositional (downwarped) plains or basins. The upwarped structures provide locations favourable for the development of oil and natural gas. In some instances such upwarping needs to be inferred from radial drainage patterns, or from the observation that rivers are skirting round an area that may prove to have been upwarped. The most recent tectonic movements may be reflected in nick points in the river thalweg (Piotrovski et al., 1972).

A systematic approach to the search for minerals and ore bodies through landform analysis is provided by geomorphological mapping (Chap. 14). In the case of placer deposits this can be carried out within the context of the drainage basins likely to contain them (Chap. 1), though past changes in river patterns must be taken into account. The most valuable gold placer deposits of the Sierra Nevada are to be found predominantly within gravels associated with prevolcanic, early Tertiary, or Eocene drainage channels which do not coincide in position with the present river channels. As a result the problem of locating the richest placer deposits is one of reconstructing the early Tertiary drainage pattern (Thornbury, 1954).

10.4. Soils

(a) Introduction

The whole subject of the relation between soil and geomorphology is so large that nothing less than a book can do justice to it. This brief section seeks only to indicate some of the reasons why a study of soils cannot be undertaken without also making a critical assessment of the geomorphology of their site.

Soils form the most ubiquitous natural resource available to man. Soil is the essential element in all agriculture in that it forms the basis for plant growth. From the soil plant roots receive mechanical support, water, the minerals required for growth, and oxygen. The physical condition of the soil determines its ability to supply water to plants. This is largely a function of available pore spaces. Soil chemical properties, on the other hand, determine the capacity of soils to supply nutrients. Both physical and chemical properties determine the nature of plant root extension, adequate extension being essential to healthy plant growth. A good physical state usually indicates that a soil is also in a good chemical condition.

The main physical properties of the soil that matter to agriculture are texture, structure, depth of horizons, total depth of the soil, soil consistency, and temperature. Field assessment of physical properties such as texture, structure, and consistency may be made by reference to the *Soil Survey Manual* (Soil Survey Staff, 1951) of the U.S. Department of Agriculture. These physical properties determine water movement through the soil, weathering processes, and the translocation of colloids and minerals. A soil acquires these properties as a result of the weathering of parent materials and the denudational (i.e. erosional and depositional) processes that have affected it since its formation began. The influence of parent materials may also be seen in the mineral composition of the soil, though this is particularly vulnerable to weathering processes. Parent materials containing a large proportion of aluminosilicate minerals weather readily into clays. Parent materials rich in calcium, magnesium, or potassium (i.e. alkaline rocks) tend to weather more slowly than acid rocks. Intense chemical weathering may cause minerals to be leached out of the soil as weathering proceeds. Thus, although a parent material may have a high calcium content (to take but one example) this leaching process may leave a soil with a low calcium content. That is why liming is sometimes necessary for agricultural soils developed on calcareous bedrock. As time proceeds, however, the depth of weathering tends to increase and climate may begin to impose characteristics upon the soil that over-ride the earlier dominant influence of parent materials. Similarly the influence of vegetation may increase with time. The direct influence of parent materials may also become obscured by the mixing of materials, as where a low site receives material from hillslopes above.

It is generally assumed that climate may be a dominant influence over parent materials in tropical areas where chemical weathering is intense and may produce a weathered profile tens of metres deep. Nevertheless tropical soils may still be dependent upon site for some of their characteristics (e.g. Moss, 1965; McFarlane, 1971).

Applied geomorphology is concerned not only with landform but also with earth materials and denudational and weathering processes. Soils thus provide a key element in geomorphological investigations. In addition, a geomorphologist's interests lie not only in the soil itself, but also in its environmental setting, its relationship to landform, parent materials, and the processes that are likely to affect its development. This interest also continues into an assessment of the soil's potential vulnerability to erosion by water (Chap. 2) or by wind (Chap. 3). The links between the geomorphologist's interest in soils and those of the agriculturalist are therefore clear. Through his work the geomorphologist has much to contribute to a proper understanding of the agricultural assets of a soil. An analysis of soil genesis and an evaluation of soil environment provide useful information to those whose task it is to manage agricultural land.

(b) The geomorphological context of soils

Soils represent the interaction between processes of weathering and rock materials. Their accumulation or removal is a function of both time and processes of denudation. They are therefore closely related to the geomorphological history of the site they occupy. For example, sandy-textured soils with a granular structure may be directly related to their situation on an alluvial terrace. Likewise the soils of floodplains, alluvial fans, screes, pediment surfaces, or deeply-weathered etch-plains may be predictable in terms of site geomorphology. Sometimes soils have been shown to vary with the age of the land surface on which they occur (Ruhe, 1956; Mulcahy, 1960). Conversely, however, the soil can be thought of as an indicator of geomorphological history. Thus a thin skeletal soil or a truncated soil profile generally indicates that material has recently been removed from the site. On the other hand a buried soil profile shows that deposition has taken place. A soil which displays a full profile development is indicative of a relative geomorphological balance between the processes of weathering and denudation. To the extent, therefore, that soils are determined by geomorphological processes, the recognition and analysis of those processes has a bearing on soil investigations for resource management.

The influence of relief on soil-profile formation, and therefore soil type, can be direct in that steep slopes tend to develop thin soils, which also tend to be less weathered than those on gentler slopes in the same area. Relief also influences the movement of water. Sites which are freely drained develop eluvial soils, with the tendency for material to move out of the soil either in suspension or solution. Conversely where water tends to accumulate it also brings in material to form illuvial soils, or even an organic soil. The amount of water available to a soil usually increases down-slope (see the discussion in Chap. 2 on throughflow), providing a down-slope sequence of soils related to hydrological conditions of the slope. Valley-floor sites or lowland depressions may remain wet for most of the year, giving waterlogged conditions that strongly influence soil-profile development. Such hydrological sequences are particularly common in areas of hummocky glacial deposits, and in certain areas of stabilized sand dunes (Fitzpatrick, 1971). Thus, a progressive change down-slope from a freely drained site (Fig. 10.5) at the top of a slope to a waterlogged site at the bottom can give rise to a soil sequence from podzol through gleys to peat. The podzol is identifiable because of its eluvial upper horizons, while the gley has illuvial characteristics and the peat arises from the accumulation of organic material.

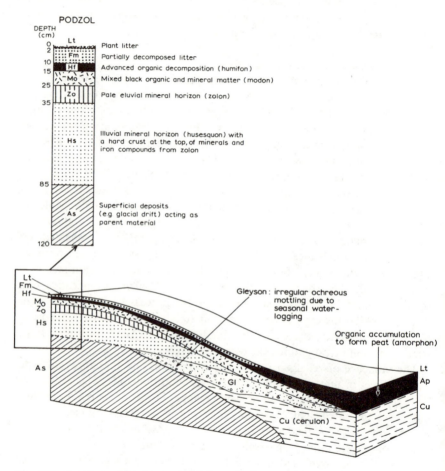

FIG. 10.5. A soil sequence, from podzol through gleys to peat, associated with slope and drainage (after Fitzpatrick, 1971)

The influence of slope angle and distance down-slope on soil properties formed the basis of a study by Furley (1971) of four transects over convexo-concave grass-covered slopes underlain by Upper Cretaceous Chalk near

Oxford (England). The study was based on a minimum of 34 soil pits on each transect, and on physical and chemical analyses of the surface horizons in each soil pit. The slopes were divided into two sections; an upper one, where net erosion is greater than net deposition, and a lower one where deposition predominates. Multiple regression analysis showed that the combined effect of slope angle and distance down-slope could account for much of the variation in the soil properties. For example, in the upper parts of the slope, soil acidity (pH), the coarser soil fraction, carbonate content, exchangeable sodium, potassium, and calcium, all increased with angle of slope, while the percentage of organic carbon, percentage of total nitrogen, moisture content, and the fine soil fractions all decreased as angle of slope increased. This effect of slope steepness was much less pronounced on the lower slopes, which showed greater variability in soil properties. On the lower slopes distance down-slope and position on the slope appeared to be the more important controls.

(c) Mapping soil resources

In many areas landform and soils are closely related, not only because the former may reflect processes that influence soil formation (Chap. 11) or soil erosion (Chaps. 2 and 3), but because they may also be related to the parent materials upon which the soils have developed.

In 1935 Milne reported for parts of East Africa a recurring and definable relationship between soil site and soil type. He suggested, perhaps somewhat optimistically, that a predictable relationship could be established, in any one area, between a soil and its position down a valley-side profile. Indeed all similar valley-side positions might be expected to have the same soil type. Milne (1935) called this relationship a soil catena. This principle has been very important in terms of soil mapping chiefly because once a definite relationship between soil and site has been established it is much easier to delimit provisional soil boundaries by mapping sites than by digging soil pits. This principle is common to many reconnaissance soil maps derived from aerial photographs, and it is echoed in the principles of the land-systems approach to be described in Chapter 13.

Common examples of catenas are quoted by Young (1972). In temperate latitudes the down-slope sequence often to be found is brown earth/gleyed brown earth/gley. In upland Britain there is often a repeating sequence of hill peat/thin iron-pan podzol/iron podzol/shallow acid brown soil, becoming deeper down-slope/gleyed acid brown soil/gley/basin peat. In some locations not all members of this catenary sequence are present.

The soil catena, therefore, is an expression of the direct relationship between geomorphological and soil-forming processes. The forces that produce the geomorphologist's 'materials' create the pedologist's 'soils'.

The most useful general-purpose model of site–process relationships in terms of predicting soil characteristics is that developed by Dalrymple *et al.* (1968). These authors recognized nine recurring landform units (Fig. 1.4) which have definable surface and subsurface processes. The model provides a ready framework into which areas may be subdivided and from which primary inferences may be made about the processes influencing soil development. All nine units of this model will not be present on every hillside, but it should be possible to classify any hillside in terms of these units. More complex hillsides may have one or more of these units repeated down their profile. Equally valuable is the mapping of each unit laterally along the slope as an initial estimate of the likely extent of associated soil types. Such a model is especially useful during soil mapping at the reconnaissance level.

Whenever a reconnaissance survey of soils is required, especially over large or remote areas, recourse normally has to be made to the interpretation of aerial photographs (Lueder, 1959; American Society of Photogrammetry, 1960). The important elements of a landscape that can be directly interpreted from aerial photographs and which have a direct bearing on soil mapping include the following geomorphological properties (Buringh and Vink, 1965):

(i) Elements having a positive and direct correlation with soils (e.g. waterlogged sites, patterned ground);
(ii) Elements related to terrain morphology (e.g. land type, relief form, slope, drainage pattern, watershed pattern, rivers);
(iii) Elements related to special aspects of terrain (e.g. gully dissection (form and pattern), lithology, tonal or colour variations).

A common starting-point for soil mapping from aerial photographs is a classification of the relief of the area to provide a framework for the soil map legend (Goosen, 1966, 1967). The units mapped may resemble those of Figure 1.4 or they may be like those of a land-systems analysis (Chap. 13), and they may be adapted, through subsequent fieldwork and laboratory analysis, into soil-mapping units. Such a classification can also save much time and expense during subsequent fieldwork as it can form the basis of a stratified random sampling programme of soil-pit investigations. A limited number of soil pits are required as only the within-unit variability needs to be tested in order to confirm boundaries. It is also easier to check the reality of provisional soil boundaries than it is to make a primary definition of them during a field survey. This approach, however, frequently leads to a classification of soils by their site characteristics rather than on the basis of their own intrinsic properties.

10.5. Conclusion

Since a close relationship exists between site and sand and gravel deposits, mineral deposits, and soils, it is often possible to evaluate these resources through the geomorphology of their sites. This evaluation may be applied at a reconnaissance stage in order to discover the likelihood of a particular type of material resource being present in an area. It may also be undertaken, in the case of soils, in order to discern the physical controls on their continuing development. Deciphering the geomorphological history of a deposit or site may form an essential component of the evaluation process if predictions are required of the composition of the materials and of their likely internal variability.

11

THE DESTRUCTION OF NATURAL
MATERIALS BY WEATHERING

11.1. Introduction

Rock is a material of fundamental importance to many aspects of human activity—it is the parent of soils upon which food production depends, it is widely used in buildings, roads, and other engineering works, and it forms the natural foundation for most engineering structures. All rocks may be subjected to attack by natural *weathering* processes, and most tend to be altered and destroyed by them, especially if the rocks were formed under conditions different from those of the present environment. The nature and effects of such processes are therefore of more than passing interest to scientists and engineers who study rocks and use them as raw materials in their work.

Several different groups of scientists have long-standing interests in the weathering of rocks. Geomorphologists have studied weathering processes theoretically, in the field, and in the laboratory, mainly with a view to interpreting weathering landforms and to understanding the supply of materials to erosional systems. Recent geomorphological surveys include Ollier's *Weathering* (1969), and Carroll's *Rock Weathering* (1970). The building industry has been concerned for decades with weathering processes that destroy natural and artificial building materials, with determining the durability of the raw materials they use, and with preventing their decay (e.g. Schaffer, 1932; Crowder, 1965; Simpson and Horrobin, 1970). Schaffer's enduring classic *The Weathering of Natural Building Stones* (1932) has recently been reissued by H.M. Stationery Office (London). Research in this field has usually been focused on organizations such as the Building Research Establishment (United Kingdom), The Division of Building Research (National Research Council, Canada) and the Division of Building Research (Commonwealth Scientific and Industrial Research Organization, Australia).

To the construction engineer, the degree of weathering and its relation to important engineering properties of materials such as strength, particle size, and permeability are important in terms of the selection of materials for construction, and the appraisal of sites for foundation design (e.g. Hosking and Tubey, 1969). Here the interests of engineering are closely allied to those of soil mechanics and links between the two fields have been securely forged (e.g. Scott and Schoustra, 1968; Terzaghi and Peck, 1948). Fundamental work in this area owes much to Karl Terzaghi and others at Harvard University and the Massachusetts Institute of Technology.

Weathering processes are also of consequence to soil scientists who study soil from an agricultural point of view and especially to pedologists working in soil chemistry, for an understanding of weathering processes contributes to the knowledge of soil development and hence to land management (e.g. Cruikshank, 1972; see also Chap. 2, 3, and 10). The Department of Agriculture in the United States, The Soil Survey of England and Wales, and the C.S.I.R.O. in Australia are amongst the organizations active in this area of study.

Despite their common interest in rock weathering, the cross-fertilization of ideas between these groups appears to have been surprisingly limited. Take the example of rock disintegration by salt crystallization. Geomorphologists have devised tests to simulate in the laboratory the effect of salt crystallization under different climatic conditions (e.g. Birot, 1954; Goudie, Cooke, and Evans, 1970); building-research workers employ a variety of tests to determine the resistance of rocks to salt crystallization that are usually descended from the studies of Brard in the 1820s (e.g. Schaffer, 1959; Honeyborne, personal communication, 1972); and different 'standard tests' are used by engineers to determine the 'soundness' of aggregates (e.g. in Britain: B.S. 1438 (1948); in America: ASTMS C88 (1971); in Australia: AS-A77; Minty, 1965). These tests have much in common, but the ideas and experience of the different groups do not seem to have been extensively exchanged.

One consequence of the fact that weathering interests such very different groups is that the relevant literature is widely disseminated, and some of it is not generally available (e.g. the notes and unpublished reports prepared by Honeyborne and others at the Building Research Establishment (England) used in this chapter). Thus it is difficult for a review from a single standpoint to do justice to the literature. In this chapter, no attempt will be made to be comprehensive. Rather, some of the scattered literature will be integrated and the applied context of weathering studies will be examined through the exploration of three important themes: the nature of the weathering system in general (Sect. 11.2), the disintegration by crystallization processes of natural building stones (Sect. 11.3); and, more briefly, the consequences of decomposition by chemical processes in engineering problems (Sect. 11.4).

11.2. The Weathering System

(a) Introduction

From most practical points of view there are two main categories of weathering processes—*disintegration* processes and *decomposition* processes. The first group causes rock breakdown without significant alteration of minerals; the second causes chemical alteration of the rock. Table 11.1 lists the principal processes. One point must be emphasized at the outset of this discussion—in most situations, rock disintegration and decomposition occur together, and the results of weathering normally reflect the combined effects of several different processes. It is often difficult or impossible to be certain in any one locality of the processes involved and of their relative importance. Indeed, quantitative evaluation of the relative importance of different weathering processes is a subject that has been grossly neglected.

TABLE 11.1

Classification of weathering processes

I. *Processes of Disintegration*

  a *Crystallization processes*
    Salt weathering
    Frost weathering
  b. *Temperature-change processes*
    Insolation weathering
  c. *Weathering by wetting and drying*
  d. *Organic processes*
    Root wedging
    Colloidal plucking

II. *Processes of Decomposition*

  a. Hydration and hydrolysis
  b. Oxidation and reduction
  c. Solution and carbonation
  d. Chelation
  e. Biological chemical changes

The nature and effectiveness of any weathering process or group of processes depends largely on three sets of variables: climatic conditions, material properties, and local variables (such as vegetation, animal life, exposure, building design, and groundwater). The following examination of these variables is selective; for general views, including brief descriptions of each weathering process, see Carroll (1970) and Ollier (1969).

(b) Weathering processes and climate

As most weathering processes depend mainly on climatic conditions and as these conditions vary from place to place, there have been several attempts to map spatial variations in the nature of individual processes and groups of processes in terms of climatic characteristics on a world-wide scale. Such efforts normally use quantitative parameters derived from the analysis of climatic statistics to describe the dependent weathering processes.

Temperature conditions and the availability of water are two primary controls on many weathering processes, but the precise relations between the controls and the processes are so complex that it is difficult to derive simple climatic parameters that can adequately describe them. In considering temperature, for example, information is often required on absolute values, ranges over different time periods, and the frequency and rates of temperature fluctuation about freezing point; and availability of water involves study of such features and precipitation–evaporation–runoff relations, and rainfall intensity, frequency, and duration. Despite the difficulty of generalization, Peltier (1950) used simple measures—mean annual temperature and mean annual precipitation—as a first approximation of these two controls to describe the intensity of weathering processes on a world-wide scale. Figure 11.1a shows the areas where chemical weathering is likely to be strong,

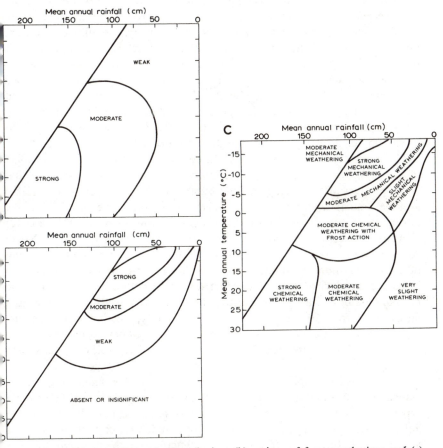

FIG. 11.1. (a) Regions of chemical weathering; (b) regions of frost weathering; and (c) general classification of weathering regions (after Peltier, 1950)

moderate, or weak. It is based on the assumptions that chemical weathering increases with the availability of water (and hence with mean annual precipitation in so far as the two are related), and that the rate of chemical reactions increases with temperature. Thus chemical weathering is strongest in hot, wet climates and weakest in cool, dry ones. Figure 11.1b shows the pattern of frost weathering. Here mean annual figures are less meaningful because the intensity of frost weathering is dependent to a considerable extent on the frequency of freezing days. If the pattern of frost weathering is equated with mechanical weathering processes (by ignoring other disintegration processes), Figures 11.1a and 11.1b can be combined to produce a definition of world weathering regions (Fig. 11.1c).

Such generalizations are inevitably problematical at the world scale, not only because the complexity of the climatic controls prejudices the use of simple climatic parameters, but also because appropriate climatic data are limited, and knowledge of the processes and their relative importance is incomplete. Slightly more meaningful and perhaps more useful regionalization can be obtained if the scale of inquiry is enlarged and more refined parameters are derived from climatic data. Two examples of studies at different scales will illustrate this point, one related to building-brick durability and disintegration processes, the other concerned with highway-foundation engineering and decomposition processes.

The American Society for Testing Materials (1971b) has related the affect of weathering on facing bricks in the United States to a *weathering index* that is the product of the average annual number of freezing-cycle days and the average annual winter rainfall in inches. A freezing-cycle day is any day during which the air temperature passes either above or below 0 °C, a crude measure of freeze–thaw frequency (for other studies of this variable in North America see Russell (1943) and Williams (1964)). Winter rainfall is the sum of the mean monthly rainfall occurring during the period between and including the normal date of the first killing frost in the autumn and the normal date of the last killing frost in the spring. In effect, by incorporating measures of frost frequency and water availability, this index records susceptibility to frost action. Its pattern in the United States, in terms of negligible ($<$100), moderate (100–500), and severe ($>$500) weathering, is shown in Figure 11.2.

A contrasted example at a larger scale relates to the engineering performance of road foundations in South Africa (Weinert, 1961 and 1965). Field experience shows that weathered basic igneous rocks, notably the Karroo dolerites, are generally 'sound' in the west and 'unsound' in the east of South Africa (Fig. 11.3). In constructing Figure 11.3, the unsound dolerites are taken to be those with a liquid limit greater than 30 and a plastic index greater than 10. 'Liquid limit' is defined in Chapter 6.6, and 'plastic index' is the difference between the liquid and plastic limit values (see Table 14.3). In terms of Figure 11.3 disintegration prevails west of the 'weathering and performance boundary' defined by these values, and decomposition prevails to the east of it.

Weinert analysed climatic data in order to determine those climatic variables that could explain the weathering contrast, assuming it to be the result of *contemporary* climatic conditions. Only a few climatic parameters were needed in order to define the weathering boundary. These were potential evaporation and total precipitation during the warmest month, and annual precipitation. An appropriate climatic index expressing these variables was

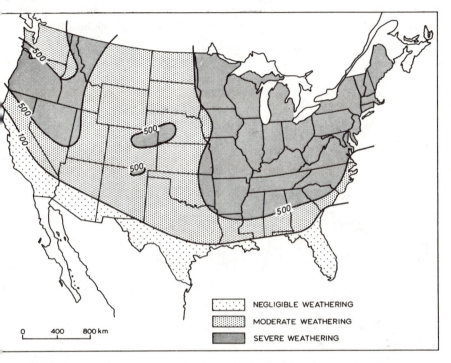

FIG. 11.2. Weathering regions in the United States based on the weathering index (after ASTM, standard C216)

derived as follows. The potential evaporation and the precipitation during the warmest month (January) was expressed by the ratio:

$$R = \frac{E_J}{P_J} \qquad (11.1)$$

where

E = potential evaporation (Meyer formula);
P = precipitation;
$_J$ = January (the warmest month).

As temperature and evaporation tend to be higher in summer, and the seasonal distribution of precipitation varies from place to place, weathering will vary according to the availability of water, and a measure of annual rainfall is required. Such an expression is:

$$D = \frac{12P_J}{P_a} \qquad (11.2)$$

where index a = annual. Summer rain occurs where $D>1$, winter rain where $D<1$, and where $D = 1$ rain is distributed equally throughout the year.

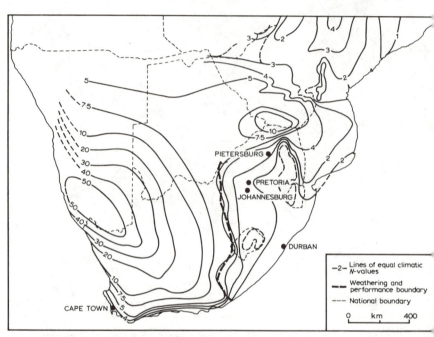

Fig. 11.3. N-values and the weathering and performance boundary in South Africa (after Weinert, 1965)

A numerical expression of the balance between the relevant climatic variables can be derived by multiplying R and D, to produce a precipitation ratio, N:

$$N = \frac{12E_J}{P_a}. \qquad (11.3)$$

N-values are shown for South Africa on Figure 11.3. $N = 5$ is the line that most closely follows the weathering boundary. Where $N>5$, disintegration is more important than decomposition and the foundation materials are relatively sound. Where $N<5$, the reverse is the case. The formation of hydrous mica is the most important product of chemical alteration where $N>6$; where N is between 2 and 5, montmorillonite is the predominant clay mineral arising from the decomposition of basic igneous rocks, and kaolinite is produced in acid igneous rocks. It should be emphasized that the N-index was derived for, and is only of proven application in, South Africa.

(c) Rocks and their properties

Before examining some processes of weathering in detail, introductory comments are required on the definition, occurrence, weatherability, and related properties of rocks.

As rock names are often loosely used, it may be useful to provide standard definitions of those rock types commonly mentioned in the remainder of this chapter, using abbreviated ASTM (1958) definitions of building stones (Table 11.2). It is also important to note the frequency of occurrence of different rock types and rock-forming minerals (Table 11.3), as well as the relative susceptibility of the latter to chemical weathering processes (Table 11.4).

TABLE 11.2

Abbreviated standard definitions of building stones

i. *Granite:* visibly granular, crystalline rock of predominantly interlocking texture, composed essentially of alkali felspars and quartz. Felspar is generally present in excess of quartz, and accessory minerals (*e.g.* micas and hornblende) are also present. *Commercial building granite* is more widely defined, and includes granite, gneiss, gneissic granite, granite gneiss, syenite, monzonite and granodiorite, and other felspathic–crystalline rocks of similar textures, containing accessory minerals and used for decorative purposes.

ii. *Limestone:* a rock of sedimentary origin (including chemically precipitated material) composed of calcium carbonate or the double carbonate of calcium and magnesium (dolomite molecule). *Dolomite* is a limestone containing an excess of 40 per cent of magnesian carbonate as the dolomite molecule, and *magnesian* (dolomitic) *limestone* contains not less than 5 nor more than 40 per cent of magnesian carbonate. *Marble* is a metamorphosed (recrystallized) limestone composed predominantly of crystalline grains of calcite or dolomite or both, having interlocking or mosaic texture.

iii. *Sandstone:* a consolidated sand in which the individual grains are composed chiefly of quartz or quartz and felspar, of fragmental (clastic) texture, and with various interstitial cementing materials, including silica, iron oxides, calcite, or clay. (In many cases, the durability of the cement is of critical importance in determining the durability of the sandstone).

iv. *Slate:* a microgranular metamorphic rock derived from argillaceous sediments, and characterized by an excellent parallel cleavage entirely independent of original bedding, by which cleavage the rocks may be split easily into slabs.

Source: ASTM (1958).

Similar sequences have been developed for crystallization processes (see below). These weathering sequences indicate the likely response of rocks to decomposition and disintegration (once their mineral content is known). Two additional aspects of rock character are also of interest from a practical point of view: (i) those properties of rocks that determine responses to weathering processes; and (ii) the changes to properties of rocks caused by weathering processes. Sometimes a property is important in both contexts.

The first group of properties are most appropriately considered in the context of individual processes, but some of general significance can be mentioned here. The ability of water and solutions to penetrate rock affects its resistance to most weathering processes and is determined primarily by *porosity* and *permeability*. The former is the proportion of voids in the rocks,

TABLE 11.3

Occurrence of rock types and minerals

Rock	Percentage of land area	Mineral	Percentage of land area
Shale	52	Felspars	30
Sandstone	15	Quartz	28
Granite	15	Clay minerals	
Limestone	7	and micas	18
Basalt	3	Calcite and	
Others	8	dolomite	9
		Iron oxide	
		minerals	4
Sedimentary	75	Pyroxene and	
Igneous and		amphibole	1
metamorphic	25	Others	10

Source: Leopold, Wolman, and Miller (1964).

TABLE 11.4

Chemical weathering sequences

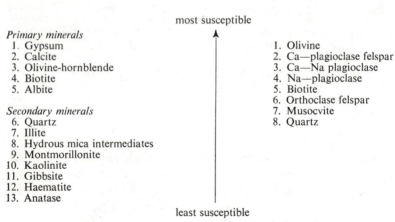

1. Weathering sequence of very fine-grained minerals

2. Susceptibility to weathering of minerals in igneous rocks

most susceptible

Primary minerals
1. Gypsum
2. Calcite
3. Olivine-hornblende
4. Biotite
5. Albite

Secondary minerals
6. Quartz
7. Illite
8. Hydrous mica intermediates
9. Montmorillonite
10. Kaolinite
11. Gibbsite
12. Haematite
13. Anatase

1. Olivine
2. Ca—plagioclase felspar
3. Ca—Na plagioclase
4. Na—plagioclase
5. Biotite
6. Orthoclase felspar
7. Musocvite
8. Quartz

least susceptible

Sources: modified from Goldrich (1938) and Jackson *et al.* (1948).

the latter is a measure of the ability of water to pass through it (and thus involves pore structure together with joints, cracks, etc.). In research on building stones, three related measures have been found to be extremely useful: *porosity*, *water absorption*, and, based on these, *saturation coefficient*.

Porosity can be measured by first saturating with air-free water a sample of rock in a vacuum, then weighing the saturated sample suspended in water ($W1$), weighing it again in air after removing excess moisture with a damp cloth ($W2$) and finally, after drying the specimen in an oven and allowing it to cool, weighing it again ($W0$). Then porosity is:

$$P = \frac{W2-W0}{W2-W1} \times 100. \qquad (11.4)$$

Water absorption, a value which depends on the method of measurement, can be found by immersing a test piece of rock in water at 15–21°C for 24 hours, removing the specimen, wiping it with a moist cloth, and weighing it (W3). Then water absorption is:

$$WA = \frac{W3-W0}{W2-W1} \times 100. \qquad (11.5)$$

The saturation coefficient, on the other hand, is a measure of the amount of water absorbed in a given time expressed as a fraction of the available pore-space. Thus, using the above definition of water absorption, the saturation coefficient is:

$$S = \frac{W3-W0}{W2-W0} \qquad (11.6)$$

The methods and definitions used in this paragraph are based on procedures adopted at the Building Research Station, England (Schaffer, 1959; Honeyborne, personal communication, 1972) and they differ from those sometimes used in soil mechanics. In an investigation of pore structure and the durability of building stones, it has been found that a distinction between large (>0·005 mm) and small (<0·005 mm) pores is valuable, because certain stones with high microporosity and saturation coefficient tend to be less durable than specimens of the same stone with lower saturation coefficient and microporosity (Honeyborne and Harris, 1958). For this purpose, microporosity was rigorously defined as the amount of water retained in the test material when it is subjected to a negative water pressure of 680 cm of water, expressed as a percentage of the total pore-space (Honeyborne, personal communication, 1972).

An illustration of the practical value of defining both saturation coefficient and microporosity is shown in a study by the Building Research Station, England, aimed at predicting the durability of Portland Stone, a limestone widely used in British buildings (Honeyborne, personal communication, 1972). Figure 11.4 shows the relations between these two variables for samples of stone of known weathering properties. Using $100S + M/2$ (where S = saturation coefficient and M = microporosity) as an index, four classes of stone can be identified from Figure 11.4:

Class *A*—index value not exceeding 79—exceptionally good service in large towns or in coastal districts.

Class *B*—index value not exceeding 95—good service in large towns or in coastal locations.

Class *C*—index value not exceeding 115—good service in large inland towns but not very satisfactory in coastal districts.

Class *D*—index value over 115—poor service everywhere except possibly in rural, inland districts.

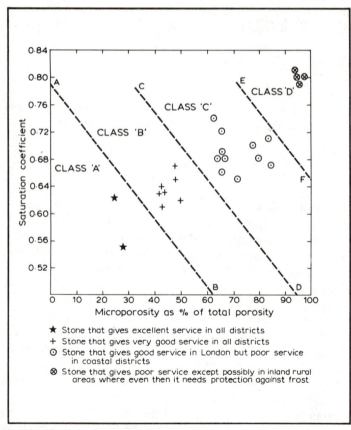

FIG. 11.4. Classification of Portland Stone on the basis of microporosity and saturation coefficient data (after the Building Research Station, England). Crown copyright

These empirical relationships say nothing directly of the weathering processes involved in attacking the stone, although in Britain crystallization processes are certainly the most important; the relations between the index and weathering are not entirely understood, but they are examined further in Section 11.3.

Several other rock properties are known to affect the durability of natural materials. Coefficients of volumetric expansion of both rocks and minerals play a significant role in determining the efficacy of insolation-weathering processes. The thermal conductivity and diffusivity of rocks is relevant in this context and in others involving crystallization processes. And tensile strength is important in determining the resistance of materials to the pressure of crystal growth in confined spaces within them. Resistance to chemical weathering is related to many other variables, such as the proportion and solubility of salts contained in a rock.

The second group of rock properties, that is to say those significantly affected by weathering, include many of those examined by the soil engineer. Mention has already been made of tensile strength, liquid limit, and plasticity index. To these can be added compressive and shear strengths (Chap. 6), particle size, permeability, infiltration capacity, and erodibility (see Chaps. 2, 3, 6, and 7). An important principle in assessing these properties is the fact that their values can change as weathering progresses.

(d) Local variables

In addition to the properties of climate and rocks, several other variables may influence the effectiveness of weathering processes at particular sites. The weathering of rocks in the landscape is considerably affected by topographic position (including aspect and slope), drainage conditions, and the nature of vegetation and animal life. In buildings, aspect, exposure, microclimate associated with the structure, and the availability of moisture with respect to features such as damp courses are similarly important. Certain parts of buildings, for example, are more likely than others to become saturated with water, such as cornices, copings, string courses, plinths, sills, and steps. Weathering may also be promoted in a building because of faulty craftsmanship or errors in the choice of materials (Schaffer, 1932). Craftsmanship faults include 'face bedding' (the placing of laminae in stone blocks vertically and parallel to the face), inadequate 'seasoning' of stones that require it, and damaging of stone by inappropriate quarrying or dressing methods. Errors of materials choice include the use of easily weathered stone in places especially susceptible to weathering, the use of unsuitable materials (such as corrodable iron in stonework) and the use together of incompatible building stones. Avoiding these mistakes requires a working knowledge not only of materials but also of their weathering characteristics.

(e) *Normal and accelerated rates of weathering*

Weathering rates, as defined by climatic parameters, vary spatially (e.g. Fig. 11.1). In circumstances where the processes have not been influenced by man these rates can be regarded as normal. There have been several attempts to determine spatial variations in the precise values of normal weathering rates, but unfortunately all of them have been hampered by lack of data. The kind of study that promises to further these efforts and at the same time relates weathering of the landscape to weathering of building stones is that by Rahn (1971). In this study, students classified the weathering of sandstone, marble, schist, or granite tombstones 100 years or more in age in the West Wilmington cemetry, Connecticut—an area without significant atmospheric pollution—into six weathering classes:

1. Unweathered;
2. Slightly weathered—faint rounding of corners or letters;
3. Moderately weathered, rough surface—letters legible;
4. Badly weathered—letters difficult to read;
5. Very badly weathered—letters almost indistinguishable;
6. Extremely weathered—no letters remaining; scaling.

Observations were corrected for the age of the tombstone by assuming (perhaps incorrectly) a constant rate of weathering and adopting a 100-year standard. The results are summarized in Figure 11.5. Sandstone tombstones have weathered most rapidly, followed by those of marble, schist, and finally granite. An examination of the topography of Connecticut shows that the average elevation (corrected to a common datum) varies with rock type

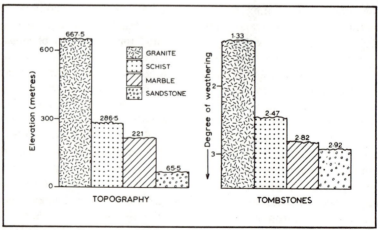

FIG. 11.5. Average elevation of topography on different rock types, and average degree of weathering of tombstones in New England (after Rahn, 1971). For explanation, see text

(Fig. 11.5). Comparison of the two measures of rock weathering shown in Figure 11.5 suggests that they adequately represent normal rates of weathering in the recent and geological past.

Man principally accelerates normal weathering rates by modifying the composition of the boundary layer of the atmosphere through air pollution. From the weathering viewpoint, sulphur dioxide (SO_2) is the most important pollutant, and its concentration tends to be higher over large urban areas than in rural districts. Sulphur gases are not normally present in the air in great quantities, but the combustion of fossil fuels may locally increase their concentrations under suitable climatic conditions. There are also suggestions that sulphur dioxide may drift many miles away from its source area, although concentrations decrease rapidly. Part of the building-stone deterioration in Sweden has been attributed to increases in sulphur dioxide pollution from sources as distant as Britain, The Ruhr, and Czechoslovakia (e.g. Anon., 1972). Sulphur dioxide, the principal sulphur pollutant, is soluble, but its solubility varies directly with atmospheric concentration and inversely with temperature. It dissolves in water to form a weak acid, sulphurous acid (H_2SO_3) which can react with calcium carbonate in limestones or in the cements of calcareous sandstones to produce relatively insoluble calcium sulphite ($CaSO_3$). This, in turn, may combine with oxygen to form more soluble calcium sulphate ($CaSO_4$) which normally crystallizes from solution as gypsum ($CaSO_4 . 2H_2O$). More seriously, SO_2 may suffer photo-oxidation, or catalytic oxidation in solution to produce sulphur trioxide (SO_3). Sulphur trioxide may combine with water to form the more aggressive sulphuric acid (H_2SO_4). This reacts with calcium carbonate to produce calcium sulphate. Similar reactions can occur with magnesium carbonate to give magnesium sulphate ($MgSO_4$). Thus sulphurous gases may contribute to rock weathering in two ways—they directly promote chemical weathering and they may lead to the formation of salts that may become involved in salt weathering. Of the two processes, the second is the more important. Both may be several times more important in urban than in rural areas.

Carbon dioxide (CO_2) is a normal constituent of the atmosphere, but its concentrations may be two or three times higher in urban than in rural areas because of the greater population densities and intensive combustion of fossil fuels in the former. Most predictions point towards an increase of CO_2 in the atmosphere in the future (Winkler, 1970). Carbon dioxide may combine with water, especially at lower temperatures, to produce weak, carbonic acid (H_2CO_3). This can dissolve calcium carbonate to produce a solution of calcium bicarbonate ($Ca(HCO_3)_2$). The solution is unstable and, on drying, calcium carbonate is redeposited, usually as a fine powder that is more susceptible to acid attack than the crystalline form. As a result, solution of limestones, dolomites, and marbles may be faster in cities than in comparable rural areas (Winkler, 1966).

(f) Approaches to the study of weathering

The two principal approaches to the study of weathering are through field surveys and laboratory tests.

Field surveys may be classified into three main types. Firstly, there are those based on the recording of degrees of disintegration or decomposition in terms of relatively simple criteria by simple methods (e.g. High and Hanna, 1970). Ideally, such studies also attempt to record changes that have taken place through time. Matthias's (1967) study of tombstone weathering rates by measurement of the depths of inscribed 1s on tombstones of different ages is an example. Other decomposition scales are mentioned in Section 11.4. A second approach is to monitor rates of debris generation or rock decay over time. An example is Rapp's (1960) analysis of rockfalls and earth-slides along railways in northern Scandinavia (Fig. 11.6). Thirdly, there have been a few attempts to monitor climatic conditions and measure rock characteristics in order to assess if weathering is taking place. Roth's (1965) study of a boulder in the Mojave Desert, California, illustrates this approach.

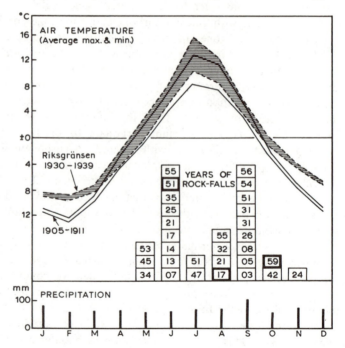

Fig. 11.6. Frequency of rock-falls and earth-slides recorded on the Riksgränsen–Narvik railway between 1902 and 1959. Squares with thick outlines indicate more than one movement in one day. Year of movement is indicated in the squares. In addition average monthly temperature and precipitation conditions are shown. Most movements coincide with those times when thawing of ice in the ground is at a maximum (June and September) (after Rapp, 1960)

Laboratory tests are normally designed to predict the durability of building stones, concrete aggregates, and other engineering materials, or to simulate the behaviour of materials under given environmental conditions. In general, two major approaches can be recognized here. Accelerated weathering tests offer one method. Samples of rock are subjected to cyclic weathering processes (e.g. freezing–thawing, wetting–drying) at much faster rates than occur naturally. Such tests have the advantages of saving time and controlling environmental conditions, but their results are open to question because the laboratory conditions are fundamentally different from specific conditions at natural weathering loci and it is impossible to ensure that by increasing cyclic frequency the effects of weathering are merely produced at an increased rate (Schaffer, 1932). Despite these qualifications, this method is widely used (e.g. Sect. 11.3). Some of the problems may be overcome by studying the behaviour of rock samples exposed to natural weather conditions. The Building Research Station's 'tray test' of frost weathering is an example (Schaffer, 1959; Honeyborne and Harris, 1958). A second approach is the empirical one of relating rock properties to known weathering characteristics and then using measurements of such properties to predict the weathering behaviour of rocks whose weathering behaviour is unknown. This method is most successful where the important rock properties and their relation to weathering processes are known and techniques for measuring them are adequate. It is illustrated by the Portland Stone study cited above.

11.3. Disintegration: Crystallization Processes and Natural Building Stones

(a) Introduction

Natural building stones are primarily weathered by chemical alteration and by mechanical processes that include insolation weathering, wetting-and-drying, various microbiological agencies (Pochon and Jaton, 1967), and salt and frost weathering. These last two weathering types are generally the most important, and they are examined in this section.

The consequences of frost and salt weathering on building stones are familiar. Winkler and Wilhelm (1970) suggested that the rapid destruction of Cleopatra's Needle when it was transferred from Egypt to the humid, polluted atmosphere of New York in 1881 was due to repeated hydration and dehydration of salts emplaced in the rock in Egypt, a process accelerated, perhaps, by frost action. Most damage to the Needle was recorded in the months of March to May, when relative humidities were high, and temperature changes were rapid.

Frost tends to fracture the stone and to generate coarse angular fragments. Buildings of a single material are not uniformly susceptible to frost action. Weathering is concentrated where the stone tends to be very wet, as in cornices and plinths (Schaffer, 1967). The effects of salts are more varied and most important. Soluble salts may crystallize at the surface in the form of an unsightly efflorescence; crystallization of salts within pores and cracks, known as cryptoflorescence, is responsible for crumbling, flaking, scaling, and blistering of rock surfaces and some artificial products such as terracotta roofing tiles (e.g. Cole, 1959). Hard skins may be formed on sandstones and limestones as a result of acid attack, and these may reduce surface evaporation and cause the deposition of soluble salts beneath them, leading to sub-surface disintegration. The surface skins may themselves suffer exfoliation (Schaffer, 1932).

(c) Frost weathering

Early studies of frost weathering, such as those by Thomas (1938), emphasized the disruptive forces associated with the phase change from water to ice. Water freezing at 0 °C in a closed system increases in volume by some 9 per cent, and a pressure of as much as 2,100 kg/cm^2 is exerted at -22 °C. This is a maximum pressure that is only reached if the system is entirely closed, air bubbles in ice and rock are scarce, and the very low temperature is attained (at this pressure, the freezing point of water is depressed to -22 °C). Nevertheless, a closed system may commonly be created when water freezes first at the surface and seals surface pores and cracks. And, although actual hydrostatic pressures may be less than the maximum, they are likely to exceed the tensile strength of many rocks. According to Reynolds (1961), for example, the tensile strength of granite is approximately 70 kg/cm^2; that of marble is 49–63 kg/cm^2; of limestone, 35 kg/cm^2; and of sandstone, 7–14 kg/cm^2. Of course, the pressure is not necessarily wholly relieved by rock shattering: for instance there could also be extrusion of ice from surface pores and cracks. The saturation coefficient has been seen by some as a useful index of this process, for if it is lower than about 0·8 (that is to say, if 20 per cent of the pore-space remains unfilled) expansion on freezing can be accommodated within the rock and frost action will not occur; if, on the other hand, the saturation coefficient is higher, disruption may result.

In many elementary reviews of weathering, disintegration as the result of phase change is still the only process attributed to frost action, but recent research has increasingly come to recognize that a second force is at work. This force is related to the *growth* of ice crystals, and it has long been recognized in the study of heaving in frozen ground (Taber, 1929; Chapter 9). Although the precise mechanism is not entirely understood, the thermodynamics of the process have been analysed by Everett (1961), and discussion of the subject in relation to building materials has been extended by Honeyborne (personal communication, 1972). It seems that when a saturated porous material freezes, ice crystals begin to form in the larger pores and water is withdrawn from the smaller pores, and mechanical disruption may accompany the growth of these crystals. The preferential growth of ice crystals in larger pores is explained as follows: 'If supercooled water in a porous system begins to freeze both in a fine pore and in a large pore, the growth of the crystal in the small pore will be limited by the pore space in which it is contained. Once a difference in size has been established between crystals in the large and small pores, the larger crystal will "feed" on the smaller until the coarse pore-space is filled' (Everett, 1961, p. 1541). Everett then goes on to develop a model which explains why, once the large pore is filled, further growth of the crystal proceeds against the constraint imposed by the walls of the pore, leading perhaps to disruption. The pressure that can build up in a large pore of radius R connected to a supply of water at the reference pressure by a capillary of radius r is, it is suggested, proportional to $(1/r - 1/R)$, and failure will occur if this pressure exceeds the strength of the porous material. The pressure therefore will ultimately depend on the pore structure of the material.

The work of Honeyborne and Harris (1958) and others has confirmed that pore structure is likely to be one of the most important rock properties governing the resistance of rock to frost action. Potts (1970) showed in his experiments that porosity and the percentage of material shattered are not significantly related statistically, and work at the Building Research Establishment (U.K.) and elsewhere tends towards the use of more refined measures of pore structure, such as direct measures of the relations between micropores and macropores, and indirect assessments based on suction (dynes/cm^2) (Honeyborne and Harris, 1958). In addition, as illustrated in Figure 11.4, measures of pore structure need to be combined with other variables, such as the saturation coefficient and the frequency of planes of weakness, in order to improve the estimation of rock durability. Experimental and field studies certainly show that mineralogically and texturally different rocks vary in their resistance to frost action (Table 11.5).

Of the relevant climatic variables, it is generally agreed that the frequency of temperature fluctuations around 0 °C is most important, and the intensity of freezing is of secondary significance (Potts, 1970; Wiman, 1963). Opinions differ on whether or not rate of freezing is important. It could be that a rapid rate of freezing may promote the development of a closed system.

Despite considerable progress in understanding the frost-weathering system through field observations and experimental work, it is perhaps surprising that it has proved rather difficult to reproduce the effects of natural freezing in the laboratory and to discriminate between frost-resistant and frost-susceptible rocks in a way accordant with rock performance in buildings (Schaffer, 1959). Durability prediction based on rock properties is quick and effective for very resistant and very susceptible materials, but it is less satisfactory for materials between the extremes. And accelerated weathering tests are less than perfect: Schaffer (1959) observed that Portland Stone failed freezing tests on two separate occasions in European laboratories,

TABLE 11.5

Relative resistance of rocks to frost action

Field studies		Experimental studies— 'Icelandic' conditions	
Ardennes	*Dartmoor*	*Potts*	*Wiman*
1.* Phyllite—pure schist	1. Metamorphosed sediments	1. Igneous rocks	1. Slate
2. Calcareous schist	2. Fine-grained granite	2. Sandstone ⎫ 3. Mudstone ⎭	2. Gneiss
3. Phyllite— quartz schist	3. Diabase	4. Shale	3. Porphyritic granite
4. Limestone	4. Elvan		4. Mica-schist
5. Grits/sand- stones	5. Tourmalinized medium-grained granite		5. Quartzite
6. Quartzite	6. Quartz-shorl		
7. Conglomerate	7. Coarse-grained granite		

* 1 = most susceptible, . . . 7 = least susceptible.

Sources: Ardennes, Dartmoor, and Potts, in Potts (1970). The Ardennes data are based on the work of J. Alexandre, the Dartmoor data on R. S. Waters's study. Wiman's work appears in the *Geografiska Annaler*, **45**, 1963, pp. 113–21. The local names for rocks are explained in these references.

despite the fact that it has stood the test of time in buildings and gravestones for many years. Several standard accelerated weathering tests are in use, such as the ASTM Standard C666–71 *Standard method of test for resistance of concrete to rapid freezing and thawing*, and the laboratory version of the 'tray test' used at the Building Research Station (Honeyborne, personal communication, 1972). One drawback of these standard tests is that they relate to a 'standard' climate. Climatic variability is important with respect to frost-weathering tests and it has been explored by several geomorphologists who have compared experimentally, for example, the effects of 'Icelandic' and 'Siberian' winters on rock disintegration (e.g. Potts, 1970; Tricart, 1956; Wiman, 1963).

(c) Salt weathering

Salts that may be responsible for the disintegration of building stone fall into three categories (Schaffer, 1932): salts present in the material before its incorporation in a building; salts formed during decomposition of the material; and those derived from external sources. Within the first category, salts may be emplaced during the deposition of many sedimentary rocks, they may be derived from sea spray in coastal quarries, or they may be produced during the manufacture of bricks and terracotta tiles. Salts may be formed by chemical decomposition associated with sulphurous gases and air pollution, as described in Section 11.2. External sources of salts in building stones include jointing materials (e.g. mortar), backing materials (e.g. bricks behind facing stones), the atmosphere, and the soil. In the case of the soil, salts may move upwards by capillary absorption of solutions to at least as far as the damp course, or to a height where the rate of evaporation and rate of absorption are balanced. The most common salts include sulphates, carbonates, and chlorides of sodium, calcium, potassium, and magnesium.

Three major processes of disintegration associated with the crypto-florescence of salt have been identified (e.g. Cooke and Smalley, 1968; Evans, 1970; Kwaad, 1970): thermal expansion of crystallized salts; hydration of salts; and the growth of salt crystals. The first of these processes has yet to be studied experimentally, but it seems likely that it may operate in areas with high diurnal temperature ranges where volumetric changes of salts might be adequate to cause disintegration. The important point is that the coefficients of thermal expansion of many common salts are higher than those of most rocks. Sodium nitrate ($NaNO_3$), sodium chloride ($NaCl$), and Potassium chloride (KCl), for instance, expand by volumetric proportions more than three times that of granite (Cooke and Smalley, 1968).

Hydration forces are certainly important (Kwaad, 1970; Winkler and Wilhelm, 1970). Pressure created by the hydration of entrapped salts can be repeated many times during a season or perhaps even during a single day. The precise mechanism of anhydrous salt formation and its translation into higher hydrates varies greatly according to local conditions. For example, when soil temperatures during the day are high, capillary-rising salt solutions may yield crystals without or low in water of crystallization; then during the night, when temperatures fall, the anhydrous salts or lower hydrates absorb water vapour from the atmosphere when the pressure of the atmospheric aqueous vapour exceeds the dissociation pressure of the hydrate involved, and higher hydrates are formed. The transformation is therefore related to relative humidity and temperature conditions in the atmosphere and to dissociation vapour pressures of salts. In addition some salts have transition temperatures above which only the anhydrous form or a lower hydrate are stable (Kwaad, 1970).

The pressure developed by hydration can be calculated using an equation that takes these and other related variables into account. One equation is that used by Winkler and Wilhelm (1970):

$$P = \frac{(nRT)}{(V_h - V_a)} 2 \cdot 3 \log \left(\frac{P_w}{P'_w}\right) \tag{11.7}$$

where

P = hydration pressure in atmospheres;
n = number of moles of water gained during hydration to the next higher hydrate;

R = gas constant;

T = absolute temperature (degrees Kelvin);

V_h = volume of hydrate $\left(\dfrac{\text{molecular wt. hydrated salt}}{\text{density (gm/cc)}}\right)$;

V_a = volume of original salt $\left(\dfrac{\text{molecular wt. original salt}}{\text{density (gm/cc)}}\right)$;

P_w = vapour pressure of water in the atmosphere (mm mercury at given temperature);

P'_w = vapour pressure of hydrated salt (mm mercury at given temperature).

This equation has been used to determine hydration pressures for the various salts shown in Figure 11.7. The figure shows that under ideal conditions hydration pressures for some salts may approach those associated with frost weathering. Kwaad (1970) pointed out that at least three conditions must be satisfied for rock disintegration to be caused by hydration: the hydration process should be accomplished in at least 12 hours, or else it cannot be completed by diurnal temperature changes; the hydrating salt must not be able to escape from the pore it is in; and hydration pressure must exceed the tensile strength of the rock.

The third process is that of crystal growth by the evaporation of the solvent in an open system. It is a process similar in some ways to that related to ice-crystal growth, and it is probably the most important cause of salt weathering (Evans, 1970). The effectiveness of the force of crystal growth has now been demonstrated experimentally many times, and there have been several attempts to explain it. Correns (1949; in Kwaad, 1970, and Evans, 1970), for example, concluded that salt crystals can continue to grow against a confining pressure when a film of solution is maintained at the salt–rock surface. The maintenance of this solution depends on the interfacial tensions at the salt–rock, salt–solution, and solution–rock interfaces: when the sum of the last two is smaller than the first, the solution can penetrate between the salt and the surrounding rock. The pressure exerted by a growing crystal depends mainly on the degree of super-saturation of the solution and may be described by the following equation:

$$P = \frac{RT}{V} \cdot \log n\left(\frac{C}{C_s}\right) \qquad (11.8)$$

where

P = pressure on the crystal;

R = gas constant;

T = absolute temperature;

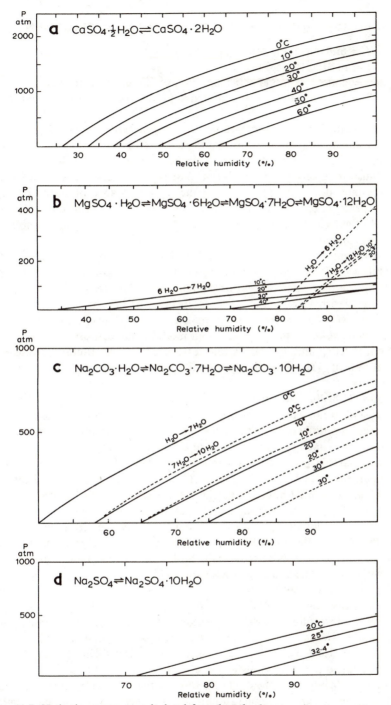

FIG. 11.7. Hydration pressures calculated for selected salts at various temperatures and relative humidities (after Winkler and Wilhelm, 1970)

V = molar volume of crystalline salt;

log n = natural logarithm;

C_s = concentration at saturation point without pressure effect;

C = concentration of a saturated solution under external pressure P.

If the major force of disintegration in salt weathering is crystal growth, a process similar in many ways to that of ice crystal growth, then tests of rock durability might be expected to be similar for both frost and salt weathering. It is thus interesting to reflect that in some of the earliest work, Brard used salt crystallization tests for assessing the frost resistance of stone. Honeyborne (personal communication, 1972) believed that the principal difference between the two processes is that salt crystallization involves higher stresses than freezing, and therefore causes greater damage.

Experimental work on salt weathering has been concerned mainly with durability prediction and monitoring of stone disintegration, and less with studying the particular process or combination of processes that is responsible for weathering. Standard tests include ASTM Standard C88–71a, *Standard method of test for soundness of aggregates by use of sodium sulfate or magnesium sulfate*, a test that has been modified in Australia as Australian Standard A77; and crystallization tests used at the Building Research Establishment (Honeyborne, personal communication, 1972; see also Section 11.2).

Figure 11.8 illustrates the important point that the response of rocks to salt crystallization varies according to the type of salt involved, and different rocks respond differently to the same salt. In these experiments, an attempt was made to simulate a desert climate (Goudie, Cooke, and Evans, 1970). The rock samples were immersed in saturated salt solutions at 17–20 °C for one hour, then dried in an oven at 60 °C for six hours, and at 30 °C for the remainder of each 24-hour cycle. The cycle was repeated 40 times, and the weight loss from specimens was recorded after each cycle. Figure 11.8b shows that sodium sulphate (Na_2SO_4) was most effective in destroying Arden Sandstone under these conditions, followed closely by magnesium sulphate ($MgSO_4$). Of the rocks tested, chalk broke up most rapidly, followed by Cotswold Limestone and Arden Sandstone (Fig. 11.8a).

The reasons for such variability are not altogether clear. Of the rock properties that may be relevant to the system, the rate of water absorption and water absorption capacity—measures related to pore structure and permeability—were found to be significantly related to rate of rock disintegration (see also Cole, 1959, and Minty, 1965). And other studies of pore structure mentioned previously are also relevant in the context of salt weathering. Minty and Monk (1966) showed that surface texture (smoothness) and surface area/volume ratio are also significant. For example, Figure 11.9 shows that in experiments on dolerite using ASTM Standard C88, percentage loss varies with the size of sample, and that rate of loss diminishes with time for fine materials and increases with time for coarse particles.

The effect of climatic variables on salt weathering has not, apparently, been widely investigated. For crystal growth, the solvent needs to be evaporated or cooled. Evaporative conditions are probably most common, and they depend on several climatic conditions such as temperature, humidity, and wind velocity. For hydration, temperature and humidity are of importance; and for thermal expansion, short-period temperature change controls the process. Figure 11.10 shows one example of the kind of influence climatic variables may have on the nature of salt weathering—the disintegration of dolerite samples by seawater is much more rapid in the higher temperature range (Minty, 1965).

The third group of variables influencing the salt-weathering system relates to the nature of the salts themselves. Figure 11.8b shows the effects of different salts on one rock, and suggests that sodium sulphate (Na_2SO_4) causes most rapid decay. Near to the sea, NaCl may be extremely damaging. And NaCl and $CaSO_4$ together may under certain circumstances be more damaging than both of them working separately (Honeyborne, personal communication, 1972). The effectiveness of salts may also vary with climatic conditions. For example, because the solubility of NaCl changes little with temperature whereas the solubility of Na_2SO_4 sharply increases with rising temperature, temperature change affects NaCl far less than Na_2SO_4. Thus it could be that crystallization of Na_2SO_4 on a fall of temperature affects a greater volume of salt per unit time than crystallization of NaCl by evaporation (Kwaad, 1970), and this type of contrast may help to explain the greater weathering by Na_2SO_4.

(d) Prevention of crystallization weathering on buildings

The efflorescence of salts and rock disintegration by salts and frost are problems sufficiently serious to justify the development of preventative measures. These measures fall into three categories (Schaffer, 1932): design and aspect; washing; and surface treatments. Sometimes the effects of frost can be prevented by omitting from the design of a building features liable to become saturated frequently, such as cornices, or at least using material for these features that is not susceptible to frost weathering. Another obvious example of a design measure is the damp course, which prevents the penetration of saline solutions above its level.

Washing of buildings with water (water *without* cleaners or detergents) at regular intervals is still probably the best way of keeping stonework clean. Washing removes loose dirt and dust, it removes in solution some of the products of chemical reactions between air, moisture, and the stone (such as $CaSO_4$), it assists the removal of other salts, and it maintains the natural

(a)

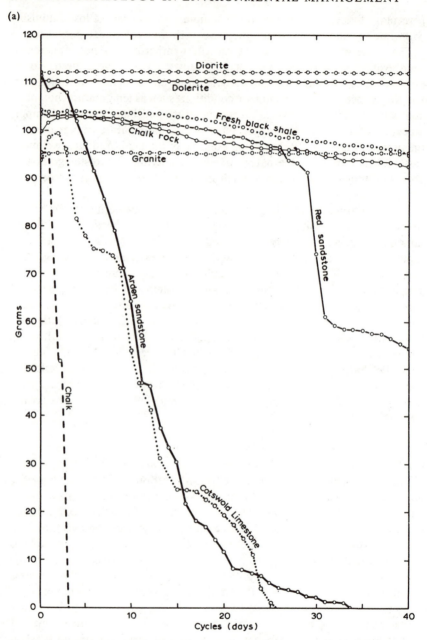

Fig. 11.8. (a) Change in weight of different rock types on treatment with sodium sulphate; (b) (p. 297) change in weight of Arden Sandstone samples on treatment with different salts (after Goudie, Cooke, and Evans, 1970). For explanation, see text

colour and texture of the stone. Particularly dirty buildings may require preliminary cleaning with a steam jet or by scrubbing (Schaffer, 1932).

A third approach is to apply artificial coverings to the surface of buildings in order to seal the surface and prevent or restrict the access of solutions. This approach has been attempted for many years and has yet to be fully vindicated. 'Preservative methods' include: those which seek to waterproof the stone (e.g. paints); those which seek to fill pores with substances not readily attacked by weathering agencies (e.g. silicates); processes that form inert materials by reaction with constituents of the stone (e.g. silico fluorides); processes that seek to retard decay by reaction with decomposition products of the stone; antiseptic processes; and limewash (Schaffer, 1932). The results of these methods have proved to be disappointing mainly because a water-proofing agent, while preventing access of solutions, also prevents the escape of solutions already within the material, so that sub-surface disintegration

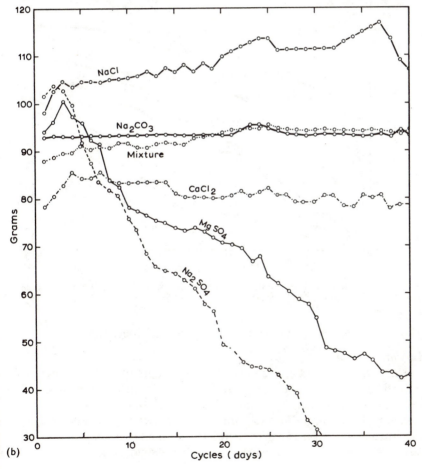

(b)

may result. Also, skin properties tend to be different from those of the rock beneath and new stresses may be established.

Clarke and Ashurst (1972) have described experiments on the effect of surface treatments designed to consolidate and/or make water-repellent natural stones in buildings at twenty-four sites in the United Kingdom. The preparations used were limewater, 2 per cent, 5 per cent, and 1 per cent, followed by 5 per cent solutions of silicones; a 5 per cent solution of siliconate, a silicon ester; a silicon ester followed by a silicone; and a liquid containing a silicone resin and a silicon ester in a petroleum solvent. Observations of the buildings over a period of years led to a clear conclusion: 'these experiments have shown that none of the treatments has had any overall beneficial effect in retarding decay on any of the sites'.

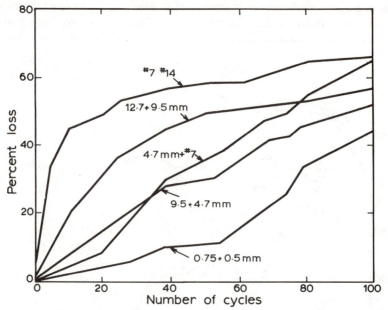

FIG. 11.9. Relationship between particle loss by salt weathering and time for dolerite samples of different-sized particles (after Minty and Monk, 1966; numbers 7 and 14 refer to sieve sizes)

11.4. Decomposition: Weathering and Engineering Problems

(a) Introduction

The engineer has extensive interests in the nature of rocks and especially in their weathering characteristics. Like those concerned with building research, he requires information on durability of rocks in order to allow the selection of appropriate materials for dams, breakwaters, and other engineering structures and for concrete aggregates. Durability is essentially a short-term problem, and it is a characteristic that should be known before the materials are used. As it has been examined in the previous section, it requires no further elaboration here.

Another aspect of weathering of interest, especially to the foundation engineer, is the nature of surface and subsurface materials present at a site or in an area where development is to take place. The engineer is mainly concerned with the engineering and soil-mechanics properties of the foundation materials (e.g. Woods, 1960, section 8). These properties reflect many things, but in particular they are often influenced by chemical weathering processes. For instance, permeability and infiltration capacity may be reduced as weathering proceeds. The weathering processes tend to alter the original properties of the material—often, but not always, making it less suitable for engineering purposes. The engineer is less concerned with studying the nature and progress of weathering processes than he is with recording the material properties and their distribution in both space and depth. His understanding of these may be illuminated through a study of geomorphological change.

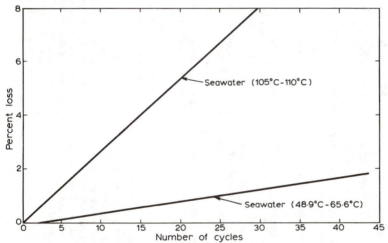

FIG. 11.10. Relationship between particle loss by salt weathering and time at different temperatures for $\frac{3}{4}-\frac{1}{2}$ inch dolerite samples (after Minty, 1965)

Fookes, Dearman, and Franklin (1971) suggested that for many field investigations of mechanical performance of rocks at engineering sites, three main characteristics will provide adequate supplementary information to basic geological descriptions: weathering grade, fracture-spacing index, and strength index. As far as weathering is concerned the aim is to classify degrees of weathering in the field using simple but precise criteria, to determine either in the field or the laboratory the engineering properties of the classes identified, and to plot the areal and vertical distribution of the classes. Soil scientists, geomorphologists (e.g. Ruxton and Berry, 1957; Ruxton, 1968;

TABLE 11.6

Engineering grade classification of weathered rock

Grade	Degree of decomposition	Field recognition (after Fookes & Horswill, 1969)		Engineering properties of rocks (after Little, 1969)
		Soils (*i.e.* soft rocks)	Rocks (*i.e.* hard rocks)	
VI	Soil	The original soil is completely changed to one of new structure and composition in harmony with existing ground-surface conditions.	The rock is discoloured and is completely changed to a soil in which the original fabric of the rock is completely destroyed. There is a large volume change.	Unsuitable for important foundations. Unstable on slopes when vegetation cover is destroyed, and may erode easily unless a hard cap present. Requires selection before use as fill.
V	Completely weathered	The soil is discoloured and altered with no trace of original structures	The rock is discoloured and is changed to a soil, but the original fabric is mainly preserved. The properties of the soil depend in part on the nature of the parent rock.	Can be excavated by hand or ripping without use of explosives. Unsuitable for foundations of concrete dams or large structures. May be suitable for foundations of earth dams and for fill. Unstable in high cuttings at steep angles. New joint patterns may have formed. Requires erosion protection.
IV	Highly weathered[1]	The soil is mainly altered with occasional small lithorelicts of original soil. Little or no trace of original structures.	The rock is discoloured; discontinuities may be open and have discoloured surfaces and the original fabric of the rock near the discontinuities is altered; alteration penetrates deeply inwards, but corestones are still present.	Similar to grade V. Unlikely to be suitable for foundations of concrete dams. Erratic presence of boulders makes it an unreliable foundation for large structures.

III	Moderately weathered[1]	The soil is composed of large discoloured lithorelicts of original soil separated by altered material. Alteration penetrates inwards from the surfaces of discontinuities.	The rock is discoloured; discontinuities may be open and surfaces will have greater discolouration with the alteration penetrating inwards; the intact rock is noticeably weaker, as determined in the field, than the fresh rock.	Excavated with difficulty without use of explosives. Mostly crushes under bulldozer tracks. Suitable for foundations of small concrete structures and rockfill dams. May be suitable for semipervious fill. Stability in cuttings depends on structural features, especially joint attitudes.
II	Slightly weathered	The material is composed of angular blocks of fresh soil, which may or may not be discoloured. Some altered material starting to penetrate inwards from discontinuities separating blocks.	The rock may be slightly discoloured; particularly adjacent to discontinuities which may be open and have slightly discoloured surfaces; the intact rock is not noticeably weaker than the fresh rock.	Requires explosives for excavation. Suitable for concrete dam foundations. Highly permeable through open joints. Often more permeable than the zones above or below. Questionable as concrete aggregate.
I	Fresh rock	The parent soil shows no discolouration, loss of strength, or other effects due to weathering.	The parent rock shows no discolouration, loss of strength, or any other effects due to weathering.	Staining indicates water percolation along joints; individual pieces may be loosened by blasting or stress relief and support may be required in tunnels and shafts.

[1] The ratio of original soil or rock to altered material should be estimated where possible.

Source: Fookes, Dearman, and Franklin (1971).

TABLE 11.7

A summary table on rock weathering, climate and related foundation engineering features

Climatic zone—predominant weathering process	Periglacial zone—disintegration	True temperate zone—decomposition and disintegration	Arid zone—disintegration	Humid tropical zone—decomposition
Classification characteristics	(a) Susceptibility to frost weathering and shattering generally increased with increased grain size of parent material (b) Hard rocks break down to produce well-graded material, often gap-graded (c) Soft rocks show little change during initial stages of weathering but chemical breakdown increases plasticity and decreases grain size	(c) Grading appears unaffected to any large degree except in late stages (b) Decomposition processes may alter plasticity especially in late stages	(a) Good sorting by transporting agents produces poorly-graded material; becomes more fine-grained with distance from parent material (b) Evaporites can cause aggregation	(a) Tests often dependant on pretreatment techniques; large scatter of results makes comparison difficult (b) Generally well-graded in initial stages of weathering, becoming finer-grained and poorly-graded
Strength	(a) Reduction in strength with weathering, but residual strength not obtained (b) Mode of failure changes from brittle to plastic	(a) Marked loss in strength does not occur until late stages of weathering	(a) Generally high and directly related to grading	(a) Published work indicates consistent and isotropic (b) Increases with increased grain size of parent material (c) Laterization increases strength owing to cementing
Permeability	(a) Increases in early stage but generally reduced with advanced disintegration	(a) Generally as expected from index properties	(a) Generally high for weathered soils near parent material (b) Away from parent material, lower permeability (c) Local variation frequent	(a) Variable, but generally as expected from index properties; often low
Compressibility and consolidation	(a) Soil becomes more compressible with increased weathering (b) Chemical weathering removes effects of preconsolidation	(a) Generally as expected from index properties (b) Chemical weathering removes effects of preconsolidation	(a) Generally as expected from index properties (b) Low for soils near parent material (c) Metastable structure of loess can cause dramatic consolidation (d) Sand dunes of variable density result in changing characteristics	(a) Chemical weathering removes all effects of preconsolidation (b) Some soils metastable from leaching processes (c) Coefficients of consolidation vary over wide range but are often extremely high (d) Compressibility values are average Cementing (e.g. laterites) may reduce compressibility
Soil stabilization	(a) Normal procedures of stabilization satisfactory, but presence of organic matter can reduce the rate of gain of strength	(a) Small thicknesses of weathered material generally make removal more economical than stabilization	(a) Addition of coarse-grained material to produce a well-graded soil	(a) Satisfactory results for most purposes obtained by few per cent of lime or cement or oil with additives

Engineering uses	(a) Scree slopes of disintegrated material, often suitable for road stones	(a) Limited, but provides suitable top-soil for plant growth, for stabilization, or for improving visual appearance	(b) the addition of lime or cement (c) presence of organic matter can reduce the rate of gain of strength (a) Provides satisfactory borrow material for roads and runways; destroy metastable structure of aeolian deposits by wetting prior to rolling (b) Surface crust of calcareous and saline soils provide good bearing layers (c) Vertical cuts in loess stand almost indefinitely (d) Generally good foundation materials; deep water level has little effect on most engineering constructions (c) Scree slopes of disintegrated material make fair to good road stones	(b) Firing to high temperatures; suitable for black clays (c) Presence of organic matter can reduce the rate of gain of strength (a) Generally makes satisfactory base courses for roads and runways; often requires stabilizing for wearing surfaces (b) Fired black clay can be used as low-grade aggregate (c) Satisfactory material for slopes; use near-vertical cuts, as flatter slopes erode badly (d) Some soils compact to very low density but make acceptable fill
Engineering problems encountered	(a) Weathering of soft rocks associated with decrease of strength and increase of both primary and secondary consolidation (b) Continual freeze-thaw cycles can result in graded bedding (c) Glacially induced rock fractures parallel to topography (d) Successive deformation results when foundations become water-logged during thaw, together with the formation of solifluction lobes	(a) Weathering of soft rocks associated with decrease in strength and increase of both primary and secondary consolidation	(a) Unless destroyed, the metastable structure of aeolian deposits can cause catastrophic collapse (b) Infilling of excavations by blown sand (c) Often the salts resulting from evaporation of mineral bearing waters necessitates the use of sulphate resistant cement	(a) When used as a wearing surface can corrugate owing to loss of fines (b) Compaction difficult to attain if high Proctor densities specified (c) Seasonal climatic variations can cause swelling and shrinking to great depths especially the montmorillonitic black soils (d) Weathered products are often minerals of unusual engineering behaviour (e.g., halloysite or montmorillonite) (e) Strength may decrease with depth—particularly when laterization has occurred (f) Softening on contact with water necessitates satisfactory drainage

Ollier, 1965), and engineering geologists (Little, 1969; Fookes and Horswill, 1969) have all attempted to classify weathering profiles and degrees of weathering for their own purposes. Two examples are shown in Table 11.6. Most of these schemes have several common features. Firstly, they are based on criteria that are simple to use in the field. Secondly, they all recognize a small number of often arbitrarily defined categories (usually no more than six). Thirdly, all schemes describe the alteration of material down a profile to fresh bedrock, which usually forms the first class; the boundary between weathered and unweathered material in depth is called the *weathering front*. The form of the weathering front is extremely varied, reflecting the relations between climatic conditions, groundwater, and water table, the nature of rock lithology and structure, and changes in the weathering processes that have affected the area through time. In general, weathering is deepest where materials are most susceptible to weathering, where they are porous and permeable, where surface erosion is limited, and where climatic conditions favour chemical decomposition. Some of the deepest weathering profiles have been described in tropical areas, such as in northern Nigeria where weathering in places extends to some 100 m in depth (Thomas, 1966)

Fourthly, the criteria used in the classification schemes vary greatly from those based on qualitative impressions to those requiring precise measurement. Some schemes merely refer to 'degree of coloration'; others record colour in terms of hue, value, and chroma, as shown on Munsell Color Charts. The approximate proportion of 'residual' (unaltered) material in a zone is commonly used as a criterion, and some indication of the extent to which rock structure and texture are preserved is usually recorded. Also often mentioned is the extent of mineral alteration, and particularly the degree to which clay minerals are developed, although this is difficult to determine precisely in the field.

(b) Case Study: decomposition of rocks on Dartmoor, England

Fookes, Dearman, and Franklin (1971) described a case study of engineering aspects of different rock types on Dartmoor using weathering grade (as shown in Table 11.6), a fracture-spacing index, and a strength index. The fracture-spacing index, I_f, is defined as the average size of drill-core or outcrop material (in metres) between fractures. The scale is logarithmic:

	Fracture-spacing index, I_f (m)
Extremely high	>2
Very high	0·6–2
High	0·2–0·6
Medium	0·06–0·2
Low	0·02–0·06
Very low	0·006–0·02
Extremely low	<0·006

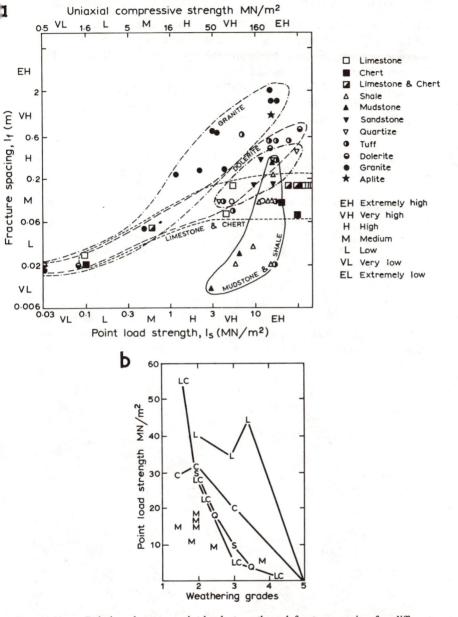

Fig. 11.11. a. Relations between point-load strength and fracture spacing for different Dartmoor rocks of varying weathering grades. Trend lines showing the path of the progress of weathering relative to I_f and I_s have been drawn for each rock type except tuff
b. Relations between point-load strength and weathering grades for selected Dartmoor rocks (after Fookes, Dearman, and Franklin, 1971). Letters are the initials of rock types listed in a.

The strength classification is based on point-load strength measurements in the field using a portable tester. The point-load strength index is $I_s = P/D^2$, where P = force required to break the specimen and D = the distance between platen contacts on this instrument. This simple measure can be shown to be closely related to uniaxial compressive strength and it also gives a measure of tensile strength. The scale is:

	$I_s(MN/m^2)$
Extremely high strength	>10
Very high strength	3–10
High strength	1–3
Medium strength	0·3–1
Low strength	0·1–0·3
Very low strength	0·03–0·1
Extremely low strength	<0·03.

The relations between point-load strength, fracture spacing, and weathering trend for rocks from Dartmoor are shown in Figure 11.11a. The three variables can be plotted separately for, say, a quarry face (Fig. 11.12). One practical value of such plots is in assessing excavation conditions, a property shown in Figure 11.12. Relations between the variables vary considerably

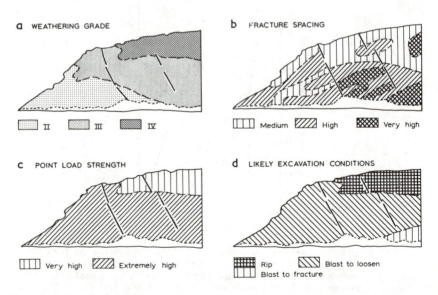

FIG. 11.12. Weathering grade, fracture spacing, point-load strength, and likely excavation conditions of a quarry face of cherts and limestones, Meldon Quarries, Dartmoor (after Fookes, Dearman, and Franklin, 1971)

with rock types (e.g. Fig. 11.11b), although in general, as would be expected, point-load strength is inversely related to weathering grade.

This study shows the relations between weathering characteristics and other rock properties and rock type in one locality. Clearly, however, engineering properties of rocks can be expected to vary with climate. Table 11.7 is an attempt to relate rock weathering, climate, and foundation engineering features (Saunders and Fookes, 1970). This generalization is valuable as a rough guide to the relations between variables for project-planning purposes, but it does not, of course, do away with the need for site studies.

11.5. Conclusion

Many geomorphologists, building-research workers, pedologists, and engineers share a common interest in the processes of weathering, the performance of materials subjected to them, and the changes in rock and soil properties arising from them. This survey has sampled the extensive literature relating to these research interests in a variety of disciplines. It has attempted to show that recent work has significantly illuminated understanding of weathering processes and their relations to climate and rock characteristics, and has defined field methods for describing the consequences of weathering. The implications of such work for environmental management are numerous and diverse, but in general the benefits arise mainly from the fact that progress towards understanding processes and their consequences promotes precise prediction, especially of such things as weathering intensity and rock performance, by means of easily used climatic surrogates and simply measured properties of materials.

12
LANDFORMS AND TECHNIQUES OF SCENIC EVALUATION

12.1. Aesthetics and Landforms

The importance of landscape in Western society for recreation, spiritual nourishment, and posterity has long been recognized, and it is reflected in the designation of 'green belts', 'national parks', 'wilderness areas', 'nature monuments', 'protected landscapes', and 'areas of outstanding natural beauty'. In recent years, however, the numbers with a taste for scenery, a desire to protect it, and the ability to visit it have grown rapidly, and the demand for recreational landscape has correspondingly increased. As a result, those responsible for environmental management have had to add a new dimension to the complex process of resource evaluation, and they have had to come to terms with several fundamental problems.

Of these problems, two are critical. Firstly, planners face the difficulties of defining the landscapes people enjoy. The emotional responses of individuals or groups to natural or rural scenes are immensely complicated and varied, and the psychology of environmental perception is a field mined with difficulties (e.g. Lowenthal, 1967). And even if it is possible to determine the ways in which particular cultural groups respond to views as a whole, the problem remains of determining those components of views that chiefly stimulate the responses. In addition, the planner has to remember that perception is something which changes with time.

Secondly, the planner must face the problem of describing the aesthetic value of scenery in terms that can be compared realistically with the more easily quantified claims of other resource users, such as those of mining interests, farmers, and urban developers. Even if particular landscape preferences are known and understood, and usually they are not, the formidable problem remains of presenting the results in a way that can validly be used in the planning process. Who can say, or who can deny, that a babbling brook is more valuable than a hydro-electric power station? If such questions could be fairly answered, the value of a particular site for recreational purposes would still depend in part on other considerations, such as accessibility, and local provision of services. In the light of such problems, it is scarcely surprising that various attempts to appraise precisely the aesthetic value of landscape have been roundly criticized; but the fundamental need for such appraisal remains.

One further thing is clear: in the appraisal of landscape, the form of the ground and the nature of geomorphological processes are normally regarded as being important ingredients. It is therefore possible that geomorphologists may have a useful role to play in contemporary efforts to evaluate landscapes for planning purposes.

In studying the aesthetic qualities of landforms, one is concerned also with an area of intensely personal emotions, and it could be argued that the trained student of landforms, far from being an objective observer, is too prejudiced to present impartial views. His attention in a view, for example, may often be directed to those aspects of landforms that related to processes of change. To the non-geomorphological visitor, on the other hand, such considerations may be unimportant; to him, the forms may express deeply-felt emotions arising largely from his particular cultural background. Perhaps this important reservation on the geomorphologist's role in judging landform aesthetics can be effectively illustrated by an artistic analogy from John Ruskin's *Modern Painters*.

Figure 12.1 (top) shows Ruskin's sketch of the Faido Pass near St. Gothard in Switzerland. Deliberately, only those features seen by the artist as being of 'some continual importance' are included. The sketch would grace any geomorphologist's notebook, for it is simple, clear, precise, and relatively unprejudiced; it records only observable phenomena in their correct locations, and it makes few assumptions of geomorphological processes. From the point of view of scenic evaluation, the sketch is valuable but incomplete. As Ruskin (Cook and Wedderburn, 1904) observed, the drawing takes no account of the fact that the scene is approached through one of the narrowest and most sublime ravines in the Alps, and after the traveller has become familiar with the nature of the St. Gothard scene. Thus to the traveller, or to the amateur appraiser of the view, the Pass of Faido may present a quite different aspect.

The confused stones, which by themselves would be almost without any claim upon his thoughts, become exponents of the fury of the river by which he has journeyed all day long; the defile beyond, not in itself narrow or terrible, is regarded nevertheless with awe, because it is imagined to resemble the gorge that has just been traversed above; and, although no very elevated mountains immediately overhang it, the scene is felt to belong to, and arise in its essential character out of the strength of those mightier mountains in the unseen north (Ruskin, in Cook and Wedderburn, 1904, p. 35.)

How, then, can the artist describe the qualities of 'fury', 'awe', and 'strength', and other attributes he subjectively identifies in the scene? One answer to this question is given in a powerful and remarkably different sketch based on the same scene by Turner (reproduced in etched outline by Ruskin, Fig. 12.1, bottom), in which astute observations of landform and geological structure, transformation of scale, and modification of location are skilfully used to convey an emotional portrait of the scene.

In terms of practical solutions to the problem of landscape appraisal

FIG. 12.1. The Pass of Faido (top) as sketched by John Ruskin, and (bottom) as reproduced in etched outline by Ruskin from a drawing by Joseph Turner.

facing the planner, several techniques have been proposed, and they fall between two extremes. At one extreme, measurable components of landscape deemed to be representative of scenic quality are evaluated in quantitative or semi-quantitative terms. At the other extreme, emphasis is placed on the perception by 'consumers' of scenic quality; 'comprehensive' consumer responses are classified, perhaps on numerical scales, and individual components of scene are not in themselves considered. Some techniques are only concerned with the evaluation of sites or particular views; others attempt to provide areal classification of terrain. Techniques may vary according to the scale of inquiry.

As Craik (1972) has observed, the results derived from these techniques yield several advantages to environmental decision-makers. They focus attention on important criteria involving such important planning issues as landscape preservation and the routing of power lines. A regional description of landscape quality provides a useful context for making choices among specific sites and for evaluating land-use proposals. In addition, quantitative information may endow environmental factors with a status comparable to that of economic and social factors, although there is a danger that such status is quite spurious. And if the surveys are carried out from time to time, the rate of change of environmental quality can be monitored.

In the following discussion, a small sample of techniques is examined; other techniques, in particular those without a significant geomorphological component, are reviewed by the Countryside Commission for Scotland (1971) and by Clout (1973). Some relevant studies not detailed below include those by Lowenthal and Prince (1964, 1965); Shafer *et al.* (1969); and Morisawa (1971). Studies in England of local interest include those by Coppock (1959) and Jackson (1971). More general studies are provided by Medhurst (1968), Steinitz (1970), and Weddle (1969).

12.2. Approaches to Appraisal

(a) *Analysis of landscape components*

(*i*) *Leopold's method: site evaluation.* A proposal to build a dam in Hell's Canyon on the Snake River, Idaho, was strongly opposed by conservation interests. Leopold (1969a, 1969b) attempted to quantify aesthetic features of the proposed site, together with those of eleven other possible dam sites in Idaho and sites of renowned beauty elsewhere in the United States, in order to determine objectively the aesthetic quality of the Hell's Canyon site in the context of information from other sites. Leopold's approach was founded on the assumptions that there is some benefit to society from the existence of 'unchanged' landscapes, that a unique landscape has more value to society than a common one, and that the unique qualities enhancing landscape value

TABLE 12.1

Factors representing the aesthetic qualities of a site

Factor Number	Descriptive Categories	Evaluation numbers				
		1	2	3	4	5
	Physical Factors					
1	River width (m) (at low flow)	<1	1–3	3–9	9–30	>30
2	Depth (m) (low flow)	<0·15	0·15–0·30	0·3–0·6	0·6–1·52	>1·52
3	Velocity (m per sec.) (flow)	<0·15	0·15–0·30	0·3–0·6	0·6–1·52	>1·52
4	Stream depth (ft.)	<0·30	0·30–0·60	0·6–1·2	1·22–2·44	>2·44
5	Flow variability	Little variation		Normal		Ephemeral or large variation
6	River pattern	Torrent	Pool and riffle	Without riffles	Meander	Braided
7	Valley height/width	≦1	2–5	5–10	11–14	≧15
8	Stream bed material	Clay or silt	Sand	Sand and gravel	Gravel	Cobbles or larger
9	Bed slope (m/m)	<0·0005	0·0005–0·001	0·001–0·005	0·005–0·01	>0·01
10	Drainage area (sq. km)	<2·59	2·59–25·9	25·9–259	259–2589	>2589
11	Stream order	≦2	3	4	5	≧6
12	Erosion of banks	Stable		Slumping		Eroding large-scale deposition
13	Sediment deposition in bed	Stable				
14	Width of valley flat (m)	<30·5	30·5–91	91–152	132–305	>305
	Biological and Water-Quality Factors					
15	Water colour	Clear colourless		Green tints		Brown
16	Turbidity (parts per million)	<25	25–150	150–1000	1000–5000	>5000
17	Floating material	None	Vegetation	Foamy	Oily	Variety
18	Water condition (general)	Poor		Good		Excellent
19	Algae Amount	Absent				Infested
20	Algae Type	Green	Blue-green	Diatom	Floating green	None
21	Larger plants Amount	Absent				Infested
22	Larger plants Kind	None	Unknown rooted	Elodea, duck weed	Water lily	Cattail
23	River fauna	None				Large variety
24	Pollution evidence	None				Evident

No.	Factor					
25	Land flora Valley	Open	Open w. grass, trees	Brushy	Wooded	Trees and brush
26	Hillside	Open	Open w. grass, trees	Brushy	Wooded	Trees and brush
27	Diversity	Small				Great
28	Condition	Good				Overused

Human-use and Interest Factors

No.	Factor					
29	Rubbish and litter Metal (no. per 30.5 m of river)	<2	2–5	5–10	10–50	>50
30	Paper	<2	2–5	5–10	10–50	>50
31	Other	<2	2–5	5–10	10–50	>50
32	Material removable	Easily removed				Difficult removal
33	Artificial controls (dams, etc.)	Free and natural				Controlled
34	Accessibility Individual	Wilderness				Urban or paved access
35	Mass use	Wilderness				Urban or paved access
36	Local scene	Diverse views and scenes				Closed or without diversity
37	Vistas	Vistas of far places				Closed or no vistas
38	View confinement	Open or no obstructions				Closed by hills, cliffs or trees
39	Land use	Wilderness	Grazed	Lumbering	Forest, mixed recreation	Urbanized
40	Utilities	Scene unobstructed by power lines				Scene obstructed by utilities
41	Degree of change	Original				Materially altered
42	Recovery potential	Natural recovery				Natural recovery unlikely
43	Urbanization	No buildings				Many buildings
44	Special views	None				Unusual interest
45	Historic features	None				Many
46	Misfits	None				Many

Key: < less than,   > greater than,   ≦ less than or equal to,   / divided by

Source: Leopold, 1969b (with metric conversion).

are those having some aesthetic, scenic, or human-interest connotations. All these assumptions, of course, are open to question.

Three types of factors were selected to represent the aesthetic qualities of a site: physical factors, biological factors, and human-use and interest factors. A total of 46 factors were identified (Table 12.1). The value of each factor at any one site was determined on a scale of 1 to 5: in some cases, the evaluation number was based on precise measurement (e.g. stream width), in others it is based on qualitative assessment (e.g. water condition), as shown in Table 12.1. The physical factors are, in general, the easiest to measure precisely. It should be emphasized that the list, though long, by no means comprehensively covers the factors that might be considered to compose the beauty of a site: for instance, important features such as smell, weather, and illumination are omitted. And there is rather little substantive evidence that the variables are in fact, either important, or equally important, in personal perception of scenic beauty by most Americans.

Site 'uniqueness' can be defined as the reciprocal of the number of sites sharing a particular evaluation number for a factor. For example, if all of 12 sites share the same number, the 'uniqueness ratio' for each site will be 1/12 or 0·08; equally, if only one site of twelve had a particular number, it would be 'relatively unique' and would have a uniqueness ratio of 1. The

TABLE 12.2

Summary totals of uniqueness ratios of aesthetic factor values, Hell's Canyon region

Site in Idaho	Aesthetic factors			Total
	Physical	Biological	Human interest	
1. Wood River, nr. Ketchum	3·06	2·92	5·09	11·07
2. Salmon River, nr. Stanley	3·73	2·66	4·61	11·00
3. Middle Fork, Salmon River, at Dagger Falls	3·53	2·81	5·53	11·87
4. South Fork, Salmon River, nr. Warm Lake	4·69	5·15	4·09	13·93
5. Hell's Canyon, Snake River	3·20	4·41	8·48	16·09
6. Weiser River, at Evergreen Forest Camp	3·75	3·67	3·75	11·17
7. Little Salmon River, nr. New Meadows	8·88	9·74	4.48	23·10
8. Little Salmon River, nr. Boulder Creek	3·26	4·24	6·28	13·78
9. Salmon River, nr. Riggins	3·28	2·65	4·32	10·25
10. Salmon River, at Carey Falls	3·30	3·96	7·05	14·31
11. French Creek, nr. Salmon River	3·43	3·06	5.46	11·95
12. North Fork, Payette River	2·84	3·70	3·67	10·21

Source: Leopold (1969b).

uniqueness ratios for each site can be summed, and the total can be compared with the totals for other sites. By simply adding factor uniqueness ratios, each factor is given equal weight. At the same time, the total values do not indicate if a site is uniquely aesthetic or unaesthetic. In the case of the twelve Idaho sites, for instance, Hell's Canyon (5) was unique in a positive (attractive) sense, and a site on the Little Salmon River (7) was unique in a negative (unattractive) sense (Table 12.2).

The data can be exploited further. For example, *valley character* and *river character* can be examined as two variables considered by some to be the most important in influencing a viewer's impression of a river-valley site.

Valley character may be, Leopold suggested, a function of *landscape scale*, the availability of distant views (*landscape interest*), and the degree of urbanization. The ranking of sites in terms of these variables on a scale of valley character can be achieved as follows (Fig. 12.2). The landscape scale is derived by plotting the height of mountains against valley width for each site, and projecting these points orthogonally on to a line sloping at 45 degrees across the graph (thus giving equal weight to both variables). The values on this landscape scale are then plotted against *scenic outlook* (degree of view confinement, or availability of distant vistas), and the resulting points are projected, as before, at 45 degrees on to a new scale of landscape interest. Landscape interest values are then plotted against degree of urbanization, and the points are projected on to a *scale of valley character*. The data plotted on Figure 12.2 are for the twelve Idaho sites—and the Hell's Canyon (5) stands out as the most wild and spectacular location.

River character can be determined in a similar way (Fig. 12.3). In general, grandeur of rivers can be attributed to size, apparent speed of flow, and the extent to which the water surface is broken. Thus river width and depth data are plotted and projected on to a scale of river size, and this variable is related to the presence or absence of rapids, riffles, and falls, to produce a *scale of river character*.

The data on valley character and river character can be combined to produce a ranking of sites in terms of all the variables that go to make up the two scales. Figure 12.4 shows the information of the twelve Idaho sites and for a selection of sites elsewhere in the United States. In terms of this sample of sites, Hell's Canyon is clearly outstanding, both locally and nationally.

Leopold's technique is attractive because it is largely based on many fairly easily measured factors, and it provides a clear, semi-quantitative result. But it has several drawbacks. Firstly, it relates only to sites, and it is difficult to see how it could easily be extended to *areas*. Secondly, some important components of scene, especially transitory ones, are omitted. And finally, although many agree with it, the scheme is based on the assumption that the criteria selected are important, and are equally important.

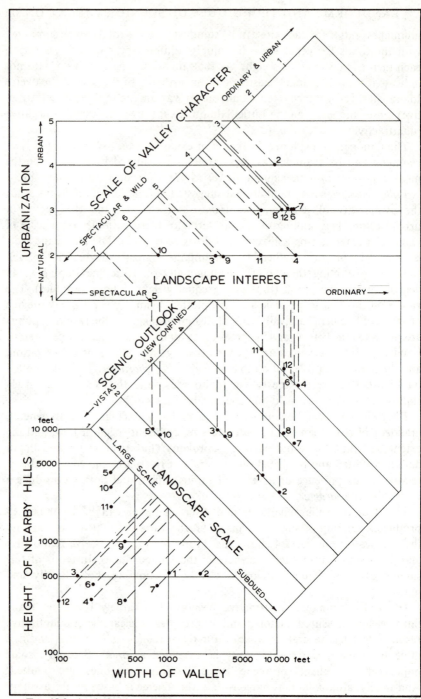

Fig. 12.2. Analysis of valley character at 12 sites in Idaho (after Leopold, 1969a)

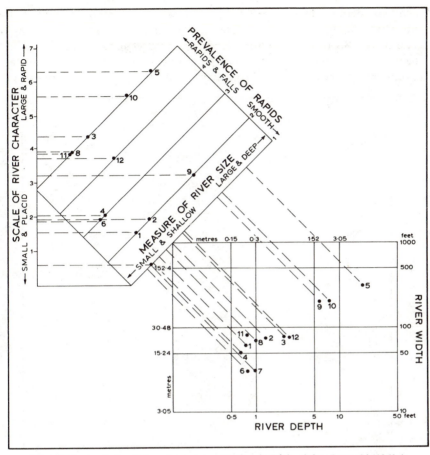

FIG. 12.3. Analysis of river character at 12 sites in Idaho (after Leopold, 1969a)

(*ii*) *Linton's method: areal differentiation.* Linton (1968) developed a technique for the assessment of scenic resources in Scotland that provided an areal classification of landscape. He argued that the appraisal should be founded on the elements of scenery that influence our reactions to it, the spatial variations of the elements should be mapped, and the several categories mapped should be arranged in a hierarchy of value. In Linton's view, those elements of scenery that contribute fundamentally to its quality can be classified into two major categories: the form of the ground (*landform landscapes*), and the use made of the land by man (*land-use landscapes*).

Landform landscapes could be quantitatively described in terms of relative relief (differences in height between the highest and lowest points within unit areas) or a similar measure, using topographic maps as a data source. But, in Linton's view, this would be a fairly laborious process. Instead, using his own considerable experience of Scottish scenery, Linton substituted a subjective appraisal of relief features shown on 10-miles-to-the-inch maps (1: 663,600). Six types of landform landscape were recognized, each based on the qualitative assessment of such criteria as relative relief, steepness of slope, abruptness of accidentation, frequency and depth of dissecting valleys in

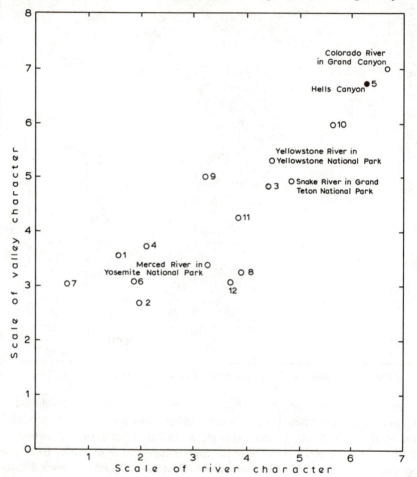

FIG. 12.4. Relations between river character and valley character at 12 sites in Idaho, and four other sites in the United States (after Leopold, 1969a)

upland regions, and isolation of hill masses from their neighbours. Each type was given a point rating, the rating being based on the personal evaluation of the relative attractiveness of the six types. The scenic quality of landscape was generally reckoned to increase with relief, slope, etc., although the intervals between ratings were not always considered to be the same. Linton also

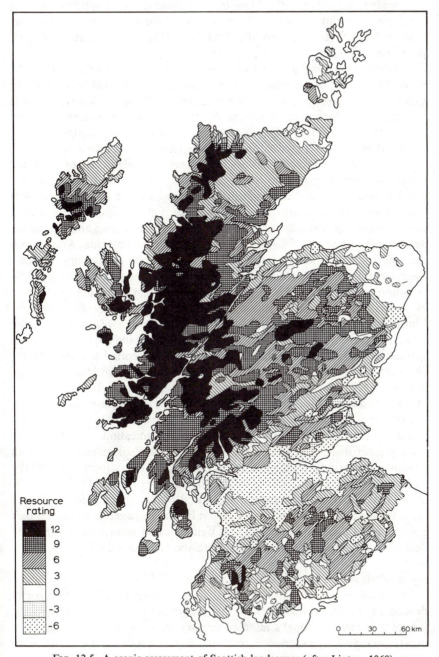

Resource
rating

12
9
6
3
0
-3
-6

0   30   60 km

FIG. 12.5. A scenic assessment of Scottish landscapes (after Linton, 1968)

appreciated that other natural features contributed to the attractiveness of landform landscapes; the most important of these is water, which was evaluated but not built into the final map (Fig. 12.5) because of scale problems.

A fundamental problem in the areal classification of landscapes on the basis of landform characteristics is that of defining the boundaries of units. In Linton's technique, the boundaries are determined subjectively, and the same is true of the technique proposed by the Countryside Commission for Scotland (1971). Perhaps the experience of land-systems mapping, to be discussed in Chapter 13, could usefully be applied to this problem.

Linton identified several land-use landscapes, delimited their extent using information on topographic maps, and gave each a point rating. In the classification, urbanized and industrial landscapes lie at one extreme (-5) and are generally assumed to be, for example, ugly, dull, or depressing; at the other extreme, there are wild, lonely landscapes, too steep and remote for development ($+6$) and considered to be of major scenic value. Armed with the maps of landform and land-use landscapes, a composite map of scenic assessment based on the combination of ratings can be produced (Fig. 12.5).

Linton's technique clearly produces, relatively quickly and cheaply, a map of scenic value that is suitable for planning purposes. But as applied by Linton, the method has several limitations. In the first place, and despite claims to the contrary, it is subjective: significant landscape elements are subjectively selected, delineated, and rated, and perceptual responses to the elements are assumed. Secondly, it takes no account of coastal scenery, of individually striking features (such as the parallel roads of Glen Roy), or of water features. This limitation could perhaps be overcome by the judicious use of bonus points for areas with especially attractive sites. Thirdly, landscape and land-use ratings are equated without justification.

At present there is little evidence that 'consumers' of scenery agree with Linton's appraisal. But the results certainly appear to conform substantially with the qualitative earlier survey of the Highlands by Murray (1962) in so far as the highland areas 'of outstanding natural beauty' designated by Murray are rated highly on Linton's map. Could it be that this agreement may in part reflect the fact that both surveyors were well-educated, astute, and experienced field observers, uncommonly familiar with the Scottish scene?

(b) Analysis of personal responses to scenery

The 'total experience' of an individual in viewing a scene is only partly dependent on the components of the landscape; it is also related to the interplay in the individual's mind of sensory, psychological, and sequential experiences. Sensory experiences relate to sight, of course, and also to sound, smell, feel, and taste. Psychological experiences might include such emotions as sentiment, fear, curiosity, surprise, and veneration. And sequential experiences comprise changes occurring in the scene or in the location of the spectator. Thus the response of an individual is the result of numerous variables, most of which are incapable of realistic assessment in isolation. At the same time, groups of individuals respond to scenic stimuli in different ways, and the variety of responses should be considered in evaluating landscape for scenic purposes.

Fines (1968) developed a survey technique in which subjective responses to view are classified on a single, comprehensive, and predetermined scale. The scale was produced using the responses to a set of photographs by one group (of a number studied) comprising people 'with considerable training and experience in design'. It is a partly geometric scale, with 32 categories (Table 12.3).

TABLE 12.3

Scale of Landscape Values

0–1	Unsightly
1–2	Undistinguished
2–4	Pleasant
4–8	Distinguished
8–16	Superb
16–32	Spectacular

Source: Fines (1968).

The field survey involves the rating of views in terms of their overall beauty on this scale. Subsequently, the rating values of views are converted into land-surface values, 'the criterion being the value of a particular tract of land to the totality of views in which it features' (Fines, 1968, p. 45). The precise method of transformation is not clear. Figure 12.6 shows the Landscape Evaluation Map of East Sussex produced by a single observer using this technique.

An extension of the Landscape Evaluation Map is the summary description of the scenic value of a given region by a Landscape Value Profile. The profile is based on measuring the percentage areas within a given region in each value category, and plotting the percentages as a frequency distribution (Fig. 12.7).

It may be profitable to attempt an interpretation of the Landscape Evaluation Map in terms of those landscape features that contribute most to the individual's perception of views. In East Sussex it seems clear that geology and geomorphology play an important role. For example, the chalklands of the South Downs (excluding the urbanized coastal belt) have a mean value of 8·3 (superb) with a standard deviation of 2·4. And along the north-facing escarpment of the Downs, in the Cuckmere Valley and along the cliffs east of Seaford, values rise to between 10 and 12. These high values, associated with

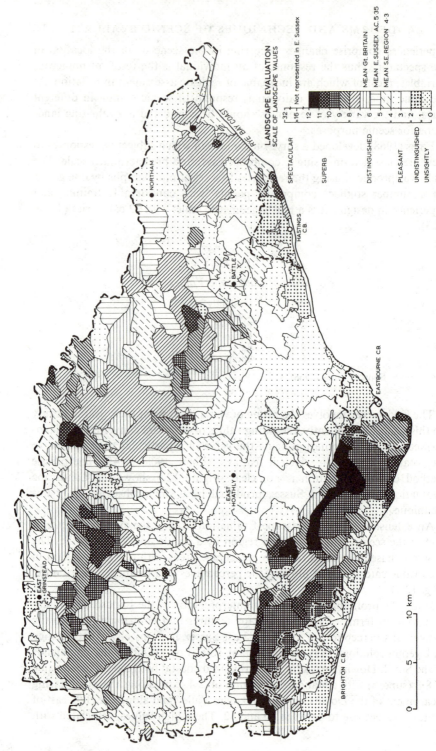

LANDSCAPE EVALUATION
SCALE OF LANDSCAPE VALUES

SPECTACULAR	32
	16 − Not represented in E. Sussex
	12
SUPERB	11
	10
	9
	8
DISTINGUISHED	7
	6 − MEAN Gt. BRITAIN
	5 − MEAN E. SUSSEX A.C. 5·35
PLEASANT	4 − MEAN S.E. REGION 4·3
	3
	2
UNDISTINGUISHED	1
UNSIGHTLY	0

NORTHIAM

RYE BAY COAST

HASTINGS C.B.

BATTLE

EASTBOURNE C.B.

EAST HOATHLY

EAST GRINSTEAD

HASSOCKS

BRIGHTON C.B.

0   5   10 km

FIG. 12.6. Landscape Evaluation Map of East Sussex (after Fines, 1968)

areas of higher land and steeper slopes, contrast with the low ratings (mean 3·3) for parts of the Clay Vale with its relatively subdued topography.

The Landscape Evaluation Map is valuable in formulating planning policy and appraising conflicting claims for the use of land. But the techniques have been severely criticized, notably by Brancher (1969). (See also Fines, 1969.) Some of the more serious criticisms include the facts that the scale, critical to the whole exercise, is based on a small set of photographs assessed by only a small and possibly biased sample of respondents, and it has not been rigorously tested. The reason for the partly geometric scale is unclear. The Sussex survey was the work of a single observer who was apparently attempting to represent the responses of most, or the most important 'consumers' of Sussex scenery. The relations between the Sussex landscape and the 'test' photographs is not stated. Finally the procedure for transforming site data into spatial information is not explained.

Ultimately the technique must be judged in terms of its success in reflecting the communities' view of the Sussex landscape, and such judgement is at present impossible. It would be interesting to know, however, how Fines' map would compare with a map of East Sussex compiled using Linton's technique.

(c) Other techniques

The preceding discussion has examined in detail some techniques of landscape evaluation in which gemorphological considerations explicitly or implicitly play an important role. There are many other approaches to this extremely difficult problem. Particularly stimulating is the analysis of forest landscapes by Litton (1968), and the more general survey of the aesthetic dimensions of landscapes by Litton and Craik (1969). In these studies, the factors that are identified as being of greatest importance are form, space, distance, observer position, time variability, and sequence. Of particular interest to the geomorphologist is the form factor which includes the form of the ground, of course, together with, at a smaller scale, the form of individual objects and groups of objects, such as trees and woods. Some of the variables deemed to contribute to distinctions between forms include isolation, relative size or scale, contour distinction or silhouette, and surface variance. It is interesting to observe that these variables are all familiar to the geomorphologist and could be subjected to quantitative analysis in terms of morphometric parameters.

12.3. Conclusion

Within the range of relatively precise techniques that have been used to evaluate scenery, those based on the analysis of landscape attributes ring must true to the geomorphologist, for it is in them that there is usually a need

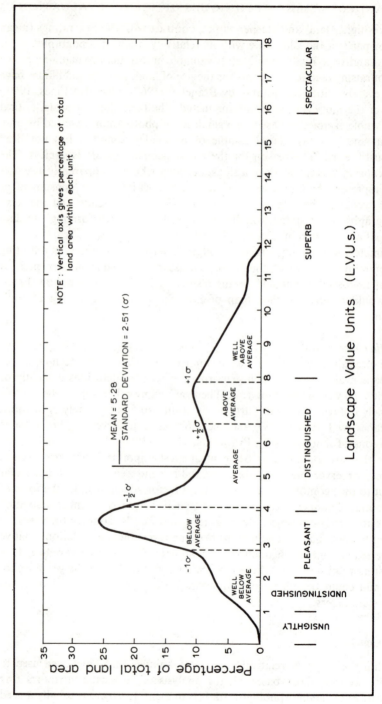

NOTE : Vertical axis gives percentage of total land area within each unit

MEAN = 5·28
STANDARD DEVIATION = 2·51 (σ)

Fig. 12.7. Landscape Value Profile of East Sussex (after Fines, 1968)

for geomorphological data. As it happens, such techniques are also attractive to planners because they apparently provide useful, quantitative, and relatively quick classifications of scenery. But the unwary geomorphologist should be warned of the dangers in most of these techniques.

He should beware of all claims to objectivity. The physical variables selected, and the values placed on them, are cultural appraisals. At best they may be adequate indirect measures of the attitudes to landscape of a particular cultural group at a particular time; at worst they reflect the unverified guesses by professionals of what they believe most people value, or should value, in landscape. It is a truism, but nevertheless a fundamental one, that perception of landscapes varies within and between social and cultural groups, and also varies with time. The 'spectacular', the 'unique', and the 'natural' may be prized by Anglo–Americans and many Europeans today, but they may be less prized by American Indians; and certainly they have not always been as important generally as they now are to some.

Furthermore, the geomorphologist (and, indeed, anyone concerned with landscape aesthetics) should beware of the apparent precision bestowed on the assessment of landscape quality by quantitative techniques. Even if the techniques are mathematically sound—and they sometimes are not—the appreciation of a scene by an observer or group of observers is so complex that quantification may merely tend falsely to transform shades of grey into a spectrum of attractive, clearly differentiated, but misleading colours.

But the planner's problem remains: decisions concerning landscape development have to be made, and the claims of conservationists and others interested in preserving or protecting attractive countryside have a right to be heard. The aesthetic arguments need to be presented precisely, if not quantitatively. Perhaps the most sensible approach to the problem is through the examination of personal preferences and prejudices. If the techniques requiring analysis of landscape components are to carry conviction, especially with social scientists, then it is essential to demonstrate that physical attributes of landscape can provide a realistic set of substitute variables for true perception variables embodied in landscape attitudes.

13

LAND-SYSTEMS MAPPING

13.1. Introduction

In the discussions on slope stability (Chap. 6), coasts (Chap. 8), and permafrost terrain (Chap. 9) it was indicated that a general method of relief classification could prove to be very useful in the delimitation of areas containing similar landform properties. For example, areas liable to landsliding because of defined physical conditions can be distinguished from adjacent more stable areas. Some coastlines may be best managed by recognizing their subdivision into units within which there are definable relations between beach form, materials, and coastal processes, even though there is an interaction between neighbouring units. Terrain evaluation at the reconnaissance level in, for example, permafrost areas often takes the form of a mapped subdivision of the ground into 'like' units, based on air-photo interpretation. These 'like' units tend to have relatively little internal variation in their geomorphological properties, though they may each be quite different from other neighbouring units.

In each of these examples it is recognized that a scheme for classifying the earth's surface based on an integrated assessment of relief form, materials, and processes is of value in environmental management. For this purpose a formalized scheme known as the land-systems approach is already in existance.

The land-systems mapping procedure involves subdividing the country into areas that have within them common physical attributes that are different from those of adjacent areas. Land systems may range in size from only tens of km² up to some hundreds of km². Within any one land system there is usually a recurring pattern of topography, soils, and vegetation (Christian and Stewart, 1952). This threefold association is justified in the following way:

The topography and soils are dependent on the nature of the underlying rocks (i.e. geology), the erosional and depositional processes that have produced the present topography (i.e. geomorphology), and the climate under which these processes have operated. Thus the land system is a scientific classification of country based on topography, soils and vegetation correlated with geology, geomorphology and climate (Stewart and Perry, 1953, p. 55.)

A land-systems map defines those areas within which certain predictable combinations of surface forms and their associated soils and vegetation are likely to be found.

In practice, whether the boundaries are being drawn in the field or from aerial photographs, the simplest criterion for distinguishing between land systems is surface relief (Bawden, 1967). Secondly, the interpretation of soils, geology, and depositional and erosional processes comes most readily through landform analysis. That is why geomorphology is a key element in land-systems mapping, or in any type of integrated resource survey, especially at the rapid reconnaissance level (Verstappen, 1966).

The land-systems approach was used as far back as 1933 by Veatch, who was concerned with a classification of the agricultural land of Michigan. He also recognized an important aspect of any evaluation exercise in land planning, which is that the economic significance of the land type will vary with changes in economic conditions and scientific discoveries. Similar types of land classification were also considered by Bourne (1931, for forestry stock-taking), Wooldridge (1932), Unstead (1933), and Linton (1951). The significance of these and other early works is discussed by Christian and Stewart (1968), and by Wright (1972b) who also considered more recent studies in the whole context of regional resource surveys.

Land-systems mapping was developed by the Commonwealth Scientific and Industrial Research Organization in Australia (CSIRO) as a rapid method of reconnaissance survey of otherwise unmapped or poorly mapped areas (Christian, 1957; Mabbutt and Stewart, 1963). The primary aim was to establish a classification of the suitability of land for agricultural purposes. The first area, the Katherine–Darwin region, was mapped in 1946, and the published report followed six years later (Christian and Stewart, 1952). Since then many more areas have been mapped as this method of reconnaissance survey continues to prove its value. Other uses of the system in Australia have included the evaluation of terrain for engineering purposes (Aitchison and Grant, 1967). The single most comprehensive volume on the land-systems approach was also published in Australia (Stewart, 1968).

The method has been adopted by the Land Resources Division (LRD) of the Overseas Development Administration, Foreign and Commonwealth Office (formerly the Ministry of Overseas Development) in Britain. Reports have been produced for parts or all of such countries as Bechuanaland, Tanzania, Nigeria, Lesotho, Botswana and Gambia (Ministry of Overseas Development, 1970).

Similar mapping techniques were put forward by Hunt (1950) as applicable to military purposes for assessing the difficulties of passing across different types of ground. As a recognition of this application support came from the Military Engineering Experimental Establishment (MEXE) for terrain classification techniques to be developed in the Soil Science Laboratory at Oxford (Webster and Beckett, 1970). This led to one major publication *Terrain classification and data storage: Uganda Land Systems* (Ollier, et al., 1969), and several smaller reports.

In the U.S.S.R. a hierarchy of landform divisions is recognized in which units that are repeated in a regular pattern form a 'landscape' (Solentsev, 1962), which is equivalent to the land system. 'Landscapes' in the Russian approach are considered to be the surface expression of a homogeneous geological–geomorphological foundation. As with the CSIRO approach these 'landscapes' are also conceived as ecological units, and form the basis for land-use planning.

The land-systems mapping approach has also been used in South Africa where it has proved to be useful in road engineering, both for route alignment planning, for identifying sources of construction material, and for land assessment in urban land planning. In Japan this type of mapping has been used in assessing geological structure, soils, land classification for land-use improvement, flood prevention, and for population mapping in terms of the density per landform subdivision (Nakano, 1963). An approach similar to that of land-systems mapping has been used for land-use planning in Illinois, by the Illinois State Geological Survey (Hackett and McComas, 1969).

13.2. Presentation of Land-Systems Data

In the CSIRO reports land systems are shown on a general map covering the whole of a survey area (ranging in size from 5,000 km² in New Guinea to 350,000 km² in the arid zone of Australia). The reports themselves carry a series of block diagrams to illustrate the general character of each land system. On the block diagram separate component parts of the land system, known as *land units* (*land facets*), are identified by numbers (Fig. 13.1 and Table 13.1).

The *land unit* has a simple form, usually occurs on a single rock type or superficial deposit, and carries soils which if not the same across the whole unit then at least vary in a consistent manner across the land unit (facet). Examples of a land unit would be an alluvial fan, a levee, a bedrock cliff, or a group of sand dunes. Land units are often recognized on the basis of their geomorphology.

A land unit may be composed of two or more separate *elements*. For example, a levee has both a crest and flanking side-slopes, each of which form separate elements. Some land units may also show local deviations, in one property or another (e.g. soils), from the general character of the unit. These local deviations are isolated as *variants*, and are rarely easy to distinguish upon aerial photographs. Their identification requires field investigation (Webster and Beckett, 1970).

Since the land system is composed of a group of land units, frequently up to ten, which form recurrent patterns across an area it resembles a morphological region as defined by Linton (1951). Land-system boundaries tend to coincide with some discernible geological or geomorphological boundaries.

Figure 13.1 illustrates the differing nature of two adjacent land systems in the Alice Springs area of Australia (Perry, 1962). The land-systems map

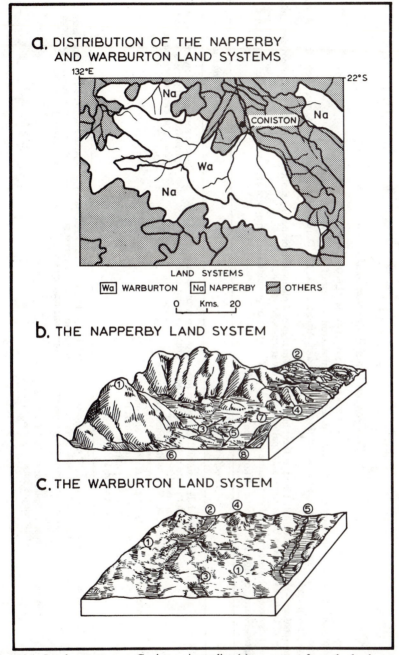

FIG. 13.1. Land systems near Coniston, Australia: (a) an extract from the land systems map, (b) and (c) block diagrams representing the landforms of the two land systems and identifying the component land units (after Perry, 1962)

TABLE 13.1

The Napperby and Warburton land systems

NAPPERBY LAND SYSTEM (2,600 sq km)

Granite hills and plains with lower rugged country in a strip from Aileron to west of Mt. Doreen homestead.

Geology—Massive granite and gneiss, some schist. Pre-Cambrian age, Arunta block, Mt. Doreen–Reynolds Range; Lower Proterozoic, Warramunga geosyncline.

Geomorphology—Erosional weathered land surface: hills up to 150 m high and plains with branching shallow valleys; less extensive rugged ridges with relief up to 15 m, and a dense rectangular pattern of narrow steep-sided valleys.

Water Resources—Isolated alluvial or fracture aquifers may yield supplies of groundwater. There are areas suitable for surface catchments.

Climate—Nearest comparable climatic station is Tea Tree Well.

Unit	Area	Landform	Soil	Plant Community
1	Large	Granite hills: tors and domes up to 150 m high; bare rock summits and rectilinear boulder-covered hill slopes, 40–60%, with minor gullies; short colluvial aprons, 5–10%	Outcrop with pockets of shallow, gritty, or stony soils	Sparse shrubs and low trees over sparse forbs and grasses, *Triodia spicata*, or *Plectrachne pungens* (spinifex)
2	Medium	Closely-set gneiss ridges and quartz reefs: up to 15 m high; short rocky slopes, 10–35%; narrow intervening valleys		
3	Medium	Interfluves: up to 7 m high and 0.8 km wide; flattish or convex crests, and concave marginal slopes attaining 2%	Mainly red earths, locally red clayey sands and texture-contrast soils, stony soils near hills	Sparse low trees over short grasses and forbs or *Eragrostis eriopoda* (woollybutt)
4	Medium	Erosional plains: up to 1.6 km in extent, slopes generally less than 1%.		

5	Small	Drainage floors: 180–365 m wide, longitudinal gradients about 1 in 200	Mainly texture-contrast soils, locally alluvial soils and red earths	Eremophila spp.—Hakea leucoptera over short grasses and forbs; minor Kochia aphylla (cotton-bush)
6	Small	Alluvial fans: ill-defined distributary drainage; gradients above 1 in 200	Alluvial brown sands and red clayey sands	Sparse low trees over short grasses and forbs or Aristida browniana (kerosene grass)
7	Small	Rounded drainage heads: up to 180 m wide and 1·5 m deep on the flanks of unit 3	Red earths	Dense A. aneura (mulga) over short grasses and forbs
8	Very Small	Channels: up to 45 m wide and 1·5 m deep and braiding locally	Bed-loads mainly coarse grit	E. camaldulensis (red gum)—A. estrophiolata (ironwood) over Chloris acicularis (curly windmill grass)

WARBURTON LAND SYSTEM (1,554 sq km)

Sparsely-timbered granite plains in the north and north-west of the area, mainly near Coniston homestead.

Geology—Quaternary soils and Tertiary 'deep weathering profile'. Overlying Pre-Cambrian granite, schist, and gneiss.

Geomorphology—Erosional weathered land surface: peneplain of selectively weathered rocks; open sub-rectangular pattern of shallow valleys.

Water Resources—Prospects generally poor but isolated fracture aquifers may yield supplies of groundwater. There are areas suitable for surface catchment.

Climate—Nearest comparable climatic station is Tea Tree Well.

Unit	Area	Landform	Soil	Plant Community
1	Very Large	Interfluves: up to 0·8 km wide and 4·5 m high; flat or slightly rounded stony crests with minor rock out-crops; short concave marginal slopes attaining 1%	Mainly red earths, locally red clayey sands and stony soils including texture-contrast soils	Sparse low trees. *A. aneura* (mulga) or *A. kempeana* (witchetty bush) over short grasses and forbs or *Eragrostis eriopoda* (woollybutt)
2	Medium	Drainage floors: up to 270 m wide; flat, unchannelled central tracts and gently sloping margins	Presumably red earths	*A. aneura* (mulga) over short grasses and forbs or *Eragrostis eriopoda* (woollybutt)
3	Small	Short valley heads shallowly entrenched on the flanks of unit 1: unchannelled floors up to 460 m wide liable to shallow gullying; longitudinal gradients above 1 in 500	Texture-contrast soils	*Eremophila* spp.—*Hakea leucoptera* over short grasses and forbs or *Bassia* spp; or minor *Kochia aphylla* (cotton-bush)
4	Very Small	Small hills and quartz reef ridges: up to 9 m high	Outcrop with pockets of shallow, gritty, or stony soil	Sparse shrubs and low trees over *Triodia pungens* (spinifex), *T. spicata* (spinifex), or sparse forbs and grasses. Far north: *E. brevifolia* (snappy gum) over *Triodia pungens* (spinifex) or *Plectrachne pungens* (spinifex)
5	Very Small	Channels: wide, shallow and braiding		*E. camaldulensis* (red gum)—*A. estrophiolata* (ironwood) over *Chloris acicularis* (curly windmill grass)

Source: Perry (1962) (with metric conversion).

produced for the report (Fig. 13.1a) does not identify the position of the separate land units. The block diagrams are of no particular view within the land systems. They are provided only as an indication of the general character of the system and of the relationship of one land unit to another. The distinction between the land units shown in Figure 13.1, as is generally the case, is made on geomorphological grounds. The land units are also listed on tables (Table 13.1) which carry information concerning their geology, geomorphology, water resources, climate, slope form and steepness, soils, and vegetation.

This mode of presentation has been modified over the years and the 1972 report on the 'Lands of the Aitape–Ambunti area, Papua, New Guinea' (Haantjens, 1972) dispenses with block diagrams and tables. Instead it supplies stereoscopic pairs of aerial photographs of each land system, with a synoptic description of the main system characteristics. The detailed account of each land system includes an assessment of its agricultural and engineering possibilities and limitations, thus moving further towards a specific application of the inventory made. The coloured maps that accompany the report include not only the standard and necessary land-systems map, but also maps of vegetation, land altitude and sea depth, drainage divisions, transport, administration, population distribution, agricultural land-use capability, forest resources and land-use intensity, terrain access categories, and the associations of soil groups. As such, therefore, this report has a greater interest in the potential applied value of its content than may be true of the earlier reports. In addition, it follows the advice of Haantjens (1968) in that it provides more quantitative data about the land-system attributes than had been the case in earlier reports.

It is generally implicit in the assumptions of a land-systems survey that land units are somewhat homogenous in terms of their physical properties. A land unit should be sufficiently uniform for a prudent arable farmer to manage the whole extent of one unit in the same way (Beckett and Webster, 1965). This uniformity has been found to exist for agriculturally significant soil properties, plant populations, and the physiognomy of vegetation, as well as for the engineering properties of soil, and the morphometric and hydrological characteristics within similar land units (Brink, Partridge, and Mathews, 1970).

Not only can homogeneity within land units be demonstrated, but significant differences often occur between adjacent land units. For example, a study of the Kyalami land system, in South Africa (Brink, Partridge, and Mathews, 1970) showed a number of land units (facets) (Fig. 13.2). The engineering properties of the soils were measured, and these revealed the differences between the units. For example, Table 13.2 shows that for units 1 (variant 1) and 2 (variant 1) the differences are appreciable, although the two units are separated only by a convex change of slope. The junction between unit 2 (variant 1) and unit 7, on the other hand, is bold, and the differences in the engineering properties are hardly surprising.

In practice the homogeneity of land unit characteristics is likely to vary with the environment. Most variation is likely to occur in areas where the parent materials for soils have a large and sometimes unpredictable variation. With glacial material, sands and clays may lie next to each other in a complex pattern without separate units being defined for each. Similarly, large unpredictable variations may occur in areas of dense population or where man has

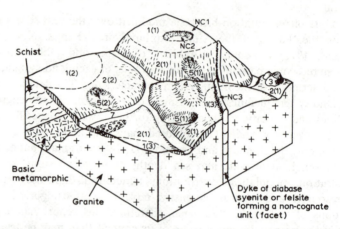

FIG. 13.2. Block diagram to illustrate the components of the Kyalami land system (Source: Brink, Partridge, and Mathews, 1970)

for a long time modified the land. For example, although the internal variability within one land unit near Oxford (Webster and Beckett, 1970) was found to be small for soil–water tension and physical properties of the soil (including clay content, plastic limit and liquid limit, see Chap. 6), much greater variability was found to exist for the chemical properties of the topsoil. This was ascribed to the fact that the area has a long history of diverse farm management and fertilizer practices. Under less advanced farming systems in areas of low population density, or in areas of much more recent agricultural development, the variations in topsoil chemical composition may be much less. However, in tropical areas with a high density of population, such as in parts of Nigeria, relations between physical conditions and land use (vegetation) are obscured by patterns imposed by socio-economic factors (Moss, 1968).

The reports published by the Land Resources Division of the Overseas Development Administration (LRD), such as that for Lesotho (Bawden and Carroll, 1968), begin with general descriptions of the physiography, geology, geomorphology, climate, vegetation, soils, drainage, and water supplies, as well as a summary of human aspects (e.g. population, communications, and

TABLE 13.2

The engineering properties of some of the soils in the Kyalami Land System

Unit	Variant	Form	Soils and hydrology	Statistic	Soil constants		
					Liquid limit	Plasticity index	Linear shrinkage
1	1	Hillcrest, 0–2°, width 450–1800 m	Residual sandy clay with collapsing grain structure on granite (9–23 m). Ferrallitic soil. Above groundwater influence except at depth	$\overline{X}$	44·1	22·6	9·78
				S	11·0	10·88	4·27
				V	0·25	0·48	0·44
	2	Convex side slope 2–12°, width 450–910 m	Hillwash of silty sand derived from granite (0·3–0·9 m) on granite schist or basic metamorphic rocks. Occasionally saturated	$\overline{X}$	17·3	4·90	2·04
				S	8·98	3·70	1·04
				V	0·52	0·76	0·51
7	—	Alluvial floodplain 1–3°, width 45–230 m	Expansive alluvial clays and sands (3–6 m) on granite schists and basic metamorphic rocks. Mineral hydromorphic soil. High water table (data for black alluvium)	$\overline{X}$	58·5	37·1	13·7
				S	14·7	13·67	10·31
				V	0·25	0·37	0·75

$\overline{X}$ = mean,   S = standard deviation,   V = coefficient of variation

Source: Brink, Partridge, and Mathews (1970) (with metric conversion).

land use). Details are then supplied of the land systems, each of which is identified on maps. These details briefly refer to the physiography, soils, vegetation, and land use. The report continues with a classification of the area, based on the land systems, into groups according to their agricultural potential.

In the Uganda land-systems report produced by MEXE (Ollier et al., 1969) each land system is illustrated by aerial photographs. These enable a stereoscopic view to be made of a part of the land system. In addition the boundaries between the land units are marked ʻon these photographs, enabling the actual relationship between one land unit and another to be better appreciated. The MEXE group, realizing the importance of effectively communicating the information to the map user, also devised a strong visual aid (module) for each land system, which contained all the information necessary in order to recognize the constituent parts of a land system (Webster and Beckett, 1970). This module consists of an envelope carrying verbal descriptions of the land system as a whole and all the land units present, together with diagrams to aid recognition. The envelope also contains selected, annotated stereo-pairs of air photographs, ground photographs, and land-unit (facet) maps of small parts of the land system.

South African land-systems studies differ from those elsewhere mainly because recent practice has been to include the identification of land units on the land-systems map (NIRR, 1971). This arises from the practical needs of route-planning, for which the technique is frequently employed. Planning the alignment of a road has to be carried out at a more detailed scale than is possible from a land-systems map. More specific predictions have to be made about the *site* across which the road is to pass than is provided by a knowledge of the general character of a land system. Not until the land units are mapped is the required level of detail even approached, and then its practical usefulness depends on the information contained in the accompanying tables. An example of the type of analysis carried out in South Africa is contained in Brink, Partridge, and Mathews (1970).

The fact that the CSIRO, LRD, and MEXE reports do not present comprehensive maps of land units is a reflection of the fact that at the general reconnaissance level, as opposed to the site recognition level, the land-unit map is inappropriate. The expense involved in mapping land units of large areas (e.g. whole countries) would be very large, not only in terms of map production, which in order to be meaningful would have to be on at least a 1:50,000 scale; but also in terms of the manpower required in the primary air-photo analysis and the necessary field checking.

Through land-systems mapping the general physical characteristics of a territory can be rapidly determined, and even at this general reconnaissance level it is frequently apparent that for a specific purpose some parts of the territory are of no value and need no further investigation. Land-unit mapping can then be carried out in the remaining areas if, for the purpose in hand, more detailed information is required about specific sites. In this way the mapping programme can be user-orientated in terms of the scale of map production. It is vital, of course, that if it is to be of any practical value the information listed in the land-unit table should be of the kind required by the user. Thus, although the classification system is general, the land-unit information may need to be very specific.

13.3. Methods of Land-System Mapping

The basic method used to identify separate land systems is by analysing relief as portrayed on aerial photographs (Webster, 1963; Bawden, 1967; Ollier et al., 1967, Webster and Beckett, 1970). Initially these photographs are laid out as a large mosaic upon which boundaries are drawn between areas having distinct but differing photographic images. These are predominantly related either to the relief character of the ground or to its vegetation, but in essence it is an integrated expression of the geological, geomorphological, soil, vegetation, and land-use properties of the region. As such it is a particularly useful approach at the reconnaissance stage of a survey when an over-all view of the area is required. The validity of the boundaries drawn in this way are tested by examining selected pairs of photographs under a stereoscope. A boundary is usually accepted on the basis of there being a distinct difference in the landform assemblages on either side of the line. Once the boundaries have been established then each area is examined, by means of aerial photographs, in order to assess the character of the land units it contains. The best photo scale for identifying land units is between 1:20,000 and 1:30,000, though sometimes a scale of 1:40,000 is used (Brunt, 1967), while photo mosaics reproduced at a scale of about 1:125,000 are generally the most useful for identifying land-system boundaries (Bawden, 1965; Webster and Beckett, 1970).

The character of land units can very largely be determined from good stereo-pairs of photographs, with an optimum scale of about 1:20,000 depending upon the complexity of the terrain. Field work is necessary, however, in order to sample soils and identify bedrock characteristics and plant types. How much preliminary work can be done from the photographs depends very largely on the skill of the interpreter. Monochromatic aerial photographs can indicate a great deal about land form, drainage, soils, bedrock, vegetation and land use, through the patterns and textural and tonal differences which they display (Am. Soc. Photogrammetry, 1960; Lueder, 1959). Advances in multi-spectral imagery make possible the gathering of more information about land units than can be done from monochromatic film alone (Colwell, 1968). For example, false-colour photography provides much information about vegetation, land use, exposures of bedrock and gravel, and it can discriminate efficiently between land and water; infra-red line-scan imagery is useful for detecting thermal differences and may therefore be useful in identifying drainage patterns and soils having different moisture contents (Cooke and Harris, 1970).

The preliminary definition of land-system boundaries is checked in the field by following selected traverse lines. At the same time information is accumulated, on a stratified random sampling basis, concerning the soils, vegetation, hydrology, and slope characteristics of the land units within each land system. Particular attention may be given to one parameter or one area, if a particular development project warrants it. For example, in the Bechuanaland (Botswana) survey (Bawden, 1965) little attention was given to those areas which were quickly recognized on the basis of the reconnaissance survey as being of no potential agricultural value.

Field sampling, as under the CSIRO programmes, is usually carried out by a team which may include a geologist, pedologist, botanist or forester, geomorphologist, and possibly a hydrologist and any other persons required for the purpose of the survey. The time taken for any one project depends on the purpose of the mapping programme, the complexity of the terrain, and the size of the area to be covered. The CSIRO programmes have covered, in a single project, areas as large as 350,000 km², in the Australian arid zone, down to 5,000 km² in the humid lowland tropics of New Guinea. The time taken for a project involving only one field season is from 15 to 18 months. This is divided into five phases (Christian and Stewart, 1968):

a. Pre-field work (3 months)—collecting existing information, preliminary aerial-photograph interpretation, selection of sampling site, and planning of field itinerary.
b. Fieldwork (3 months).
c. Final aerial-photograph interpretation (3 months).
d. Evaluation of field data (3–4 months).
e. Report preparation (3–5 months).

13.4. Land-Systems Mapping for Agriculture

Land-systems mapping has been applied more in agriculture and engineering projects than in any other. It is pertinent therefore to examine its applicability in at least one of these two fields of environmental management.

(a) Land-systems mapping and land-capability classifications

Many of the CSIRO and LRD mapping projects (Sect. 13.1 and 13.2) were undertaken to assess the suitability of land for agricultural development. In the United States the Soil Conservation Service of the Department of Agriculture formalized a land-capability classification for agriculture which resembles the land-systems method. The United States system recognizes a threefold hierarchy from smaller to larger groupings of (i) capability units, (ii) capability subclasses, and (iii) capability classes (Klingebiel and Montgomery, 1961).

The capability unit resembles the land unit of a land system in that the soils in a capability unit are sufficiently uniform (a) to produce similar kinds of cultivated crops and pasture plants with similar management practices, (b) to require similar conservation treatment and management under the same kind and condition of vegetative cover, and (c) to have comparable potential productivity. The capability unit is described as a grouping of soils that are nearly alike, and is based on an interpretation of soil data. The relevance of geomorphology can only be assessed when the basis for describing the soils is known. If site relief and drainage conditions are included in the soil classification then geomorphology becomes intimately involved (Chaps. 2, 3, and 10). On the other hand if the soils are defined more on the basis of their internal physical and chemical properties then the relevance of geomorphology to the classification may be more remote.

At the subclass and class level of the United States capability classification, however, there is a great dependence on relief and drainage characteristics. Subclasses are defined according to their limitations for agricultural use and according to the hazards to which they are exposed. Four general kinds of limitations are recognized: (i) erosion hazard, (ii) wetness, (iii) rooting-zone limitations, and (iv) climate. Of these the first three are closely related to the geomorphology of an area. Subclasses are composed of groups of capability units.

At the most general level there are eight capability classes, which, under the United States system of notation, are numbered from I to VIII according to a scale of increasing severity of soil damage or limitations. Soils in class I to class IV under good management are capable of producing adapted plants, such as forest trees or range plants, and the common cultivated field crops and pasture plants, but some conservation practices become increasingly necessary through classes II–IV. The range of uses decreases through classes V–VII, and soils in class VIII do not return on-site benefits for inputs of management for crops, grasses, or trees without major reclamation. The particular nature of hazards and limitations is summarized in Table 13.3. Locally soil salinity may produce a limitation additional to those listed in this table. Relief characterisitics provide an important element in ascribing land to a particular capability classification. Indeed, an initial classification of land on the basis of slope steepness and its liability to erosion or flooding can provide a first approximation of the eventual land-classification class based on soil characteristics as well. This is on the assumption, of course, that climate alone has not already determined the allocation to a particular class. In the United States an estimated 2 per cent of the land falls in class I, classes

TABLE 13.3
Land classification for agriculture, United States

The table is organised under three main headings — **SOILS** (Texture, Structure, Rooting depth, Drainage, Stoniness), **RELIEF** (Slope, Liability to erosion, Liability to flooding) and **CLIMATE** (General character). A × indicates the condition applicable to each class. The Texture column is marked **NOT SPECIFIED** (spanning Good / Fairly good / Poor for all classes).

Class (physical limitations)	Texture	Structure	Rooting depth	Drainage	Stoniness	Slope	Liability to erosion	Liability to flooding	General character (Climate)
I None or minor	NOT SPECIFIED	Good (×)	Good (×)	Good (×)	—	Flat/gentle (×)	Moderate (×)	None (×)	Favourable (×)
II Some		Fairly good (×)	Moderate (×)	Imperfect (×)	—	Flat/gentle (×)	Small (×)	Some (×)	Slightly unfavourable (×)
III Severe		Poor (×)	Restricted (×)	Poor (×)	—	Moderate (×)	Severe (×)	Frequent (×)	Slightly unfavourable (×); Unfavourable (×)
IV Very severe (i)		—	—	Very poor (×)	Many (×)	Steep (×)	Severe (×)	Frequent (×)	Moderately severe (×)
V Very severe (ii)		—	Poor (×)	Very poor (×)	Many (×)	Steep (×)	Severe (×)	Frequent (×)	Moderately severe (×)
VI Very severe (iii)		—	—	Very poor (×)	Many (×)	Flat/gentle (×)	—	Frequent (×)	Severe (×)
VII Extremely severe		—	Poor (×)	Very poor (×)	Many (×)	Steep (×); Very steep (×)	Severe (×)	—	Severe (×); Very severe (×)
VIII Absolute		—	Poor (×)	—	—	—	Severe (×)	—	Very severe (×)

Source: Compiled from information by Klingbiel and Montgomery, (1961)

II, III, VI, and VII occupy about 20 per cent each, class IV 12 per cent, class V only 3 per cent, and class VIII occupies 2 per cent of the United States mainland area (U.S. Dept. Agric., 1962; see also Young, 1973).

In Britain a similar land-capability classification system has been proposed (Bibby and Mackney, 1969). Here land is graded into seven classes also based upon the site relief, climate, and soil characteristics (Table 13.4). Apart from climate all of the criteria used once more concern landform, materials, and process, and as such are geomorphological in character. At a more precise pedological level the chemical properties of the soil may become significant in terms of limitations on its use for agriculture, but at the primary level of classification it is the physical properties of both soils and site that matter.

The mapping of land units, as in a land-systems survey, potentially appears to be of value at the reconnaissance level of a land-capability classification. These units can then be combined into capability subclasses, or classes, according to their susceptibility to erosion or their inherent limitations for agricultural use. These combinations would not equate with land systems, however, for land systems may contain land units of differing slope steepness and liability to erosion.

(b) Case study: Land-unit mapping and land classification for agriculture, Tideswell, Derbyshire, England

The land-classification system described by Bibby and Mackney (1969) has been applied to a limestone upland area near Tideswell, Derbyshire (Johnson, 1971). In general this is an undulating upland dissected by deep and steep-sided river valleys, occasionally with flat alluvial valley-floors. Altitudes range from 150 m O.D. in the valley-floors to 400 m on the higher parts of the limestone plateau. Near-vertical outcrops of massive Carboniferous Limestone occur on some of the steeper valley-sides and scree slopes (mainly grassed over) are common within the deeper valleys. There is no upland surface drainage, though isolated areas of ill-drained land occur where clays prevent rapid infiltration to the limestone bedrock. Plateau-top slopes are locally steep (up to 25 degrees), but generally the upland slopes are less than 6 degrees.

A land classification of this area is of considerable interest because it is close to the upper limit for agriculture in the Peak District. This becomes apparent in the classification as minor changes in aspect introduce micro-climatic effects which change the classification category of otherwise continuous or similar slopes.

A portion of the land-use capability map of the Tideswell area is reproduced in Figure 13.3a, together with the classification index (refer to Table 13.4). Climate provides an important limiting factor on the capability of the land for agriculture, but within this general context it is the factors of slope steepness, soil materials, and drainage which, either singly or in combination,

TABLE 13.4
Land classification for agriculture, Britain

Class physical limitations	Texture: Good	Texture: Fairly good	Texture: Poor	Structure: Good	Structure: Fairly good	Structure: Poor	Rooting depth: Good	Rooting depth: Moderate	Rooting depth: Restricted	Rooting depth: Poor	Drainage: Good	Drainage: Moderate	Drainage: Imperfect	Drainage: Poor	Drainage: Very poor	Stoniness: None	Stoniness: Many	Slope: Flat/gentle	Slope: Moderate	Slope: Steep	Slope: Very steep	Erosion: None	Erosion: Slight	Erosion: Severe	Flooding: None	Flooding: Some	Flooding: Frequent	Climate: Favourable	Climate: Slightly unfavourable	Climate: Unfavourable	Climate: Moderately severe	Climate: Severe	Climate: Very severe
1 None or minor	x			x			x				x							x										x					
2 Some		x			x			x				x							x				x					x	x				
3 Moderately severe			x			x			x				x	x						x			x							x	x		
4 Severe (i)										x				x	x		x		x				x			x					x	x	
5 Severe (ii)														x	x					x				x			x				x	x	
6 Severe (iii)										x					x		x				x			x			x					x	
7 Extremely severe										x					x		x				x			x									x

N.B. In order to be assigned to the next poorest class an area must show one or more of the detrimental characteristics typical of that class.

Source: Compiled from information by Bibby and Mackney. (1969)

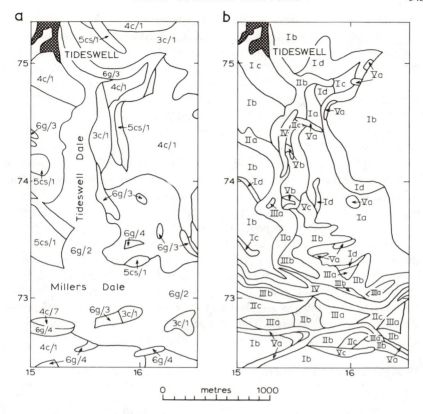

FIG. 13.3. A comparison of the land-use capability classification of a part of the Tideswell area, Derbyshire, England (map (a)), and a land-unit subdivision of the same area (map (b)) (Source for map (a): Johnson, 1971)

Land capability class			Land unit
3	c/1	with moderately severe climate	I Plateau top
4	c/1	with severe climate	(a) under 300 m—moderately severe climate
5	cs/1	with severe climate and shallow soils	(b) over 300 m—severe climate
			(c) aspect NW–N–NE—severe climate
6	g/2	with steep slopes	(d) locally steeper slopes
6	g/3	with shallow soils and limestone outcrops	II Valley side (a) Rock outcrops
			(b) Steep
6	g/4	with unfavourable soil pattern, high in trace elements	(c) Very steep
			III Valley bench (a) High on valley-side
			(b) Low on valley-side
			IV Flat valley floor
			V Human interference dominant
			(a) Hummocky ground (mineral rakes)
			(b) Quarry
			(c) Rough ground

provide the key to the class of land-use capability. Fieldwork was carried out, by means of soil-profile analysis and hand augering, in order to compile a soils map of the area. The land-use capability classification arises from combining the soils information with agronomic, climatic, and topographic information (Johnson, 1971). It is admitted that some of the boundaries were arrived at subjectively, and that some arbitrary decisions were made.

If the purpose in mapping this area had been solely to define land-use capability then many of the boundaries and class characteristics could have been defined through land-unit mapping. Figure 13.3b shows the land units of the same area as Figure 13.3a as derived from an air-photo interpretation with field checking. A comparison between these two maps shows a considerable degree of accordance in the boundary positions but also with much more detail on the land-unit map. A comparison of the keys to these figures indicates that many of the physical properties of the land-capability groups can be derived from a land-unit description produced as a basis for land-use capability classification. The significant advantage of the land-unit approach is that it involves less effort and a lower cost than standard field mapping techniques. In addition, many of the land-unit boundaries coincide with soil boundaries, since relief is an important factor in soil formation (and profile characteristics) in this area. Thus it also provides a valuable guide to soil boundary locations at the reconnaissance stage in mapping soils.

13.5. Advantages and Disadvantages of the Land-Systems Approach

(a) *Some criticisms*

Notwithstanding the apparent success of the land-systems approach as applied to agricultural land evaluation, it has been criticized (Thomas, 1969; Moss, 1969). The criticism is made on both conceptual and practical grounds. It is accepted that the approach provides much information about the land but it does little or nothing to measure or evaluate the important complex of functional relationships between soils, climate, plants, and animals. This emphasizes the static, cataloguing approach of the land-systems method, and reinforces the need for an alternative *dynamic* approach.

The accuracy of some land-system surveys has been questioned by Wright (1972a) on the grounds that little account is taken of local variations, and that classification may be based on imprecise criteria and strongly influenced by the more evident contrasts in air-photo patterns. In some cases Wright observed that this leads to final subdivisions into land units which differ both in kind and in their order of magnitude.

(b) *Alternative approaches*

(i) *The biocenological approach.* For agricultural purposes Moss (1969) proposed a *biocenological* approach, by which agricultural development is thought of in the context of the biological system within which it is to take place. This follows from the fact that the agricultural production operates according to the ecological laws in the chosen location. Moss argued that an approach based upon established ecological principle and theory, with reference to the contemporary interactions and influences between the different classes of feature to be found at the surface of the earth, offers a more valid, and therefore a more relevant approach to the problem of land assessment, particularly in areas such as tropical Africa. In addition, Moss suggested that the operation of the ecological system is frequently independent of much of the data collected during a land-systems survey (e.g. geology, slope), but more dependent upon present-day human land use, in the broadest sense of the term. It is not denied by Moss that plant and agricultural data can be grouped on a landform basis in any particular situation, as in the land-systems approach, but his point is that meaningful relationships, either functional or causal, cannot be universally demonstrated between the features thus classified and grouped. Since the biocenological approach focuses upon plant–soil relationships in relation to agricultural practices, a different balance of data, a distinct order of presentation, and a separate view of slope, soils, and climate are inevitable. In particular soil genesis is of little importance, and denudation chronology and stratigraphy are of almost no importance at all (Moss, 1969).

In the context of land development for agriculture the arguments put forwards by Moss are valid. In practice his approach may be difficult to apply in countries or areas where there is almost no organized background information about environmental and ecological conditions. Where this is the case a land-systems reconnaissance survey, concentrating on appropriate data collection, remains a valuable basis for later and more dynamically orientated work. Indeed the organizations which have been most involved in the development of land-systems surveys have moved in this direction (Bawden, personal communication, 1972). In an earlier study Moss developed the concept of a *habitat-site*. This he defined as:

a distinct area of the land-surface with a defined range of slope angle and a particular slope position, which possesses defined morphological limiting factors with respect to plant growth. Each groups of sites thus possesses similar limitations with respect to agricultural development, and is also a *recognisable* distinct unit of the land-surface (Moss, 1968, p. 124.)

This brings the basis for the mapping carried out by Moss very close to that of land-unit mapping, as in the land-systems technique. The starting-point for a biocenological survey (i.e. land use/vegetation patterns) nevertheless remains somewhat different, and the emphasis by Moss (1968, 1969) on a recognition of the interaction between ecological and socio-economic systems, of which the land-use pattern is an expression, is important.

(*ii*) *The site-analysis approach of Wright.* Instead of the process of subdividing an area beginning with the general (land systems) and progressing down to the particular (land units) Wright (1972a) advocated a method whereby details are defined for site units before these are grouped together to form the equivalent of land systems. In this context Wright's site-analysis procedure is based on slope-profiling along transects chosen so as to include all the variability within the area being investigated. This variability is observed by preliminary air-photo analysis and field reconnaissance. Along the profiles measured the principal survey stations are spaced 15–30 m apart, with intermediate measurements made to embrace any evident slope discontinuities or micro-relief. The profile is then classified into straight or curved sections identified between marked breaks or changes of slope. Units separated by a uniformly gradational change are defined as one site unless this change is marked by a relative discontinuity in the physical properties of the soil. In addition to the record of surface form, each profile carries details of the soil and vegetational or land-use characteristics within each measured section (Fig. 13.4). The principal terrain characteristics of landform, vegetation, and soils are confirmed laterally away from each cross-section for up to

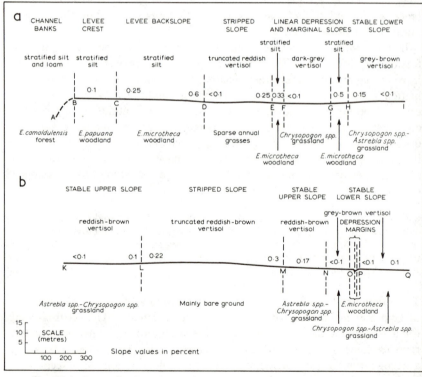

FIG. 13.4. An application of the site-analysis approach (after Wright, 1972a)

about 100 m. This information is compared with the air-photo image to enable lateral extrapolations to be made between the surveyed profiles.

On this basis distinct site-types can be identified, each comprising several individuals with closely similar landform, soil, and vegetation characteristics. Recurring members of a few site-types occurring in one area can be grouped, just as land units can be grouped into land systems.

This more precise approach to land classification has been applied in the Fitzroy floodplains and the Daly River basin, Australia; in the Nagarparkar peninsula, West Pakistan, and in Murcia province, south-east Spain. In each case it provided a useful framework for examining agricultural land use, especially in relationship to defined properties of relief amplitude, gradient, ground shape, and soil texture, depth, and debris cover (Wright, 1972a).

(c) *Land systems and data banks*

With the development of advanced filing methods, especially computerized data banks, land units have become recognized as potentially convenient categories for holding landscape information in store (Brink *et al.*, 1966, 1968). This principle is used in South Africa as the basis for a data-bank system, to which additional information is continually added and from which information can be gained when a second project is to cover the same ground as that already mapped for an earlier purpose (Brink and Partridge, 1967; NIRR, 1971).

(d) *Advantages of the land-systems approach*

Land-systems mapping results in a map that is easy to assimilate. It only carries boundaries and a brief reference code superimposed on a location map. The visual representation of each land system, by means of a block diagram, is both appealing and easy to understand. The table of information concerning each land unit can be problem-orientated, and only information directly relevant to the problem need be measured and recorded. The land-systems concept is applicable on a wide range of scales. If land-systems boundaries only are to be shown then small scales are feasible, such as 1:1 million, as used for the land-systems maps of the Alice Springs area (Perry, 1962) and Uganda (Ollier, *et al.*, 1969). Much of the mapping can be done from aerial photographs, and once completed the field work is limited to making sample investigations within examples of each type of land unit. This is, therefore, an economical and comparatively rapid method of acquiring information about the land resources of an area.

Whatever the criticisms raised against the land-systems approach it has one paramount advantage, that of simplicity, over any other method yet devised for relaying terrain information from the field scientist to the client environmental manager, in whatever capacity he may operate. The method of classification and the manner of data presentation can be readily understood by those who need the information for planning and decision-making purposes. Without this, no analysis, however correct, can be usefully applied to land mangement and the wise use of resources.

13.6. Case Study: Soil Conservation in Lesotho

The western lowland of Lesotho is composed predominantly of sedimentary rocks (grits, sandstones, and shale), or deep accumulations of the weathering products from these beds. Annual rainfall in these lowlands is 635–900 mm, and is seasonally distributed with high-intensity storms between October and April. The rest of the year tends to be dry. There is an almost complete absence of trees, probably as a result of overpopulation, overgrazing, and overburning.

During a survey of the whole of Lesotho, this lowland was subdivided into twelve land systems (Fig. 13.5) according to distinctive combinations of physiography, soils, and vegetation (Table 13.5) (Bawden and Caroll, 1968). Many of these could be seen to include extensive areas of gullying. Some were prone to extensive wind erosion during the dry season. For example, the Caledon lowlands, bordering the Caledon River and within which Maseru is situated, possess claypan soils with very erodible subsurface horizons that gully easily, especially since they tend to occupy the lower sites within the land system. The hilltop sites carry fersiallitic soils and are less susceptible to water erosion. However, they have a loose sandy topsoil which is easily removed by strong winds. Their siting on the higher parts of the land system is therefore a disadvantage.

In similar manner the report (Bawden and Carroll, 1968) identified the other land systems in which soil erosion is a problem. Its conclusion refers to the need for introducing stricter conservation measures in these areas, not least through education of the local inhabitants. It is clearly stated that there is a lack of understanding among the indigenous population of the necessity for devising and maintaining soil-conservation measures. Erosion is accelerated by ploughing alongside dongas (gullies) and by overstocking. The communications barrier between the land evaluator and the local farmer needs therefore to be broken for at least a part of this land-system survey to be effective. Much applied work faces the same, or a similar, problem. By means of the survey, however, at least the land planner knows which are the danger areas with respect to soil erosion, and if necessary it is for these areas that conservation legislation can be introduced.

13.7. Conclusion

Land-systems mapping is a land classification approach to the landforms and materials of an area. The subdivision into land units is primarily based on

The Lowland Region Land Systems of Lesotho

(See Fig. 13.5; the original land system numbering is retained)

Land system	Elevation m above sea level	Landform and geology	Soils	Vegetation
13 Lowlands escarpment	1500–1800	Steep escarpment of Red Beds, with Cave Sandstone cap. Also remnant plateaux of sedimentary rocks	Lithosols, Raw mineral	*Themeda-Cymbopogon-Eragrostis* grassland
14 Caledon lowlands	1500	Undulating to rolling dissected plains of Red Beds and Molteno rocks. Alluvial terraces along Caledon River	Claypan, Fersialitic Juvenile soils on Recent Alluvium	*Themeda-Cymbopogon-Eragrostis* in part transitional to Highland Sourveld
15 Phatiatsana lowlands	1500–1800	Broad floodplain surrounded by rolling dissected plains of sedimentary rocks	Vertisols, Claypan, Fersiallitic	*Themeda-Cymbopogon-Eragrostis* in part transitional to Highland Sourveld
16 Central lowlands	1500–1800	Rolling dissected plains of Red Bed and Molteno rocks	Claypan, Fersiallitic	*Themeda-Cymbopogon-Eragrostis*
17 Makhalong lowland	1500–1800	Undulating dissected plains of Red Beds and Molteno rocks	Claypan, Fersiallitic	*Themeda-Cymbopogon-Eragrostis*
18 Red Beds plains	1500–1800	Undulating plains of Red Beds rocks	Fersiallitic	*Themeda-Cymbopogon-Eragrostis*, transitional to Highland Sourveld
19 Molteno plains	1500	Gently undulating pediplain on Molteno rocks with occasional small dolerite hills	Claypan	*Themeda-Cymbopogon-Eragrostis* grassland
20 Dissected Molteno plains	1500	Dissected gently undulating pediplain on Molteno rocks	Claypan, Lithosols	*Themeda-Cymbopogon-Eragrostis* grassland
21 Little Caledon valley	1500–1800	Steep-sided valley with gently undulating pedimented floor	Vertisols, Lithosols, Claypan	*Themeda-Cymbopogon-Eragrostis* grassland
22 Southern Beaufort plains	1500	Gently undulating pediplain on Beaufort rocks. Alluvial terraces along Caledon River	Claypan, Vertisols, Juvenile soils on Recent Alluvium	*Themeda-Cymbopogon-Eragrostis* grassland
23 Northern Beaufort plains	1500	Level to gently undulating pediplain on Beaufort rocks	Vertisols, Claypan	*Themeda-Cymbopogon-Eragrostis* grassland
24 Dolerite hills and plains	1500–1800	Steep rounded dolerite hills and gently undulating pediments	Vertisols, Lithosols	*Themeda-dominated* grassland

Source: after Bawden and Carroll (1968) (with metric conversion).

surface form, and this may prove to be highly indicative of the nature of both the underlying geology and the mantle of soils. Problem-oriented classifications (e.g. soil types, land-use potential, engineering difficulties) can be based on a land-systems and land-unit appraisal. The purist may well question the accuracy obtained by this approach. The pragmatist will reply that it

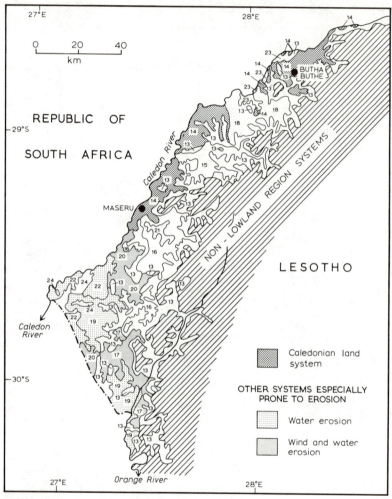

FIG. 13.5. The land systems of lowland Lesotho. Those especially prone to erosion are identified (Source: Bawden, M. G. and Carroll, D. M., 1968, *The Land Resources of Lesotho*, Land Resource Study No. 3, Directorate of Overseas Surveys (Tolworth, England)

has been shown to work, not least because it is an approach that allows data to be presented in a form that can be understood by policy-makers and environmental managers who are not specialists in geomorphology.

Proficiency in land-systems mapping does not come easily. It requires a full command of geomorphology, and familiarity with topics such as those introduced in the foregoing chapters. In addition experience in geology, soil science, hydrology, and civil engineering are added qualifications that might be expected of anyone attempting to apply land-systems mapping to a particular environmental management problem. Because such qualifications are seldom found in one person the approach has frequently been employed through a team of specialists in these fields working together. This has been the basis of much of the work carried out by the CSIRO in Australia.

The land-systems approach is based on the identification of regional boundaries that are frequently independent of watersheds (catchment divides). Each land system, however, may contain a number of drainage basins. It is interesting therefore to compare the ideas expressed in Chapter 1, about the drainage basin being a useful unit for environmental management, with the philosophy behind the land-systems approach. Which method is the most appropriate in evaluating a particular area or problem depends on the nature of the problem. The two approaches can, of course, be combined, with the land-systems analysis providing the general framework for systematic basin studies. In this way, for example, predictive equations could be developed for the relation between stream discharge and more easily measured morphometric characteristics of the drainage basins within each of the land systems (see Chap. 6). Separate equations might be found for each land system and then applied only in the case of a land system having characteristics similar to that for which the regression equation was first developed.

Land units are detailed components of land systems which have been found to be useful in evaluating land for both agricultural and engineering purposes. Yet another method of evaluating land at this more detailed level is that of geomorphological mapping. Indeed, it can be made to supply information at any level of detail demanded by the problem. Geomorphological mapping is the subject of the next chapter.

14

GEOMORPHOLOGICAL MAPPING

14.1. Introduction

In this consideration of geomorphology in environmental management stress has been laid upon several different physical systems, such as those of erosion by water (Chaps. 1, 2, 4, and 5), wind (Chap. 3), and weathering (Chap. 11). Materials, that is to say bedrock and regolith, have been included whenever relevant, such as in the analysis of coastal systems (Chap. 8) and landsliding (Chap. 6). They are of course fundamental in an appreciation of the material resources of an area (Chap. 10). Landform is important in many systems, for it reflects a balance between past processes and the physical properties of the materials of which that form is composed. Landform in part also influences future processes. For example, slope steepness may be critical in determining the potential stability of a particular slope (Chap. 6).

Some chapters have included a consideration of spatial components of landforms and geomorphological processes. The study of distributions is an important art in geomorphology. It leads to the recognition of properties that vary across an area, or are always found in association with each other. In addition, location and spatial patterns may be as important in a planning project as the height and width dimensions of a particular feature. In short, a valuable aid in many environmental management problems is a map of the relevant landforms, materials, and processes. To this end much work has been done in finding efficient methods of, and sets of symbols for use in, geomorphological mapping. A set of basic mapping symbols is provided in this chapter to which others may be added as the area or problem demands. They have been compiled with special reference to the I.T.C. system (Verstappen and Zuidam, 1968) and the I.G.U. legend (Demek, 1972), and the colours suggested are those of the I.T.C. system. No single perfect mapping technique has been devised, nor can it ever be, but several examples of past practices (e.g. Tricart, 1965) can guide in the design and adoption of methods best suited to the problem in hand. In this discussion reference will be made to earlier chapters, indicating how they may be used to isolate some of the more important characteristics of landform, materials, and processes worthy of particular attention in a geomorphological mapping approach to the land appraisal stages of environmental management.

Geomorphological mapping can be thought of in two quite distinct ways. It is a means by which the geomorphological characteristics of an area may be recorded in map form. Such a map may have little direct practical value in environmental management for it will contain much that is not relevant to a particular problem. However, a second approach is to convert the geomorphological map into a purpose-oriented map carrying only that information which is relevant for management needs. This may be done either by presenting only a portion of the original information or by using the geomorphological map as the base map to which is added supplementary information necessary for making it into such a map.

There are many examples of the practical applications of geomorphological mapping (Table 14.1). Some of these come from Europe, where the post-1950 extension of towns has led to the occupation of lands previously thought to be unsuitable for urban development. It was realized that before development plans could be made specific information was required concerning landform and materials to be found at these sites and the processes

TABLE 14.1

Applications of geomorphological mapping in planning and economic development

I. *Land Use:*

Territorial planning
Regional area planning
Conservation of the natural and cultural landscape

II. *Agriculture and Forestry*

Potential utilization
Soil conservation
Soil erosion control
Reclamation of destroyed or new areas
Soil reclamation
Drainage and irrigation

III. *Underground and Surface Civil Engineering*

Reconstruction and replanning of settlements, especially of towns
Designing of industrial buildings
Communications (roads, railways, canals, harbours)
Hydro-engineering:
  reservoirs and dams
  regulation of rivers
  natural and artificial waterways
  irrigation canals
  harbour construction
  shore protection
  fishing projects

IV. *Prospecting and Exploitation of Mineral Resources*

Prospecting
Geological survey
Exploitation
Mining
Potential and actual damage done by mining
Reclamation of abandoned open-cast mines
Landslip areas and regions of subsidence due to mining
Reclamation of areas destroyed by mining and waste dumps

Source: Demek (1972).

working upon them. It is in just such site investigations that geomorphological mapping comes into its own. Many of the relevant publications are in Hungarian or Polish; an extensive bibliography is provided in Doornkamp (1971).

Geomorphological mapping as it is known today was begun in Poland where it has been used since the early 1950s, and usually in the practical context of economic planning (e.g. Galon, 1962; Klimaszewski, 1956, 1961a, 1961b). Considerable use has been made of geomorphological mapping by the International Institute of Aerial Survey and Earth Sciences (I.T.C.) at Enschede, The Netherlands, especially in land-use planning in the Far East (e.g. Sumatra), for conservation purposes in Southern Italy (see Verstappen and Zuidam, 1968) and Northern Spain, and for use in irrigation or hydro-logical studies, as on the Indus Plain (Verstappen, 1970). Members of the Centre de Géographie Appliquée at the University of Strasbourg have also used geomorphological mapping as a basis for applied geomorphological studies. For example, they acquired data for the design of a land reclamation scheme on the Senegal River delta by this means (Tricart, 1959, 1961). In Venezuela a geomorphological map was used to define areas of limited value for agriculture, because of the dangers of erosion, thin or stony soils, excessive local relief, insufficient drainage, or flooding. These areas were excluded from later detailed soil mapping, thus saving the soil surveyors both time and money (Tricart, 1966). Likewise this centre at Strasbourg has been responsible for producing geomorphological maps which have been used as a basis for defining the desirable limits of urban development, for route planning (e.g. Route Nationale No. 2 through the Guil River valley in the Alps), and for assessing river dynamics in relation to flooding in the Cévennes Mountains. Geomorphological mapping as a resource assessment technique has also enjoyed the support of UNESCO (e.g. Fränzle, 1966).

14.2. The Compilation of a Geomorphological Map

A good example of a simple geomorphological map produced for planning purposes is that by Klimaszewski (1961b) reproduced as Figure 14.1. It includes information about the nature and location of features formed by denudation, fluvial action, fluvio-glacial, karst and aeolian processes, as well as those features created by man. Where relevant these are subdivided into destructional and constructional types, and their ages are shown. The resulting map is clear to read, and this is so because its compiler has been selective (e.g. representing relief by contours and not by steepness of slope). Arriving at such a map is a job requiring much patience and skill, and may be best achieved by carrying out one mapping stage at a time. For example a frequent distinction has been drawn in this book between relief form,

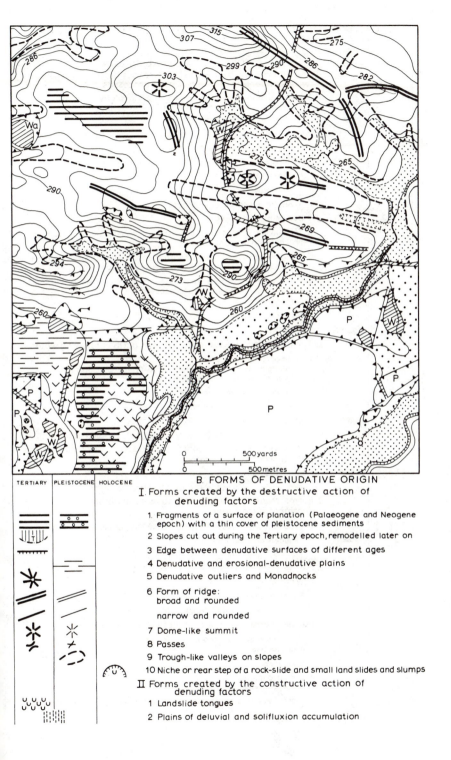

TERTIARY	PLEISTOCENE	HOLOCENE

B. FORMS OF DENUDATIVE ORIGIN

I. Forms created by the destructive action of denuding factors

1. Fragments of a surface of planation (Palaeogene and Neogene epoch) with a thin cover of pleistocene sediments

2. Slopes cut out during the Tertiary epoch, remodelled later on

3. Edge between denudative surfaces of different ages

4. Denudative and erosional-denudative plains

5. Denudative outliers and Monadnocks

6. Form of ridge:
broad and rounded

narrow and rounded

7. Dome-like summit

8. Passes

9. Trough-like valleys on slopes

10 Niche or rear step of a rock-slide and small land slides and slumps

II. Forms created by the constructive action of denuding factors

1. Landslide tongues

2. Plains of deluvial and solifluxion accumulation

PLEISTOCENE	HOLOCENE

C. FORMS OF FLUVIAL ORIGIN

I. Forms created by the destructive action of flowing water with cooperation of denudative processes

1. Edges and erosional undercuts of accumulative terraces and erosional-denudative plains with relative height

0 - 3m well preserved

badly "

3 - 6m well "

badly "

6 -12m well "

badly "

2 River beds eroded in alluvium

3 Bed of a blind creek

4 Trough-like valleys created with the cooperation of denudative processes mainly solifluxion

5 Trough-like valleys created with the cooperation of denudative processes posessing an accumulative bottom

6 Small valleys created by periodic or perennial water with cooperation of denudative processes
ravines

troughs

gorges

II Forms created by the accumulative action of flowing water

1 River accumulation plain
a. belonging to the end of the Middle-Polish Glaciation (Warta Stage)

b. belonging to the Glaciation

c higher one and lower one

2 Alluvial fan plain

D. FORMS OF FLUVIO-GLACIAL ORIGIN

I Forms created by the constructive action of glacial waters (Middle-Polish Glaciation)

1 Kames, sandr

E. FORMS OF KARST ORIGIN

I Forms created by the constructive action of the continental ice sheet

1 Ground moraine plain (Middle-Polish Glaciation)

2 Denuded and moraine walls

J FORMS OF AEOLIAN ORIGIN

I Forms created by the constructive action of wind

1 Wall-dunes

2 Fields of small dunes

I   N ANTHROPOGENIC FORMS

Forms created by the destructive action of man:

1 Quarries, clay-pits, sand-pits
a productive

b disused

2 Old strip mines
a fields of small shafts

b bore-pits (old silver, lead and zinc mines) and old lime pits

c coal open-pits

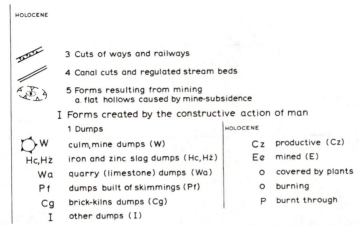

HOLOCENE

3 Cuts of ways and railways

4 Canal cuts and regulated stream beds

5 Forms resulting from mining
   a. flat hollows caused by mine-subsidence

I Forms created by the constructive action of man

1 Dumps

		HOLOCENE	
W	culm, mine dumps (W)	C z	productive (Cz)
Hc,Hż	iron and zinc slag dumps (Hc,Hż)	E ℓ	mined (E)
Wa	quarry (limestone) dumps (Wa)	o	covered by plants
Pf	dumps built of skimmings (Pf)	o	burning
Cg	brick-kilns dumps (Cg)	P	burnt through
I	other dumps (I)		

Fig. 14.1. Geomorphological map of a part of the Upper Silesian industrial district (from Klimaszewski, 1961b)

materials, and process. Each of these categories can be mapped one at a time, and in that order.

To illustrate the methods outlined below reference will be made throughout this chapter to maps compiled for an area near Johannesburg, just north of Krugersdorp, in South Africa.

(a) Mapping surface form (morphological mapping)

The classical method of landform mapping is through surveyed contours. Geomorphologists have, however, devised a technique which defines the shape of the land surface in greater detail than is normally to be found on contour maps, and in a manner more relevant to their development into geomorphological maps (Savigear, 1965). Figure 14.2 illustrates the use of this technique on the area near Johannesburg which is being taken to illustrate the mapping methods. The basis of this mapping lies in the recognition of breaks and changes of slope as they occur on the ground (Fig. 14.3). Breaks of slope, where there is a sharp junction between two portions of a hillside having differing gradients, are shown by continuous lines and changes of slope by broken lines. In each case a V-symbol is added to the line, pointing downhill, and lying on the side of the steeper slope. Thus convex and concave boundaries are readily distinguished from each other. Between these mapped slope boundaries measurements can be made of slope steepness and, if present, slope curvature. Steepness is shown by an arrow lying normal to the slope, pointing downhill, and carrying the angle of slope, in degrees. Down-slope curvature is shown by placing X on the arrow stem for a convex slope and — for a concave slope. In these cases the upper- and lower-slope angles also need to be recorded. Special symbols exist for very steep slopes. For example, a steep cliff, or free face of bedrock, is defined by a solid black symbol along a

MORPHOLOGY

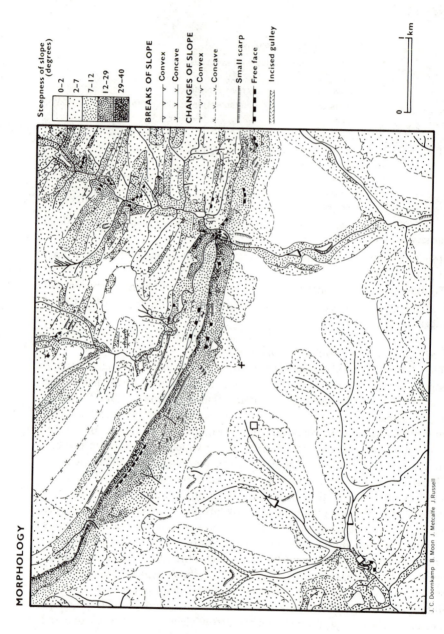

Steepness of slope
(degrees)

0-2
2-7
7-12
12-29
29-40

BREAKS OF SLOPE
Convex
Concave

CHANGES OF SLOPE
Convex
Concave

Small scarp
Free face
Incised gulley

0 ___ km

J. C. Doornkamp B. Moon J. Metcalfe J. Russell

FIG. 14.2. A morphological map of an area west of Johannesburg, South Africa (mapping carried out with J. Metcalfe, B. Moon, and R. Russell)

line marking its crest; and a short, but locally steep slope is identified by a pecked line (Fig. 14.3).

Differences in slope steepness can be emphasized by a system of selected shading or colours, according to defined slope classes. The I.G.U. Manual of Detailed Geomorphological Mapping (Demek, 1972) suggested six slope categories of:

$$0-2°, 2-5°, 5-15°, 15-35°, 35-55°, 55°+,$$

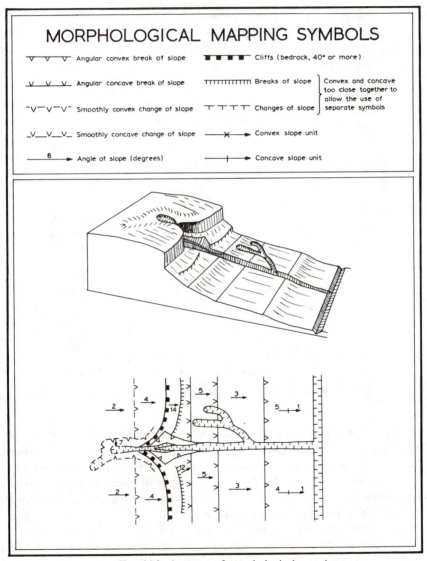

FIG. 14.3. A system of morphological mapping

though in some localities it may be appropriate to adopt a different set of categories according to local relief conditions. On the morphological map example (Fig. 14.2) slope classes have been chosen so as to bring out, through the shading used, the main morphological boundaries.

Slope steepness has been variously measured in degrees, percentage, and gradient ratios. The equivalents between these are defined by the graph in Figure 14.4.

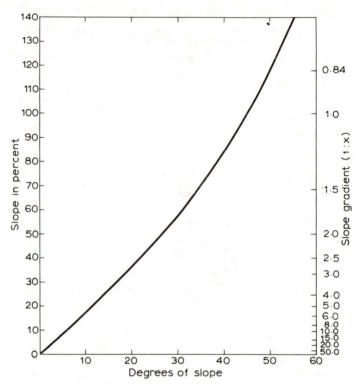

FIG. 14.4. A graph for the conversion of slope steepness from degrees to per cent or a gradient ratio

Information about slope steepness alone is of considerable importance in land management, for it can provide the critical restricting factor for some uses. Limiting gradients are especially pertinent in the case of transport and of agriculture. The limit may be set by the physical ability of equipment to cope with a slope, or it may be set by the economics of the situation whereby, for example, if roads are built without excavation on slopes above a certain gradient then heavy vehicles will lose speed, and therefore time and money. Some critical slopes are given in Table 14.2, which is based on an unpublished review of the available data by Crofts (1973). This led Crofts to suggest that suitable slope class boundaries, for practical purposes, are 2·5, 5, 10, and 20 per cent (or 1·4, 2·8, 5·7, and 11·3 degrees). Thus, if a morphological map is shaded (or better still coloured) using these class intervals it can be used directly to ascertain some critical practical limits to its direct use by man.

TABLE 14.2

Critical slope steepness for specified activities

Steepness per cent	Critical for
1	International airport runways
2	Main-line passenger and freight rail transport
	Maximum for loaded commercial vehicles without speed reduction
	Local aerodrome runways
	Free ploughing and cultivation
	Below 2 per cent—flooding and drainage problems in site development
4	Major roads
5	Agricultural machinery for weeding, seeding
	Soil erosion begins to become a problem
	Land development (constructional) difficult above 5 per cent
8	Housing, roads
	Excessive slope for general development
	Intensive camp and picnic areas
9	Absolute maximum for railways
10	Heavy agricultural machinery
	Large-scale industrial site development
15	Site development
	Standard wheeled tractor
20	Two-way ploughing
	Combine harvesting
	Housing-site development
25	Crop rotations
	Loading trailers
	Recreational paths and trails

Source: Crofts (1973).

Height differences can be included on a morphological map if it is drawn on a feint contour base. For an accurate representation of surface form the scale of mapping should be determined by the nature of the topography. In the more complicated areas, such as those which are intensely dissected, mapping might need to be at scales of 1:10,000. In areas of moderate dissection smaller scales may be sufficient to record the existing slopes. At scales less than 1:75,000, however, it is rarely possible to record the meaningful slope discontinuities.

The mapping scale chosen also depends on the thickness of the lines drawn to represent the features. The breadth of the thinnest line that can be drawn on a map in the field is about 0·5 mm, if only pecks are added the space required becomes 1·5 mm, and if an arrow is included with the appropriate curvature sign (cross or negative sign) then at least 3·00 mm, and frequently twice this space may be required. These considerations alone may determine the mapping scale. For example, at a scale of 1:10,000 1·5 mm on the map is 15 m on the ground, so no feature less than 15 m across can be individually and accurately represented on the map by the smallest symbol allowed, namely a line with pecks on it. In practice it is not possible to consider continually the effects of pencil thickness, and an appropriate mapping scale is chosen based on experience, and will be the scale that allows the representation of all features relevant to the problem in hand.

In any field mapping technique there may be an element of subjectivity about the location of boundaries, as in the case of locating breaks and changes of slope on a morphological map. In general this subjectivity decreases as slope discontinuities become sharper and as locational reference points, such as field boundaries, increase in density. Where cultural reference features are either absent on the ground or not shown on the base map, then airphotos may have to be used in field plotting.

The mapping of surface form can be taken as the first step in geomorphological mapping. The next is to make interpretations about the forms and to ascribe a genesis to them. This has to be done with full regard to both the bedrock and superficial materials of each feature (Sect. 14.2(b)) and to the past and present processes (Sect. 14.2(c)) operating in the area. Nevertheless, form itself can play an effective part in many of the geomorphological systems described in earlier chapters. For example, both the overland flow model of Horton and the saturated throughflow models (Chap. 2) require knowledge concerning slope steepness and slope length (Eqs. 2.13 and 2.17) as do the soil loss by running water formulae given as Equations 2.20 and 2.21. Soil erosion-control practices also require information about slope steepness in order to determine the efficient widths of strips under a strip-cropping system (Chap. 2). The identification of floodplains and alluvial fans is very largely based on land form (Chap. 5.2); while the accurate delimitation of a catchment and the identification of its valley network (Chap. 1) are best based on detailed morphological mapping. Slope form is important in recognizing areas of potential, present, and past landslide activity (Table 6.5, Figs. 6.7 and 6.8). In coastal areas form has been described both as a cause and as an effect in the marine systems (Chap. 8). Morphological mapping of coastal areas can define the different past influences of marine activity, and indeed it can betray both the materials and the process that have operated along a particular stretch of coastline (Sect. 8.4). Since the forms of coastal areas are subject to rapid changes (Sect. 8.2(a)) morphological mapping can be used to record the sequence of different forms. Morphological mapping can also aid in the mapping of both patterned ground under permafrost or periglacial conditions (Chap. 9), and in the identification of the extent of ground subsidence (Chap. 7). The material resources of sand and gravel

(Sect. 10.2) in glaciated areas, or along river valleys, depend upon the identification of such things as fluvio-glacial features and river terraces. This identification is frequently based on their form characteristics, while paying due regard to their material composition. The drainage characteristics of soils are an important physical attribute, and these may be dependent upon the form of the site. Soil site is best described in the spatial context by morphological mapping methods (Curtis, Doornkamp, and Gregory, 1965). Morphological mapping can also be used in the assessment of details within land units, and may help in the identification of variants (Chap. 13).

(b) Mapping materials

For any valid interpretation of the origin and development of the morphology it is necessary to know the extent and nature of the various geological materials. These fall into two basic groups, solid rock and superficial deposits (including both weathered bedrock and transported sediments). These deposits are all *soils* to the engineer, though the pedologist and agriculturalist has a more restricted use for this word, limiting it to the media within which plants grow.

The distinction between bedrock and regolith (i.e. the products of weathering and denudation) is a very important one (see Chap. 6). The boundary between the two, (the *weathering front*, Chap. 11) marks the junction of materials with normally quite different physical properties and different influences on the retention, passage, and chemical composition of water.

Mapping lithology and structure in the field is described in standard text books in geology, such as Gilluly, Waters, and Woodford (1968), and Longwell, Flint, and Sanders (1969). Mapping geology from aerial photographs is explained in Allum (1966), V. Miller (1961), Lueder (1959), and American Society of Photogrammetry (1961).

There are many distinctive properties of both bedrock and regolith that can be mapped in the field and recorded on a geomorphological map. Their relevance generally depends on the purpose of the map.

A distinction can be drawn between those characteristics which are required to be known for most purposes (e.g. bedrock lithology) and those material properties which may only be required for special purposes (e.g. slope stability analysis requires a knowledge of shear strengths, but this is not required for a problem relating to wind erosion). The former would appear on a general geomorphological map while the technical information would normally be shown on a separate special-purpose, or problem-oriented map.

Bedrock lithology is recorded in terms such as those used in Table 11.2 (i.e. granite, limestone, sandstone, and slate), but the list can be increased to include other frequently encountered rock types. These are given in Figure 14.5 together with the symbols by which they are normally recorded. Also

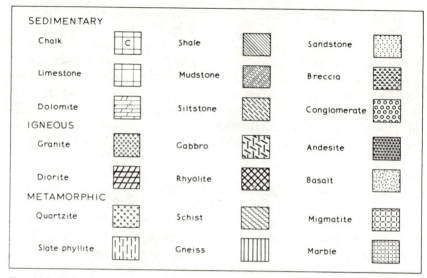

FIG. 14.5. Bedrock lithology: geomorphological mapping symbols (usually drawn in black or grey. In this case the shading patterns have been chosen from commercially available instant lettering and shading systems, e.g. the Letraset system)

the general-purpose geomorphological map would show bedrock structures, such as fold axes and fault lines (Fig. 14.6), as well as the features which are the direct result of structure (Fig. 14.7). In some areas there exist features of volcanic origin (Fig. 14.8).

It is not possible to identify every possible special-purpose application of geomorphological mapping for which separate maps of selected properties of materials would be drawn. In the context of the discussion on weathering (Chap. 11), it is clearly useful to portray not only the spatial distribution of the grades of rock decomposition as defined in Table 11.6 but also the types of superficial materials (Fig. 14.9). This has special application in many engineering problems (e.g. search for local construction materials). A more

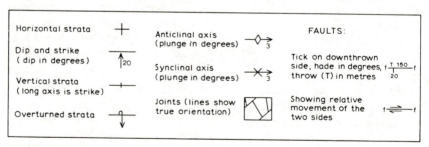

FIG. 14.6. Geological structure: geomorphological mapping symbols (usually drawn in purple)

detailed examination of the bedrock would include its subdivision on the basis of its mineral composition (Tables 11.3 and 11.4).

The depth of the weathered mantle is of relevance in many specialized studies, but the data are not easily obtained except from isolated exposures or borehole records. Porosity and permeability are physical properties of material that have a strong influence on water penetration (infiltration) and

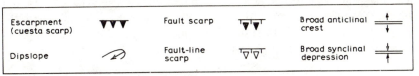

FIG. 14.7. Features resulting from bedrock structure: geomorphological mapping symbols (usually drawn in purple)

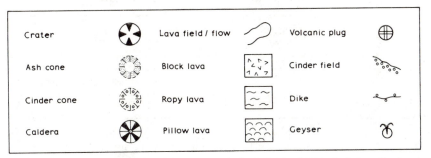

FIG. 14.8. Features of volcanic origin: geomorphological mapping symbols (usually drawn in red)

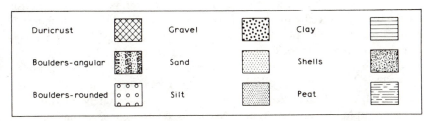

FIG. 14.9. Superficial unconsolidated materials: geomorphological mapping symbols (usually drawn in black or grey. The shading patterns have been chosen from commercially available instant lettering and shading systems, e.g. the Letraset system)

thus on weathering. This type of measure, as with soil texture, shear strengths, liquid and plastic limits (see Chap. 6), and indeed any geotechnical data concerning the mechanical properties of materials belong to very specific problems and require mapping at a detailed scale. This type of mapping takes place at the site investigation stage of particular projects, such as along proposed routes or at a foundation site. The relevant characteristics in an engineering geology sense are incorporated in a special issue of the *Quarterly*

Journal of Engineering Geology (Geological Society Engineering Group Working Party, 1972) describing a proposed legend for engineering geology maps.

In the study of the area near Johannesburg the general geomorphological map (Fig. 14.10) includes a definition of the main lithological outcrops and bedrock structures, as well as the location of the larger occurrences of superficial materials (e.g. accumulation terrace and talus slope). The soils are not shown, though steep bedrock slopes are. A comparison of this map with the morphological map (Fig. 14.2) shows that many of the breaks and changes of slope on the one coincide with feature boundaries on the other (e.g. the lower

TABLE 14.3

Field engineering estimates of plasticity index

Term	Plasticity Index	Dry strength	Field test
Non-plastic	0–3	Very low	Falls apart easily
Slightly plastic	4–15	Slight	Easily crushed with fingers
Medium plastic	15–30	Medium	Difficult to crush
Highly plastic	31+	High	Impossible to crush with fingers

boundary of the planation surface is marked by a convex change, or occasionally a convex break of slope). Indeed, the general geomorphological map can be thought of as an interpretative statement of the morphological map. In a similar way the general geomorphological map can be taken as the basis for compiling a specific map relating to the physical properties of materials in the area. For example, Figure 14.11 is a geotechnical (soil engineering) map based on the physical properties of the exposed bedrock and of the upper 0·5 m of soil. The soils were examined on a stratified random sampling basis. That is to say, the primary sampling was done on a random basis *within* each of the main geomorphological features separately identified on the geomorphological map. This was based on the concept that physical properties would display more variation between these landforms than within them (see also the discussion on land units in Chap. 13). To illustrate the results two measures are recorded in Figure 14.11, namely unconfined compressive strength (as based on the field test defined in Table 6.3 and Fig. 6.4) and a plasticity index (Table 14.3).

In only a few instances were there variations in these properties sufficient to require class boundaries within the limits of a specific landform. In addition, however, this type of special-purpose map has to be more explicit about rock type (including hardness, see Table 6.3) than the general geomorphological map, and all existing outcrops need to be shown.

GEOMORPHOLOGY

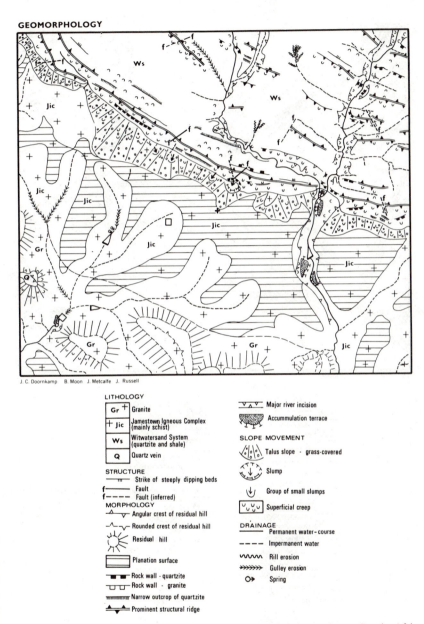

J. C. Doornkamp   B. Moon   J. Metcalfe   J. Russell

LITHOLOGY

Gr +	Granite
+ **Jic**	Jamestown Igneous Complex (mainly schist)
Ws	Witwatersand System (quartzite and shale)
Q	Quartz vein

STRUCTURE

Strike of steeply dipping beds
f——— Fault
f----- Fault (inferred)

MORPHOLOGY

Angular crest of residual hill
Rounded crest of residual hill
Residual hill
Planation surface
Rock wall - quartzite
Rock wall - granite
Narrow outcrop of quartzite
Prominent structural ridge

Major river incision
Accummulation terrace

SLOPE MOVEMENT

Talus slope - grass-covered
Slump
Group of small slumps
Superficial creep

DRAINAGE

——— Permanent water-course
----- Impermanent water
Rill erosion
Gulley erosion
O→ Spring

FIG. 14.10. Geomorphological map of an area west of Johannesburg, South Africa (mapping carried out with J. Metcalfe, B. Moon, and R. Russell)

GEOTECHNICAL

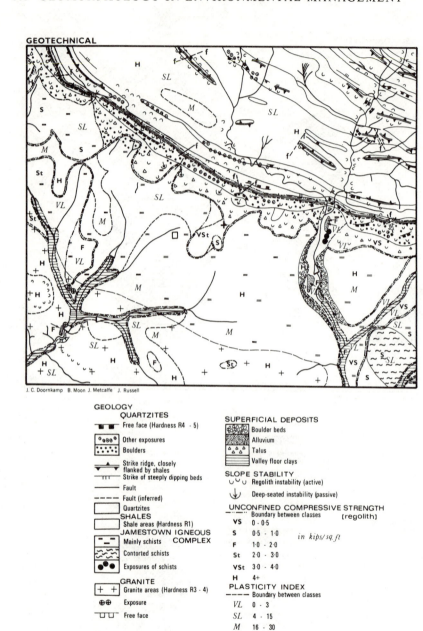

J. C. Doornkamp   B. Moon   J. Metcalfe   J. Russell

GEOLOGY

QUARTZITES
- Free face (Hardness R4 - 5)
- Other exposures
- Boulders
- Strike ridge, closely flanked by shales
- Strike of steeply dipping beds
- Fault
- Fault (inferred)
- Quartzites

SHALES
- Shale areas (Hardness R1)

JAMESTOWN IGNEOUS COMPLEX
- Mainly schists
- Contorted schists
- Exposures of schists

GRANITE
- Granite areas (Hardness R3 - 4)
- Exposure
- Free face

SUPERFICIAL DEPOSITS
- Boulder beds
- Alluvium
- Talus
- Valley floor clays

SLOPE STABILITY
- Regolith instability (active)
- Deep-seated instability (passive)

UNCONFINED COMPRESSIVE STRENGTH
- Boundary between classes   [regolith]

VS	0 - 0·5	
S	0·5 - 1·0	*in kips/sq.ft*
F	1·0 - 2·0	
St	2·0 - 3·0	
VSt	3·0 - 4·0	
H	4+	

PLASTICITY INDEX
- Boundary between classes

VL	0 - 3
SL	4 - 15
M	16 - 30

FIG. 14.11. Geotechnical (soil engineering) map of the area shown in Figure 14.10 (mapping carried out with J. Metcalfe, B. Moon, and R. Russell)

The general geomorphological map can similarly be used as a basis for compiling a map of soil characteristics of importance in agriculture (Chap. 10) (e.g. soil chemical properties, texture, soil structure, pH values). Likewise it can be used as a basis for mapping the material properties important in an analysis of potential wind erosion. These properties were defined in Chapter 3 (Sect. 3.2a and Fig. 3.6) as those of soil texture, density, its cohesive properties, abradibility, transportability, and organic matter content. The mapping of material properties in problems of soil erosion by water include those of soil infiltration capacity, permeability, thickness, and erodibility (e.g. percentage of water-stable aggregates >3 mm in diameter) (Chap. 2). Although the geotechnical map (Fig. 14.11) includes information about slope stability conditions, a project specifically concerned with the dangers of landsliding would have to include the mapping of other things such as those factors leading to a low shear strength (Table 6.2), and the material characteristics associated with large-scale instability as listed on the checklist of Table 6.5. Detailed mapping of the landslides themselves (Fig. 14.12) would require an assessment of the spatial variations in the material properties of both the soils (or regolith) and bedrock as identified in Chapter 6.

In Chapter 8 the interdependence of form, materials, and process in coastal environments was established, and shows how critical it is to know quite specifically the textural characteristics of beach materials and their distribution along a coastline. Such information is relevant not only to the geomorphological interpretation of coastline evolution, but also to the management of coastal material resources. General classifications of materials on the coast, such as rock (mapped by lithology and structure as in Figs. 14.5–14.9), boulders, shingle, sand, silt, or mud (Table 8.1) may be enough

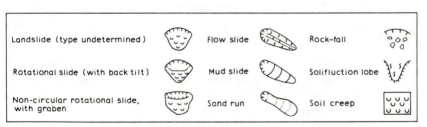

FIG. 14.12. Slope instability features: geomorphological mapping symbols (usually drawn in brown)

for many mapping purposes, but the mapped area needs to embrace all elements of the coastal system and thus should include such wind-blown features as dunes (Figs. 14.13) where they are present. Some mapping symbols are suggested in Figure 14.14 for the marine features of the coastal system.

In permafrost problem areas, Chapter 9 indicated that once again the texture and mineral composition of surface materials, together with the amount of compaction of the particles, are important physical properties to which such conditions as the material's thermal conductivity, specific heat, volumetric heat capacity, and thermal diffusivity are related. The latter, as well as water content, determines, the response to permafrost conditions. It

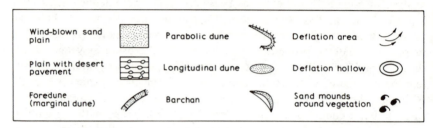

FIG. 14.13. Aeolian features: geomorphological mapping symbols (usually drawn in yellow)

can be argued, however, that it is easier to map the *results* of permafrost (i.e. the surface features), and from these to derive an assessment of the likely material properties. Indeed, it may be that by mapping the features first (Fig. 14.15) the most efficient materials sampling plan can be derived (Sect. 9.5(a) and Table 9.1). This principle is the same as that proposed in the

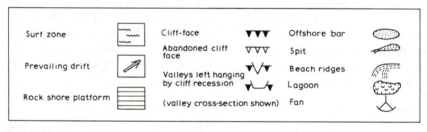

FIG. 14.14. Coastal features: geomorphological mapping symbols (usually drawn in green)

South African example where the geomorphological map (Fig. 14.10) is used to construct a stratified random sampling scheme upon which the geotechnical map (Fig. 14.11) is based.

Most subsidence problems stem from underground activities (Chap. 7) but a few, such as those due to hydrocompaction and the drainage of organic materials, are related to surface conditions, not least those of material properties. These are defined in Chapter 7, and if mapped the delimitation of their extent would be a valuable piece of information in the management and land-use planning of such areas.

The wise management of mineral resources (Chap. 10) depends on a full knowledge of their quality and extent, whether they be sands and gravels, mineral deposits, or agricultural soils. Many of the features that would normally appear on a general geomorphological map (e.g. river terraces, fluvio-glacial depositional features, marine sand and gravel forms, and screes) are well-known sources of sands and gravels. The map immediately isolates the sites worthy of field economic assessment, and defines their location and extent. As far as the search for mineral deposits is concerned this would initially be based on an examination of the appropriate relief features (Table 10.2), as defined on a general geomorphological map which could then provide the basis for a special-purpose materials map. The mapping of soils is a specialized procedure. It is clear, however, from an examination of many soil maps that soil boundaries are frequently coincident with

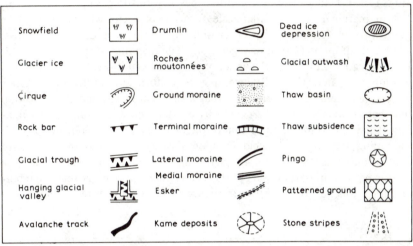

Snowfield		Drumlin		Dead ice depression	
Glacier ice		Roches moutonnées		Glacial outwash	
Cirque		Ground moraine		Thaw basin	
Rock bar		Terminal moraine		Thaw subsidence	
Glacial trough		Lateral moraine		Pingo	
Hanging glacial valley		Medial moraine Esker		Patterned ground	
Avalanche track		Kame deposits		Stone stripes	

FIG. 14.15. Forms of permafrost areas, glacial, and periglacial features: geomorphological mapping symbols (usually drawn in light blue)

either the boundaries of parent materials or of surface features, or both. The argument therefore follows that the general geomorphological map which carries information about both of these can be used as a basis for compiling a soils map. The geomorphological map can also be useful in predicting the physical properties of the soils themselves (Sect. 10.4(c)).

(c) Mapping processes

In practice geomorphological maps rarely show the distribution of processes, only the distribution of landforms resulting from defined processes. Thus on the South African example (Fig. 14.10), the general geomorphological map shows sites of slumping and groups of slumps, talus slopes, and accumulation terraces. The causative process is not identified, though gravity operating through a primary set of causes of instability may be predicted as being responsible for the slumps and the talus slopes, while fluvial deposition is likely to have created accumulation terraces close to the river within its incised section.

The identification of a feature's origin is usually based upon knowledge of its form, materials, and the past history of the area. Nevertheless, to ascribe an origin is to make an interpretation. Any interpretation is based on the ability, knowledge, and experience of the interpreter, it may therefore be open to debate.

In terms of environmental management an interpretation of the origin and processes affecting the features of the ground surface is critical to effective land planning. To return once again to the subject matter of the earlier chapters, the wind erosion system (Chap. 3) gives rise to specific landforms (Fig. 14.13). Under extreme conditions desert forms (e.g. sand dunes) occur, but in marginal areas smaller-scale forms may occur which indicate the general potential danger of wind erosion in the area (e.g. deflation hollows, sand and silt mounds around vegetation). The mapping of these (Fig. 14.13) may indicate the extent of the area potentially susceptible to this danger. Soil erosion by water (Chap. 2) gives rise to distinct surface features. Active gully systems may be the first sign that a problem exists, and their location may betray whether they are caused by overland flow or saturated through-flow (Chap. 2). Mapping the extent of the problem is a pre-requisite for both estimating the cost and deciding upon the necessary and most effective control measures (Sect. 2.6). In fact, mapping the fluvial features of any landscape is an important part of its effective management. A set of symbols useful for this purpose is shown in Figure 14.16. Such features include those

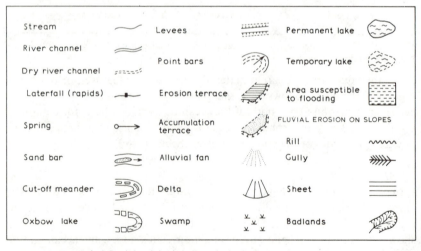

Fig. 14.16. Forms of fluvial origin: geomorphological mapping symbols (usually drawn in dark blue)

relating to river-channel morphology (Chap. 4), areas susceptible to flooding (Chap. 5), and the various other fluvial components of a drainage basin. Special-purpose hydrogeological maps can be compiled from a general geomorphological map whenever there is a special interest in, or problem relating to, this aspect of an area's resources. In Figure 14.17 the South African example is used to show how this may be done. The mapping legend in this case is based on that published by UNESCO (1970).

The background information necessary for a hydrogeological map is that of bedrock and superficial materials as well as of river characteristics, all of which are identified on the geomorphological map (Fig. 14.10). The nature of the materials determines their groundwater storage characteristics. Enquiries from local landowners indicated that the yield of wells in this area is closely related to bedrock, being much greater on the schists than on either the granites or the quartzites and shales (Fig. 14.17). The definition of watersheds (catchment boundaries) is relevant to any projects planned within the separate basins. Variations in soil drainage properties (i.e. as defined through an examination of soil texture) are closely coincident with the landforms identified on the geomorphological map. For the most part little additional fieldwork is necessary for the conversion of a geomorphological map into one showing hydrogeology. The legend for hydrogeological maps published by the UNESCO (1970) also suggests that information on groundwater hydrology, hydrochemistry, boreholes, and wells should be included where these are known in sufficient detail.

As was pointed out in Chapter 6 no one condition can be responsible for slope instability. The results of instability can be mapped (Fig. 14.12), but a prediction of those areas liable to future instability has to be based on an assessment of the spatial juxtaposition of a whole set of conditioning characteristics (Table 6.5). Prediction can be based, however, on an efficiently compiled geomorphological map.

Coastal features (Fig. 14.14) tend to form a special group in that they are generally easy to identify, except where they have been abandoned by a relative lowering of sea-level and have become degraded under subaerial processes. Thus raised beaches may become less easy to identify the older they are. Coastal processes are not always easy to map, however, though tidal ranges, the dominant wave directions and conditions of longshore drift should be mapped whenever possible.

Permafrost and periglacial features (Fig. 14.15) may be readily identified where there is little vegetation. Chapter 9 showed that such features are indicative of causative processes, and include ground hummocks formed by ground heaving, pingos which are associated with intrusive ice, polygonal patterns related to ice wedges, and flow forms due to solifluction. Such features can occur as fossil forms in temperate areas, and the importance of some of these to slope stability analysis was described in Chapter 6.

Glacial landscapes are subject to much landsliding, and also provide the sites for some sand and gravel deposits. They are also being increasingly used as tourist areas and for the development of recreational resources. In

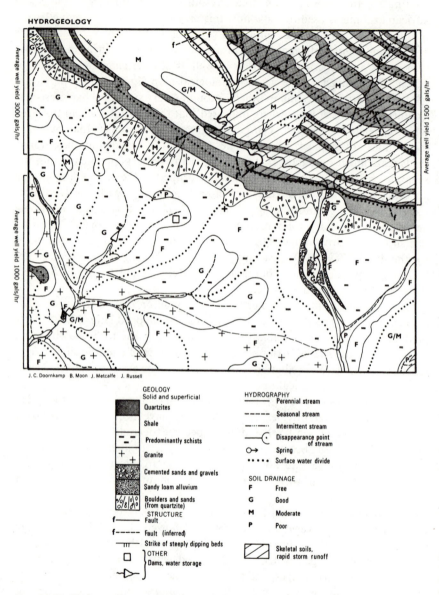

FIG. 14.17. Hydrogeology map of the area shown in Figure 14.10 (mapping carried out with J. Metcalfe, B. Moon, and R. Russell)

addition, they are frequently the setting for major civil engineering projects, such as dam construction. For all of these things advance knowledge of the terrain's attributes is important to its efficient and safe use.

To complete the basic set of legends which can be used in geomorphological mapping Figure 14.18 illustrates the symbols that may be employed

Conical karst	∩∩	Limestone pavement	Ⅲ Ⅲ Ⅲ	Swallow hole	⇐
Tower karst	ЛЛ	Clints and grikes	⋈	Cave	∩
Labyrinth karst	∧∧	Doline	⬭	Gorge	⫯⫯⫯

FIG. 14.18. Karst landscape features: geomorphological mapping symbols (usually drawn in orange)

for karst areas, while Figure 14.19 includes some useful symbols for major denudational forms. There may also be instances in which man-made features need to be recorded (Fig. 14.20). A more comprehensive set of symbols is given in Demek (1972), but this results in such a large number of symbols that they tend to overwhelm the would-be user. Other mapping systems exist, and a brief comparison between them brings out a number of critical points concerning geomorphological mapping.

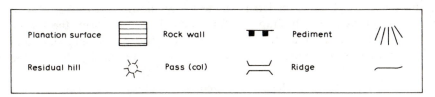

| Planation surface | ▤ | Rock wall | ▬▬ | Pediment | //\|\\ |
| Residual hill | ⋇ | Pass (col) | ⏝ | Ridge | ⌒ |

FIG. 14.19. Major features not included in previous figures: geomorphological mapping symbols (usually drawn in brown)

14.3. Different Mapping Systems

Comparability between different mapping systems has been difficult to achieve. Maps which have already been compiled show differences, according to Gellert and Scholz (1964), arising from:

(i) differences in the geomorphological characteristics of the areas mapped;
(ii) the nature of the development of geomorphology as a science in the country concerned;
(iii) the particular interests of the author of the map.

Gilewska (1967) applied four mapping systems, namely the French, Hungarian, Russian, and Polish systems, to the same area in Siberia in order to test their differences, and she demonstrates that the *French* method emphasizes the structure of the bedrock, which together with relief contours, forms the background for selected hydrographic and landform data. Although their age and origin are shown, there is little information about the steepness of landforms. Slope processes, past and present, are included.

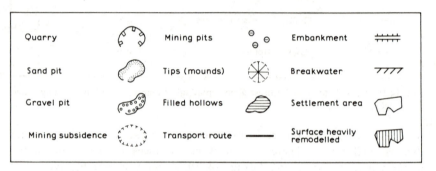

FIG. 14.20. Man-made features: geomorphological mapping symbols (usually drawn in black or grey)

The *Hungarian* system uses fewer contours. The structure of strong beds and the areas of unconsolidated sediments are shown as a background to morphogenetic and hydrographic information. This also reflects the French and Soviet concepts of insisting on the representation of changing slope and river processes, a feature which makes it difficult to classify the existing landforms.

The map produced by the *Soviet* system uses contours as a background for information concerning tectonics and lithology (including that of the waste cover and sediments), together with morphographic, morphogenetic, and age data on the landforms.

Under the *Polish* system there is a comparative lack, on the final map, of information concerning slope and river processes. The influence of structure, slope steepness, hill crest and scarp forms, and the age and origin of landforms are all shown.

Gilewska concluded this comparison with a startling, if not disturbing, conclusion that although each map brings out additional information concerning both the nature of the landforms and the successive phases of land sculpture, in her view none provided a totally satisfactory statement of the actual geomorphology. This leads to the view that 'comprehensive' maps are difficult to achieve.

14.4. Different Types of Geomorphological Maps

In view of the above it has to be recognized that there are several different types of geomorphological maps. Apart from the distinction which may be made between maps drawn for academic purposes only, and those compiled in order to attack a particular problem, there is also a distinction to be made between morphographic, morphogenetic, morphometric and morpho-chronologic maps (Verstappen, 1970).

On *morphographic* maps landforms are identified by their specific names (e.g. planation surface, drumlin), and their shapes are shown.

Morphogenetic maps define the origin and development of the landforms, with genetic descriptions provided on the map legend (e.g. sandy alluvial plain, sandy pro-glacial plain). Form and material composition also need to be shown because of their close link with processes.

On *morphometric* maps information is supplied about the dimensions of the landforms. The morphometric information includes data on such things as slope steepness and shape, amplitude of relief, heights of specific features (e.g. river terraces), and the density of the valley network.

Morphochronologic maps distinguish between landforms according to their time of initiation. Distinctions are made between recent forms and those inherited from a previous period and whose character may be related to past processes, possibly operating under a different climate. Age can be indicated by a code letter or a code number, or omitted when age is not known.

Morphographic and morphogenetic mapping proceeds by the use of symbols of the type illustrated in this chapter. In the compilation of the I.T.C. system of geomorphological mapping (Verstappen and Zuidam, 1968) which fulfils most general needs, distinct guidelines were followed (Verstappen, 1970):

(i) the mapping system should be flexible, allowing the compiler discretion in the adoption of the symbols most appropriate for the area concerned;
(ii) the maps should be as simple as possible (to counteract cartographic problems and to avoid high printing costs);
(iii) the system should be applicable for mapping at all scales;
(iv) general geomorphological maps should be supplemented by special purpose ('applied') maps;
(v) colours should be used to indicate major genetic landform units (rather than lithology or chronology as in some other mapping schemes).

The value of this last point is that in practice the colours will thereby define those areas which under a land-systems survey (Chap. 13) would normally be identified either as a land system or as a land unit. In either case compatible boundaries between these two approaches to land evaluation will become clear if this basis is adopted for the application of colours.

The colour system used by the I.T.C. is as follows (Verstappen and Zuidam, 1968):

1. Forms of structural origin — purple
2. Forms of volcanic origin — red
3. Forms of denudational origin — brown
4. Forms of fluvial origin — dark-blue
5. Forms of marine origin — green
6. Forms of glacial/periglacial origin — light-blue
7. Forms of aeolian origin — yellow
8. Forms of solution (karst) origin — orange
9. Morphometry — black
10. Lithology — black/grey
11. Chronology — black
12. Topography — black/grey

An example of the I.T.C. system of mapping is included in Verstappen and Zuidam (1968). Some contrasts occur between the I.T.C. system and others in the use of colours. Polish maps tend to use colours to indicate both land-form age and origin, while on the French system colours are used for age and lithology of materials.

14.5. Limitations and Practicalities

Good detailed published geomorphological maps are few in number considering that the technique's merits have been expounded for over twenty years. There are two main reasons for this. The first is that in order to compile a good detailed geomorphological map much fieldwork is required. Secondly, in order to be effective, a geomorphological map has to be printed in colour. This is very expensive. One of the best published maps to date is that of the Pavlovské Hills which accompanies the *Manual of Detailed Geomorphological Mapping* (Demek, 1972), as compiled at the Institute of Geography at Brno (Czechoslovakia). No one knows of course how many unpublished manuscript maps are lying in their compiler's drawers waiting for a sponsor to make their publication possible. It does not follow, of course, that all the published maps have been useful in environmental management, or that none of the manuscript maps have been used in this way. Indeed, many unpublished maps have been compiled specifically because there has been a management problem to solve.

The value of geomorphological maps has not gone unquestioned. Wright (1972b) reviewed their value for purposes of land classification and recognized in them an emphasis or specialism that reflected the compilers' interests or geomorphological philosophy. Only the maps produced at the Centre

National de Recherches Géomorphologique, in Belgium, had success-fully combined the descriptive components (morphography and morpho-metry) with the interpretative components (morphogenesis and morpho-chronology) of landform mapping (Macar *et al.*, 1961; Gullentops, 1963). On the other hand a scrutiny by Wright (1972b) of a map of the Vaunage area in France (Tricart, 1965) revealed a strong morphogenetic bias with informa-tion about the character of landforms restricted to contours and with scant information about rock type and surficial deposits. A weakness also lay in the fact that on Tricart's map different criteria were used in different parts of the area to distinguish between landforms. For example within one morpho-genetic group differentiation was according to the chemical composition of materials and in another it was based on material textures.

Not only does the clear representation of all available geomorphological information on one map produce difficult cartographic problems, it also leads to difficulties in its use, thus limiting its ultimate applied value. Most published geomorphological maps are only meaningful to other geomorpho-logists. To be of practical value the map must be clear. This means reproduc-tion at a large enough scale, and that the map needs to be problem-orientated.

General geomorphological maps have their principal value at the initial field investigation stage. They allow the person who is doing the mapping to acquire knowledge about the specific sites in the context of their geomor-phological situation. In addition, as was shown in Section 14.2, they may form the basis of a number of specialist maps to which yet further specialist knowledge may be added.

To be useful, therefore, geomorphological maps need to be made avail-able to land planners, or to the site investigator in an engineering project, in a selective form. In addition the legend has to be phrased in terms which will be readily understood by those for whom the map is intended.

14.6. Conclusion

The purpose of geomorphological maps is to portray the forms of the surface, the nature and properties of the materials of which these are com-posed, and to indicate the kind and magnitude of the processes in operation. As such they provide an integrated and comprehensive statement of land-form and drainage. They contain information of considerable potential value in land-use planning, hydrological engineering, civil engineering, soil survey-ing, and conservation.

The most useful geomorphological maps, in environmental management terms, are those which are compiled in the context of a specified problem. Such maps may be based on a general geomorphological map, but in any event they should be explicit, clear, and preferably in colour. No mapping exercise is ever complete in itself, however, for its results have to be considered in terms of the dynamic process systems of the kind described earlier in this book.

REFERENCES

AGAR, R., 1960, The post-glacial erosion of the north Yorkshire coast from the Tees Estuary to Ravenscar, *Proc. Yorks. Geol. Soc.* **32**, 409–27.

AGRICULTURAL ADVISORY COUNCIL, 1970, *Modern Farming and the Soil* (Min. Ag., Fish and Food, H.M.S.O., London), 119 pp.

AITCHISON, G. D., and K. GRANT, 1967, The P.U.C.E. programme of terrain description, evaluation and interpretation for engineering purposes, *4th Reg. Conf. for Africa on Soil Mech. and Foundation Eng.* (Cape Town) **1**, 1–8.

ALBERTSON, M. L., and D. B. SIMONS, 1964, Fluid mechanics, in V. T. Chow (ed.), *Handbook of Applied Hydrology* (McGraw-Hill, New York), 7.1–7.49.

ALLEN, A. S., 1969, Geologic settings of subsidence, in D. J. Varnes and G. Kiersch (eds.), *Reviews in Engineering Geology*, **2**, 305–42.

ALLEN, R. H., and E. L. SPOONER, 1968, *Annotated Bibliography*, Coastal Engineering Research Center, M–P 1–68, 141 pp.

ALLUM, J. A. E., 1966, *Photogeology and Regional Mapping* (Pergamon Press, Oxford), 107 pp.

ALYESKA PIPELINE SERVICE CO., 1971, *Project Description of the Trans-Alaska Pipeline System*, 64 pp.

AMERICAN SOCIETY OF PHOTOGRAMMETRY, 1960, *Manual of Photographic Interpretation* (Washington, D.C.), 868 pp.

AMERICAN SOCIETY FOR TESTING MATERIALS, 1958, Terms relating to natural building stones (C119–1950).

AMERICAN SOCIETY FOR TESTING MATERIALS, 1971a, Standard test for resistance of concrete to rapid freezing and thawing (C666).

AMERICAN SOCIETY FOR TESTING MATERIALS, 1971b, Specifications for facing brick (C216), footnote 1, 167–8.

ANDERSON, H. W., 1957, Relating sediment yield to watershed variables, *Trans. Am. Geoph. Un.* **38**, 921–4.

ANON., 1969, *Lee Valley Regional Park* (Plan of Proposals).

ANON., 1972, Sulphur pollution across national boundaries, *Ambio*, **1**, 15–20.

ARMBRUST, D. V., and J. J. DICKERSON, 1971, Temporary wind erosion control: cost and effectiveness of 34 commercial materials, *J. Soil and Water Conservation*, **26**, 154–7.

ARNETT, R. R., and A. J. CONACHER, 1973, Drainage basin expansion and the nine unit landsurface model, *Australian Geographer*, **12**, 237–49.

ASTM—see American Society for Testing Materials.

BAGNOLD, R. A., 1941, *The Physics of Blown Sand and Desert Dunes* (Methuen, London), 265 pp.

BAKER, R. F., and H. C. MARSHALL, 1958, Control and correction, in E. B. Eckel (ed.), *Landslides and Engineering Practice* (Highway Research Board, Special Report **29**), 150–88.

BARATA, F. E., 1969, Landslides in the tropical region of Rio de Janeiro, *Proc. 7th Inter. Conf. Soil Mech. and Foundation Eng.* (Mexico) **2**, 507–16.

BARNES, D. F., 1966, Geophysical methods of delineating permafrost, *Proc. Permafrost Intern. Conf.* (Nat. Academy of Sciences—National Research Council), Publ. **1287**, 349–55.

BARNES, H. H., 1967, Roughness characteristics of natural channels, *U.S. Geol. Surv. Water Supply Paper* **1849**, 213 pp.

BARRATT, M. S., *et al.*, 1970, Land use planning and mineral resources with special reference to aggregates for the construction industries, in P. T. Warren (ed.), *Geological Aspects of Development and Planning in Northern England* (Yorkshire Geol. Soc., Leeds), 48–66.

BARRY, R. G., and R. J. CHORLEY, 1968, *Atmosphere, Weather and Climate* (Methuen, London), 319 pp.

BARTELLI, L. J., A. A. KLINGEBIEL, J. V. BAIRD and M. R. HEDDLESON (eds.), 1966, *Soil Surveys and Land Use Planning* (Soil Science Soc. of Am. and Am. Soc. Agronomy, Madison), 196 pp.

BAWDEN, M. G., 1965, A reconnaissance of the land resources of Eastern Bechuanaland, *J. Appl. Ecol.* **2**, 357–65.

BAWDEN, M. G., 1967, Applications of aerial photography in land system mapping, *Photogrammetric Record*, **5**, 461–4.

BEACH EROSION BOARD, 1933, Interim Report (Washington, D.C.).

BEAVER, S. H., 1968, *The Geology of Sand and Gravel* (Sand and Gravel Assn. of Great Britain, London), 66 pp.

BECKETT, P. H. T., and R. WEBSTER, 1965, *A Classification System for Terrain* (M.E.X.E. Christchurch, England), Rep. No. **872**, 247 pp.

BENNETT, H. H., 1939, *Soil Conservation* (McGraw-Hill, New York), 993 pp.

BENSON, M. A., 1962, Factors influencing the occurrence of floods in a humid region of diverse terrain, *U.S. Geol. Surv. Water Supply Pap.* **1580**-B.

BIBBY, J. S., and D. MACKNEY, 1969, Land use capability classification (Soil Surv. England and Wales, Harpenden, U.K.), *Tech. Monogr.* **1**.

BIRD, E. C. F., 1968, *Coasts* (Australian Nat. Univ. Press, Canberra), 246 pp.

BIROT, P., 1954, Désagrégation des roches cristallines sous l'action des sels, *C. R. Ac. Sci.* **238**, 1145–6.

BISAL, F., 1960, The effect of raindrop size and impact velocity on sand splash, *Can. J. Soil Science*, **40**, 242–5.

BISAL, F., and J. HSIEH, 1966, Influence of moisture on erodibility of soil by wind, *Soil Science*, **102**, 143–6.

BLACK, R. F., 1950, Permafrost, in P. D. Trask (ed.), *Applied Sedimentation* (Wiley, New York), 247–76.

BLACK, R. F., 1954, Permafrost—a review, *Bull. Geol. Soc. Am.* **65**, 839–55.

BLASE, M. G., and J. F. TIMMONS, 1961, Soil erosion control—problems and progress, *J. Soil Water Cons.* **16**, 157–62. Reprinted in I. Burton and R. W. Kates (eds.), 1965, *Readings in Resource Management and Conservation* (Univ. Chicago Press, Chicago), 338–47.

BLENCH, T., 1969, *Mobile-bed Fluviology* (Univ. Alberta Press, Alberta), 221 pp.

BLENCH, T., 1972, Morphometric changes, in R. T. Oglesby, C. A. Carlson and J. A. McCann (eds.), *River Ecology and Man* (Academic Press, New York), 287–308.

BLUMENTHAL, K. P., 1967, Some aspects of land reclamation in the Netherlands, *Proc. 10th Conf. Coast. Eng.* 1331–59.

BORCHERT, J. R., 1971, The Dust Bowl in the 1970s, *Ann. Ass. Am. Geog.* **61**, 1–22.

BORMANN, F. H., and G. E. LIKENS, 1969, The watershed-ecosystem concept and studies of nutrient cycles in G. M. van Dyne (ed.), *The Ecosystem Concept in Natural Resource Management* (Academic Press, New York), 49–76.

BOURNE, R., 1931, Regional survey and its relation to stock-taking of the agricultural resources of the British Empire, *Oxford Forestry Memoirs*, **13**.

BOWEN, A. J., and D. L. INMAN, 1966, Budget of littoral sands in the vicinity of Point Arguello, California, *CERC Tech. Memo.* **19**, 41 pp.

BRANCHER, D. M., 1969, Critique of K. D. Fines: Landscape evaluation: a research project in East Sussex, *Regional Studies*, **3**, 91–2.

BRIDGES, E. M., 1967, Geomorphology and soils in Britain, *Soils and Fertilizers*, **30**, 317–20.

BRINK, A. B. A., and T. C. PARTRIDGE, 1967, Kyalami land system: an example of physiographic classification for the storage of terrain data, *4th Reg. Conf. for Africa on Soil Mech. and Foundation Eng.* (Cape Town), **1**, 9–14.

BRINK, A. B. A., T. C. PARTRIDGE, and G. B. MATHEWS, 1970, Airphoto interpretation in terrain evaluation, *Photo Interpretation*, **5**, 15–30.

BRINK, A. B. A., J. A. MABBUTT, R. WEBSTER, and P. H. T. BECKETT, 1966, *Report of the Working Group on Land Classification and Data Storage* (M.E.X.E., Christchurch, England), Rep. No. **940**, 97 pp.

BRINK, A. B. A., T. C. PARTRIDGE, R. WEBSTER, and A. A. B. WILLIAMS, 1968, Land classification and data storage for the engineering usage of natural materials, *Proc. 4th Conf. Australian Road Res. Board*, 1624–47.

BROWN, E. H., 1970, Man shapes the earth, *Geogr. J.* **136**, 74–85.

BROWN, R. J. E., 1965, Factors influencing discontinuous permafrost in Canada, in T. L. Péwé (ed.), *The Periglacial Environment* (McGill–Queen's U. Press, Montreal) 11–54.

BROWN, R. J. E., 1966, Influence of vegetation on permafrost, *Proc. Permafrost Inter. Conf.* (Nat. Academy of Sciences—National Research Council), Publ. **1287**, 20–5.

BROWN, R. J. E., 1967, Comparison of permafrost conditions in Canada and the U.S.S.R., *Polar Record*, **13**, 741–51.

BROWN, R. J. E., (ed.), 1969, *Proceedings of the Third Canadian Conference on Permafrost* (Nat. Res. Council Canada), Technical Mem. **96**, 187 pp.

BROWN, R. J. E., 1970, *Permafrost in Canada* (Univ. of Toronto Press, Toronto), 234 pp.

BRUNSDEN, D., and D. K. C. JONES, 1972, The morphology of degraded landslide slopes in South West Dorset, *Q. J. Eng. Geol.* **5**, 205–22.

BRUNT, M., 1967, The methods employed by the Directorate of Overseas Surveys in the assessment of land resources, *Études de Synthèse*, **6**, 3–10.

BRUUN, P., 1954, Coast erosion and the development of beach profiles, *CERC Tech. Memo.* **19**, 41 pp.

BRUUN, P., and J. B. LACKEY, 1962, Engineering aspects of sediment transport, in T. Fluhr and R. F. Legget (eds.), *Reviews in Engineering Geology* (Geol. Soc. Am.), **1**, 39–103.

BRYAN, K., 1925, The Papago Country, Arizona, *U.S. Geol. Surv. Water Supply Paper*, **499**, 436 pp.

BRYAN, R. B., 1968, The development, use and efficiency of indices of soil erodibility, *Geoderma*, **2**, 5–26.

BULL, W. B., 1964, Alluvial fans and near-surface subsidence in western Fresno County, California, *U.S. Geol. Surv. Prof. Paper*, **437**-A, 70 pp.

BULL, W. B., 1968, Alluvial fans, *J. Geol. Education*, **16**, 101–6.

BURINGH, P., and A. P. A. VINK, 1965, The importance of geology in air photo-interpretation for soil mapping, *Revertera Number, Int. Training Centre for Aerial Survey* (Delft), Ser. V, No. **33**, 16–23.

BURTON, I., and R. W. KATES, 1964, The perception of natural hazards in resource management, *Natural Resources J.* **3**, 412–41.

BURTON, I., R. W. KATES, and G. F. WHITE, 1968, The human ecology of extreme geophysical events, *Natural Hazard Research Working Paper* (University of Toronto), **1**, 33 pp.

BURTON, I., R. W. KATES, and R. R. SNEAD, 1969, *The Human Ecology of Coastal Flood Hazard in Megalopolis* Univ. Chicago, Dept. of Geography, Research Paper, **115**, 169 pp.

BURTON, I., R. W. KATES, J. R. MATHER, and R. R. SNEAD, 1969, The shores of Megalopolis: coastal occupance and adjustment to flood hazard, *Final Rep. ONR Contract No. 388–073* (Elmer, New Jersey) 603 pp.

CARLSTON, C. W., 1968, Slope-discharge relations for eight rivers in the United States, *U.S. Geol. Surv. Prof. Paper*, **600**–D, 45–7.

CARROLL, D., 1970, *Rock Weathering* (Plenum, New York), 203 pp.

CHANDLER, R. J., 1969, The effect of weathering on the shear strength properties of Keuper marl, *Géotechnique*, **19**, 321–34.

CHANDLER, R. J., 1971, Landsliding on the Jurassic escarpment near Rockingham, Northamptonshire, in D. Brunsden (compiler) *Slopes: Form and Process* (Inst. Br. Geogr. Sp. Publ. 3), 111–28.

CHANDLER, R. J., 1972, Lias Clay: weathering processes and their effect on shear strength, *Géotechnique*, **22**, 403–31.

CHEPIL, W. S., 1945, Dynamics of wind erosion: II. Initiation of soil movement, *Soil Science*, **60**, 397–411.

CHEPIL, W. S., 1951, Properties of soil which influence wind erosion: V. Mechanical stability of structure, *Soil Science*, **72**, 465–78.

CHEPIL, W. S., 1955, Factors that influence clod structure and erodibility of soil by wind: IV. Sand, Silt and Clay, *Soil Science*, **80**, 155–62.

CHEPIL, W. S., and R. A. MILNE, 1941, Wind erosion of soil in relation to roughness of surface, *Soil Science*, **52**, 417–31.

CHEPIL, W. S., and N. P. WOODRUFF, 1963, The physics of wind erosion and its control, *Advances in Agronomy*, **15**, 211–302.

CHERRY, J. A., 1965, Sand movement along a portion of the north California coast, *CERC Tech. Memo.* **14**, 125 pp.

CHIEN, N., 1956, Graphic design of alluvial channels, *Proc. Am. Soc. Civ. Eng.* **121**, 1267–80.

CHORLEY, R. J., 1959, The geomorphic significance of some Oxford soils, *Am. J. Sci.* **257**, 503–15.

CHORLEY, R. J. (ed.), 1969, *Water, Earth and Man* (Methuen, London), 588 pp.

CHORLEY, R. J. and B. A. KENNEDY, 1971, *Physical Geography: A Systems Approach* (Prentice-Hall, London), 375 pp.

CHOW, V. T. (ed.), 1964, *Handbook of Applied Hydrology* (McGraw-Hill, New York), 1418 pp.

CHRISTIAN, C. S., 1957, The concept of land units and land systems, *Proc. 9th Pacific Sci. Congr.* **20**, 74–81.

CHRISTIAN, C. S., and G. A. STEWART, 1952, Summary of general report on survey of Katherine-Darwin Region, 1946 (CSIRO, Australia), *Land Research Series*, **1**, 24 pp.

CHRISTIAN, C. S., and G. A. STEWART, 1968, Methodology of integrated survey, *Aerial Surveys and Integrated Studies, Proc. of the Toulouse Conf.* (UNESCO, Paris), 233–80.

CIRIACY-WANTRUP, S. V., 1964, Water policy, in V. T. Chow (ed.), *Handbook of Applied Hydrology* (McGraw-Hill, New York), 28.1–28.15.

CLAPPERTON, C. M., and P. HAMILTON, 1971, Peru beneath its eternal threat, *Geogr.*

Mag. **43**, 632–9.

CLARK, A. H., R. U. COOKE, C. MORTIMER, and R. H. SILLITOE, 1967, Relationships between supergene mineral alteration and geomorphology, southern Atacama Desert, Chile—an interim report, *Trans. Inst. Min. and Met.*, B (Applied Earth Science), **76**, B89–B96.

CLARKE, B. L., and J. ASHURST, 1972, *Stone Preservation Experiments* (H.M.S.O., London), 78 pp.

CLAYTON, K. M., 1971, Reality in conservation, *Geogr. Mag.* **44**, 83–4.

CLOUT, H. D., 1973, *Rural Geography* (Pergamon, Oxford), 131–7.

COASTAL ENGINEERING RESEARCH CENTER, 1966, Shore protection, planning and design, *Tech. Rep.* **4**, 3rd edition (Washington), 401 pp.

COATES, D. R., 1971, *Environmental Geomorphology* (Publications in Geomorphology, State University of New York, Binghamton), 262 pp.

COLE W. F., 1959, Some aspects of the weathering of terracotta roofing tiles, *Austr. J. App. Sci.* **10**, 346–63.

COLEMAN, ALICE, 1955, Land reclamation at a Kentish colliery, *Trans. Inst. Brit. Geog.* **21**, 117–35.

COLWELL, R. N., 1968, Remote sensing of natural resources, *Scientific American*, **218**, 54–69.

COOK, D. O., 1970, The occurrence and geological work of rip currents off southern California, *Mar. Geol.* **9**, 173–86.

COOK, E. T., and A. WEDDERBURN, 1904, *The Works of John Ruskin*, **6**, *Modern Painters* (4) (Allen, London), 485 pp.

COOKE, R. U., and D. R. HARRIS, 1970, Remote sensing of the terrestrial environment—principles and progress, *Trans. and Papers, Inst. Brit. Geogr.* **50**, 1–23.

COOKE, R. U., and I. J. SMALLEY, 1968, Salt weathering in deserts, *Nature*, **220**, 1,226–7.

COOKE, R. U., and J. R. G. TOWNSHEND, 1970, Pattern of Peru's great earthquake, *Geogr. Mag.* **42**, 765–6.

COOKE, R. U., and A. WARREN, 1973, *Geomorphology in Deserts* (Batsford, London; California U.P.), 416 pp.

COOPER, W. S., 1958, *Coastal sand dunes of Oregon and Washington*, *Geol. Soc. Am. Mem.* **72**, 169 pp.

COOPER, W. S., 1967, Coastal dunes of California, *Geol. Soc. Am. Mem.* **104**, 131 pp.

COPPOCK, J. T., 1959, The Chilterns as an area of outstanding natural beauty, *J. Town Planning Inst.* **45**, 137–41.

CORRENS, C. W., 1949, Growth and dissolution of crystals under linear pressure, in *Disc. Faraday Soc.* **5**, *Crystal Growth* (Butterworth, London), 267–71.

CORTE, A. E., 1969, Geocryology and engineering, *Reviews in Engineering Geology*, **2**, 119–85.

COUNTRYSIDE COMMISSION FOR SCOTLAND, 1971, *A Planning Classification of Scottish Landscape Resources* (C.C.S., Perth), 123 pp.

CRAIK, K. H., 1970, Environmental psychology, *New Directions in Psychology*, **4**, 1–121.

CRAIK, K. H., 1972, Psychological factors in landscape appraisal, *Environment and Behaviour*, **1**, 255–66.

CRITTENDEN, M. D., Sr., 1963, Effective viscosity of the earth derived from isostatic loading of Pleistocene Lake Bonneville, *J. Geoph. Res.* **68**, 5517–30.

CROFTS, R., 1973, Slope categories in environmental management (Department of Geography, University College London), unpublished paper.

CRORY, F. E., 1966, Pile foundations in permafrost, *Proc. Permafrost Int. Conf.*

(Nat. Academy Sciences—National Research Council), Publ. **1287**, 467–76.

CROWDER, J. R., 1965, The weathering behaviour of glass-fibre reinforced polyester sheeting, *Building Research, Misc. Papers*, **2**, 14 pp.

CROZIER, M. J., 1973, Techniques for the morphometric analysis of landslips, *Zeit. für Geom.*, **17**, 78–101.

CRUIKSHANK, J. G., 1972, *Soil Geography* (David and Charles, Newton Abbot), 240 pp.

CURTIS, L. F., J. C. DOORNKAMP, and K. GREGORY, 1965, The description of relief in the field study of soils, *J. Soil Sci.* **16**, 16–30.

CZUDEK, T., and J. DEMEK, 1970, Thermokarst in Siberia and its influence on the development of lowland relief, *Quaternary Res.* **1**, 103–20.

DALRYMPLE, J. B., R. J. BLONG, and A. J. CONACHER, 1968, A hypothetical nine unit land surface model, *Zeit. für Geom.* **12**, 60–76.

DARBYSHIRE, J., and L. DRAPER, 1963, Forecasting wind-generated sea waves, *Eng.* **195**, 482–4.

DASMANN, R. F., 1968, *Environmental Conservation* (Wiley, New York), 375 pp.

DAVIES, J. L., 1959, Wave refraction and the evolution of shoreline curves, *Geog. Studies*, **5**, 1–14.

DAVIES, J. L., 1960, Beach alignment in south Australia, *Austr. Geog.* **8**, 42–4.

DAVIES, J. L., 1964, A morphogenic approach to world coastlines, *Zeit. für Geom.*, Sp. No. 127*–42*.

DAVIES, J. L., 1972, *Geographical Variations in Coastal Development* (Oliver and Boyd, Edinburgh), 204 pp.

DAVIS, G. H., J. B. SMALL, and H. B. COUNTS, 1963, Land subsidence related to decline of artesian pressure in the Ocala Limestone at Savannah, Georgia, in P. D. Trask and G. A. Kiersch (eds.), *Engineering Geology Case Histories*, **4**, 1–8.

DAVIS, R. A., and W. T. FOX, 1971, *Beach and Nearshore Dynamics in Eastern Lake Michigan* (Tech. Rep./ONR Contract 388–092 Geographical Branch, Office of Naval Research, N 99914–69–C–0151), 145 pp.

DAVIS, R. A., and W. T. FOX, 1972, Four-dimensional model for beach and inner nearshore sedimentation, *J. Geol.* **80**, 484–93.

DAWE, H. G., 1965, Volume assessment by photogrammetric methods, in Inst. Min. and Met., *Opencast Mining, Quarrying and Alluvial Mining* (London), 493–504.

DEACON, G. E. R., 1949, Waves and swell, *Quart. J. Roy. Met. Soc.* **75**, 227–38.

DEMEK, J. (ed.), 1972, *Manual of Detailed Geomorphological Mapping* (Academia, Prague), 368 pp.

DENDY, F. E., 1968, Sedimentation in the Nation's reservoirs, *J. Soil and Water Conservation*, **23**, 135–7; reprinted in R. W. Tank, 1973, *Focus on Environmental Geology* (O.U.P., New York), 180–5.

DIXEY, F., 1962, Applied geomorphology, *S. Afr. Geogr. J.* **44**, 3–24.

DIXON, J. W., 1964, Water Resources. Part I, Planning and development, in V. T. Chow (ed.), *Handbook of Applied Hydrology* (McGraw-Hill, New York), 26.1–26.29.

DOLAN, R., 1971, Coastal landforms: crescentic and rhythmic, *Geol. Soc. Am. Bull.* **82**, 176–80.

DOLAN, R., 1973, Barrier Islands: natural and controlled, in D. R. Coates (ed.), *Coastal Geomorphology* (Publ. in Geom., State Univ. of New York, Binghamton) 263–78.

DONAHUE, R. L., J. C. SHICKLUNA, and L. S. ROBERTSON, 1971, *Soils: an Introduction to Soils and Plant Growth*, 3rd edition (Prentice Hall, New Jersey), 587 pp.

DOORNKAMP, J. C., 1971, Geomorphological mapping, in S. H. Ominde (ed.), *Studies in East African Geography and Development* (Heinemann, London and Nairobi), 9–28.

DOORNKAMP, J. C., and C. A. M. KING, 1971, *Numerical Analysis in Geomorphology: An Introduction* (Arnold, London), 327 pp.

DOUGLAS, I., 1967, Man, vegetation and the sediment yield of rivers, *Nature*, **215**, 925–8.

DOUGLAS, I., 1968, Sediment sources and causes in the humid tropics of northeast Queensland, Australia, in A. M. Harvey (ed.), *Geomorphology in a Tropical Environment* (Brit. Geom. Res. Group, Occ. Pap. **5**), 27–39.

DOUGLAS, I., 1971, Dynamic equilibrium in applied geomorphology: two case studies, *Earth Science J.* **5**, 29–35.

DOUGLAS, I., 1973, Water resources, in J. A. Dawson and J. C. Doornkamp (eds.), *Evaluating the Human Environment: Essays in Applied Geography* (Arnold, London), 57–87.

DUANE, D. B., 1969, Sand inventory program. A study of New Jersey and north New England coastal waters, *Shore and Beach*, Oct. 1969, and CERC R–2–70.

DUMONTELLE, P. B., N. C. HESTER, and R. E. COLE, 1971, Landslides along the Illinois River Valley south and west of La Salle and Peru, Illinois (Illinois State Geol. Survey, Urbana), *Env. Geol. Notes*, **48**, 16 pp.

DUNSTAN, L. M., 1966, Some aspects of planning in relation to mineral resources, *The Chartered Surveyor*, **99**, 67–73.

DURY, G. H., 1969, *Perspectives on Geomorphic Processes*, Assoc. Am. Geogr. Resource Paper, **3**, 56 pp.

DYLIK, J., 1957, Dynamic geomorphology, its nature and method, *Bull. de la Soc. des Sciences et des Lettres de Lódź*, Classe III, VIII (12), 1–42.

ECKEL, E. G. (ed.), 1958, *Landslides and Engineering Practice*, Highway Research Board, Washington, Special Report **29**; NAS–NRC Publ. **544**, 323 pp.

EDELMAN, T., 1954, Tectonic movements as resulting from the comparison of two precision levellings, *Geol. en. Mijnbouw*, **16**, 209–13.

EINSTEIN, H. A., 1972, Sedimentation (suspended solids), in R. J. Oglesby, C. A. Carlson, and J. A. McCann (eds.), *River Ecology and Man* (Academic Press, New York), 309–18.

EINSTEIN, H. A., A. G. ANDERSON, and J. W. JOHNSON, 1940, A distinction between bed-load and suspended load in natural streams, *Am. Geophys. Un. Trans.* **21**, 628–33.

ELLISON, W. D., 1947, Soil erosion studies, parts I–VII, *Agricultural Engineering*, **28**, pp. 145–6, 197–201, 245–8, 297–300, 349–51, 353, 402–5, 408, 442–4, 450.

EMBLETON, C., 1972, *Glaciers and Glacial Erosion* (MacMillan, London), 287 pp.

EMBLETON, C., and C. A. M. KING, 1968, *Glacial and Periglacial Geomorphology* (Arnold, London), 608 pp.

EMMETT, W. W., 1970, The hydraulics of overland flow on hillslopes, *U.S. Geol. Surv. Prof. Paper*, **662**–A, 68 pp.

EVANS, I., 1970, Salt crystallization and rock weathering: a review, *Rev. Géom. Dyn.* **19**, 153–77.

EVERETT, D. H., 1961, The thermodynamics of frost damage to porous solids, *Trans. Faraday Soc.* **57**, 1,541–51.

F.A.O., 1960, Soil erosion by wind and measures for its control on agricultural land, *Agricultural Development Paper*, **71**, 88 pp.

F.A.O., 1965, Soil erosion by water—some measures for its control on cultivated lands, *Agricultural Development Paper*, **81**, 284 pp.

FERRIANS, O. J., R. KACHADOORIAN, and G. W. GREENE, 1969, Permafrost and related engineering problems in Alaska, *U.S. Geol. Surv. Prof. Paper*, **678**, 37 pp.

FINES, K. D., 1968, Landscape evaluation: a research project in East Sussex, *Regional Studies*, **2**, 41–55.

FINES, K. D., 1969, Landscape evaluation: a research project in East Sussex: rejoinder to critique by D. M. Brancher, *Regional Studies*, **3**, 219.

FITZPATRICK, E. A., 1971, *Pedology: A Systematic Approach to Soil Science* (Oliver and Boyd, Edinburgh), 306 pp.

FLAWN, P. T., 1970, *Environmental Geology* (Harper and Row, New York), 313 pp.

FOLK, R. L., and W. C. WARD, 1957, Brazos river bar, a study of the significance of grain size parameters, *J. Sed. Petrol.* **27**, 3–27.

FOOKES, P. G., and P. HORSWILL, 1969, Discussion on engineering grade zones, *Proc. Conf. In Situ Testing Soils and Rocks, Inst. Civ. Eng., London*, 53–7.

FOOKES, P. G., W. R. DEARMAN, and J. A. FRANKLIN, 1971, Some engineering aspects of rock weathering with field examples from Dartmoor and elsewhere, *Quart. J. Eng. Geol.* **4**, 139–85.

FOTH, H. D. and L. M. TURK, 1972, *Fundamentals of Soil Science*, 5th edition (Wiley, New York), 454 pp.

FOURNIER, F., 1960, *Climat et érosion: la relation entre l'érosion du sol par l'eau et les précipitations atmosphériques* (P.U.F., Paris), 201 pp.

FOURNIER, F., 1972, *Aspects of Soil Conservation in the Different Climatic and Pedalogic Regions of Europe* (Council of Europe), 194 pp.

FRÄNZLE, O., 1966, Geomorphological mapping, *Nature and Resources* (UNESCO), **II(4)**, 14–16.

FRASER, J. C., 1972, Regulated discharge and the stream environment, in R. T. Oglesby, C. A. Carlson, and J. A. McCann (eds.), *River Ecology and Man* (Academic Press, New York), 263–85.

FREE, G. R., 1960, Erosion characteristics of rainfall, *Agricultural Engineering*, **41**, 447–9, 455.

FREVERT, R. K., G. O. SCHWAB, T. W. EDMINSTER, and K. K. BARNES, 1955, *Soil and Water Conservation Engineering* (Wiley, New York), 479 pp.

FROST, R. E., 1950, *Evaluation of Soils and Permafrost Conditions in the Territory of Alaska by Means of Airphotos*, Contract Report, St. Paul District Corps of Engineers, 2 vols.

FROST, R. E., J. H. McLERRAN, and R. D. LEIGHTY, 1966, Photointerpretation in the Arctic and sub-Arctic, *Proc. Permafrost Int. Conf.* (Nat. Academy Science–National Research Council), Publ. **1287**, 343–8.

FRYE, J. C., 1967, *Geological Information for Managing the Environment*, State Geol. Survey, Illinois, Env. Geol. Notes, **18**, 12 pp.

FURLEY, P. A., 1971, Relationships between slope form and soil properties developed over chalk parent materials, in D. Brunsden (compiler), *Slopes: Form and Process* (Inst. Brit. Geogr., Sp. Publ. 3), 141–63.

GABRYSCH, R. K., 1969, Land-surface subsidence in the Houston–Galveston region, Texas, *Pub. Inst. Ass. Sci. Hydr.* **88**, 43–54.

GAILLARD, D. B. W., 1904, *Wave Action* (Corps. of Engineers, U.S. Army, Washington, D.C.).

GALLIERS, J. A., 1969, Barriers, beaches and lagoons of the Ghana coast, *Brit. Geom. Res. Group. Occ. Paper*, **5**, 77–87.

GALON, R., 1962, *Instruction to the Detailed Geomorphological Map of the Polish Lowland* (Polish Acad. of Science, Geogr. Inst. Dept. of Geomorphology and Hydrography of the Polish Lowland at Toruń).

GALVIN, C. J., 1967, Longshore current velocity: a review of theory and data, *Rev. Geophys.* **5**, 287–304.

GELLERT, J. F., and E. SCHOLZ, 1964, *Katalog des Inhaltes von geomorphologischen Detailkarten aus verschieden europäischen Ländern* (Institut für Geographie, Potsdam), 119 pp.

GEOLOGICAL SOCIETY ENGINEERING GROUP WORKING PARTY, 1972, The preparation of maps and plans in terms of engineering geology, *Q. J. Eng. Geol.* **5**, 293–382.

GILBERT, G. K., 1880, *Report on the Geology of the Henry Mountains*, 2nd edition (U.S. Geol. Surv., Washington), reprinted in part in S. A. Schumm (ed.), 1972, *River Morphology* (Dowden, Hutchinson, Ross Inc., Stroudsberg, Pennsylvania), 43–77.

GILBERT, G. K. 1906, Crescentric gouges on glaciated surfaces, *Bull. Geol. Soc. Am.* **17**, 303–13.

GILEWSKA, S., 1967, Different methods of showing the relief on the detailed geomorphological maps, *Zeit. für Geom.* **11**, 481–90.

GILLULY, J., A. C. WATERS, and A. O. WOODFORD, 1968, *Principles of Geology*, 3rd edition (Freeman, San Francisco), 687 pp.

GLENN, L. C., 1911, Denudation and erosion in the Southern Appalachian Region and the Monongahela Basin, *U.S. Geol. Surv. Prof. Paper*, **72**, 137 pp.

GLYMPH, L. M., and H. N. HOLTAN, 1969, Land treatment in agricultural watershed hydrology research, in W. L. Moore and C. W. Morgan (eds.), *Effect of Watershed Changes on Streamflow* (Texas U.P., Austin), 44–68.

GODFREY, P. J., and M. M. GODFREY, 1973, Comparison of ecological and geomorphic interactions between altered and unaltered barrier island systems in North Carolina, in D. R. Coates (ed.), *Coastal Geomorphology* (Publ. in Geomorph., State Univ. New York, Binghamton), 239–58.

GOLDRICH, S. S., 1938, A study in rock-weathering, *J. Geol.* **46**, 17–58.

GOOSEN, D., 1966, The classification of landscapes as the basis for soil surveys, *Trans. 2nd Intern. Symp. Photo-Interpretation* (Paris), IV, 1.45–50.

GOOSEN, D., 1967, Aerial photo interpretation in soil survey, *F.A.O. Soils Bull.* **6**, 55 pp.

GOTTSCHALK, L. C., 1964, Reservoir sedimentation, in V. T. Chow (ed.), *Handbook of Applied Hydrology* (McGraw-Hill, New York), 17.1–17.34.

GOUDIE, A., R. U. COOKE, and I. EVANS, 1970, Experimental investigation of rock weathering by salts, *Area*, 42–8.

GRANT, U. S., 1954, Subsidence of the Wilmington oil field, California, in R. H. Jahns (ed.), *Geology of Southern California*, Chapter 10 (3), 19–24.

GREGORY, K. J., and D. E. WALLING, 1973, *Drainage Basin Form and Process, a Geomorphological Approach* (Arnold, London), 456 pp.

GUILCHER, A., and C. A. M. KING, 1961, Spits, tombolos and tidal marshes in Connemara and West Kerry, Ireland, *Proc. Roy. Ir. Acad.* **61B 17**, 283–338.

GULLENTOPS, F. B., 1963, La cartographie géomorphologique en Belgique, *Geogr. Stud.* (Warsaw), **46**, 57–8.

GUY, H. P., 1965, Residential construction and sedimentation at Kensington, Md., in Federal Inter-Agency Sedimentation Conf., Jackson, Miss., 1963, *U.S. Dept. Agric. Misc. Publ.* **970**, 30–7.

HAANTJENS, H. A., 1968, The relevance for engineering of principles, limitations and developments in land system surveys in New Guinea, *Proc. 4th Conf. Australian Road Research Board*, **4**, 1,593–1,612.

HAANTJENS, H. A. (compiler), 1972, Lands of the Aitape-Ambunti area, Papua New Guinea (CSIRO, Australia), *Land Research Series*, **30**, 243 pp.

HACK, J. T., 1957, Studies of longitudinal stream profiles in Virginia and Maryland, *U.S. Geol. Surv. Prof. Paper*, **294-B**, 97 pp.

HACK, J. T., and J. C. GOODLETT, 1960, Geomorphology and forest ecology of a mountain region in the Central Appalachians, *U.S. Geol. Surv. Prof. Paper*, **347**, 66 pp.

HACKETT, J. E., and M. R. McCOMAS, 1969, Geology for planning in McHenry County, *Illinois State Geological Survey Circular*, **438**, 31 pp.

HAMILTON, H. R., et al., 1969, *Systems Simulation for Regional Analysis: an Application to River-Basin Planning* (Massachusetts Institute of Technology Press, Cambridge, Mass.), 407 pp.

HANKE, S. H., 1972, Flood losses—will they ever stop? *J. Soil and Water Conservation*, **27**, 242-3.

HANSEN, B. L., 1966, Instruments for temperature measurements in permafrost, *Proc. Permafrost Inter. Conf.*, Nat. Academy of Sciences—National Research Council, Publ. **1287**, 356-8.

HANSEN, H. P. (ed.), 1967, *Arctic Biology* (Oregon State Univ. Press, Corvallis), 318 pp.

HARRISON, W., 1969, Empirical equations for foreshore changes over a tidal cycle, *Mar. Geol.* **7**, 529-51.

HAUGEN, R. K., and J. BROWN, 1971, Natural and man-induced disturbances of permafrost terrane, in D. R. Coates (ed.), *Environmental Geomorphology* (Publ. in Geom., State Univ. New York, Binghamton), 139-49.

HEIM, A., 1932, *Bergsturz und Menschenleben* (Fretz and Wasmuth, Zurich).

HELD, R. B., and M. CLAWSON, 1965, *Soil Conservation in Perspective* (John Hopkins Press, Baltimore), 344 pp.

HELD, R. B., M. G. BLASE, and J. F. TIMMONS, 1962, Soil erosion and some means for its control, *Agric. and Home Ec. Expl. Station, Iowa State University of Sci. and Tech., Spec. Report*, **29**, 32 pp.

HENDERSON, F. M., 1966, *Open Channel Flow* (MacMillan, London), 522 pp.

HIGGINBOTTOM, I. E., and P. G. FOOKES, 1970, Engineering aspects of periglacial features in Britain, *Q. J. Eng. Geol.* **3**, 85-117.

HIGH, C., and F. K. HANNA, 1970, A method for the direct measurement of erosion on rock surfaces, *Brit. Geom. Res. Group Tech. Bull.* **5**, 24 pp.

HITE, J. C., and J. M. STEPP, 1971, *Coastal Zone Resource Management* (Praeger, New York), 169 pp.

HJÜLSTROM, F., 1939, Transportation of detritus by moving water, in P. D. Trask (ed.), *Recent Marine Sediments* (Am. Assoc. Petrol. Geol.), 5-31.

HJÜLSTROM, F., 1949, Climatic changes and river patterns, *Geografiska Annaler*, **31**, 83-9.

H.M.S.O., 1929, *Royal Commission on Mining Subsidence, Final Report*, Cmd. 2899.

H.M.S.O., 1949, *Inter-Departmental Committee Report on Mining Subsidence, Report of Committee on Mining Subsidence*, Cmd. 7637.

H.M.S.O., 1969, *A Selection of Technical Reports Submitted to the Aberfan Tribunal* (London), 149 pp.

HOLEMAN, H. N., 1968, The sediment yield of major rivers of the world, *Water Resources Research*, **4**, 737-47.

HONEYBORNE, D. B., and P. B. HARRIS, 1958, The structure of porous building stone and its relation to weathering behaviour, *The Colston Papers*, **10**, 343-65.

HORTON, R. E., 1945, Erosional development of streams and their drainage basins: hydrophysical approach to quantitative morphology, *Bull. Geol. Soc. Am.* **56**, 275-370. This article is conveniently abbreviated and reproduced in G. H. Dury, 1970, *Rivers and River Terraces* (MacMillan, London), 117-65.

HOSKING, J. R., and L. W. TUBEY, 1969, Research on low-grade and unsound aggregates, *Road Research Laboratory Report*, LR **293**, 30 pp.

HOWE, G. M., H. O. SLAYMAKER, and D. M. HARDING, 1967, Some aspects of the flood hydrology of the upper catchments of the Severn and Wye, *Trans. Inst. Brit. Geogr.* **41**, 33–58.

HOYT, W. G., and W. B. LANGBEIN, 1955, *Floods* (Univ. Princeton Press, Princeton), 469 pp.

HUDSON, N., 1971, *Soil Conservation* (Batsford, London), 320 pp.

HUGHES, O. L., 1972, Surficial geology and land classification, Mackenzie Valley transportation corridor, *Canadian Northern Pipeline Research Conference*, (National Research Council of Canada) 17–24.

HUNT, C. B., 1950, Military geology, *Application of Geology to Engineering Practice* (Geological Soc. of America, Berkey Volume), 295–327.

HUTCHINSON, J. N., 1967, The free degradation of London Clay cliffs, *Proc. Geotech. Conf. Oslo*, **1**, 113–8.

HUTCHINSON, J. N., 1968, Mass movement, in R. W. Fairbridge (ed.), *Encyclopaedia of Geomorphology* (Reinhold, New York), 688–95.

HUTCHINSON, J. N., 1969, A reconsideration of the coastal landslides at Folkestone Warren, Kent, *Géotechnique*, **19**, 6–38.

INGLIS, C. C., 1947, Meanders and their bearing on river training, *Proc. Inst. Civ. Eng.*, Maritime and Waterways Paper No. **7**.

INGLIS, C. C., 1949, *The Behaviour and Control of Rivers and Canals*, Research Publication, Control Board Irrigation, India, No. **13**.

INSTITUTION OF CIVIL ENGINEERS, 1966, *River Flood Hydrology* (I.C.E., London), 230 pp.

INSTITUTION OF MINING AND METALLURGY, 1965, *Opencast Mining, Quarrying and Alluvial Mining* (London), 772 pp.

ISAAC, P. C. G., 1967, *River Management* (Maclaren, London), 258 pp.

JACKS, G. V., and R. O. WHYTE, 1939, *The Rape of the Earth—a world survey of soil erosion* (Faber, London), 313 pp.

JACKSON, K., 1971, Notts/Derbys: a sub-regional landscape survey, *J. Town Planning Inst.* **57**, 203–4.

JACKSON, M. L., S. A. TYLER, A. L. WILLIS, G. A. BOURBEAU, and P. R. PENNINGTON, 1948, Weathering sequence of clay size minerals in soils and sediments, *J. Phys. and Coll. Chem.* **52**, 1237–60.

JAMES, L. B., 1968, Failure of Baldwin Hills Reservoir, Los Angeles, California, in G. A. Kiersch (ed.), *Engineering Geology Case Histories*, **6**, 1–11.

JENNINGS, J. E., 1966, Building on dolomites in the Transvaal, *The Civil Engineer in South Africa* (South African Institution of Civil Engineers, Johannesburg), 41–62.

JENNY, H., 1941, *Factors of Soil Formation* (McGraw-Hill, New York), 281 pp.

JOHNSON, D. W., 1919, *Shore Processes and Shoreline Development* (Wiley, New York), 584 pp.

JOHNSON, P. A., 1971, *Soils in Derbyshire*, Soil Survey of England and Wales, Harpenden, Soil Survey Record, **4**, 100 pp.

JOHNSON, W. D., 1904, The profile of maturity in alpine glacial erosion, *J. Geol.* **12**, 569–78.

JOHNSTON, G. H., 1966a, Engineering site investigations in permafrost areas, *Proc. Permafrost Int. Conf.* (Nat. Academy Sciences—National Research Council), Publ. **1287**, 371–4.

JOHNSTON, G. H., 1966b, Pile construction in permafrost, *Proc. Permafrost Int. Conf.* (Nat. Academy Sciences—National Research Council), Publ. **1287**, 477–84.

JOLLIFFE, I. P., 1964, An experiment designed to compare relative rates of movement of different size of beach pebbles, *Proc. Geol. Assoc.* **75**, 67–86.

JONG, J. D. de, 1960, The morphological evolution of the Dutch coast, *Geol. en Mijnbouw*, **39**, 638–43.

KACHURIN, S. P., 1962, Thermokarst within the territory of the U.S.S.R., *Biul. Peryglac.* **11**, 49–55.

KATES, R. W., 1962, *Hazard and Choice Perception in Flood Plain Management* Univ. Chicago, Dept. of Geography, Research Paper, **78**, 157 pp.

KATES, R. W., 1965, *Industrial Flood Losses*, Univ. Chicago, Dept. of Geography, Research Paper, **98**, 76 pp.

KATES, R. W., and G. F. WHITE, 1964, Flood hazard evaluation, in G. F. White (ed.), 1964, *Choice of Adjustment to Floods*, Univ. Chicago, Dept. of Geography, Research Paper, **93**, 135–47.

KELLAWAY, G. A., and J. H. TAYLOR, 1968, The influence of landslipping on the development of the city of Bath, England, *Proc. 23rd Intern. Geol. Congr.* **12**, 65–76.

KING, C. A. M., 1953, The relationship between wave incidence, wind direction and beach changes at Marsden Bay, Co. Durham, *Trans. Inst. Brit. Geogr.* **19**, 13–23.

KING, C. A. M., 1964, The character of the offshore zone and its relationship to the foreshore near Gibraltar Point, Lincolnshire, *East Midland Geographer*, **3**, 230–43.

KING, C. A. M., 1965, Some observations on the beaches of the west coast of County Donegal, *Irish Geog.* **5**, 40–50.

KING, C. A. M., 1971, Geometrical forms in geomorphology, *Int. Journ. Math. Educ. Sci. Technol.* **2**, 153–69.

KING, C. A. M., 1972, *Beaches and Coasts*, 2nd edition (Arnold, London), 570 pp.

KING, C. A. M., 1973, Processes of coastal accretion in south Lincolnshire, in D. R. Coates (ed.), *Coastal Geomorphology* (Publ. in Geomorph., State Univ. New York, Binghamton), 73–98.

KING, C. A. M., and P. M. MATHER, 1972, Spectral analysis applied to the study of time series from the beach environment, *Mar. Geol.* **13**, 123–42.

KING, C. A. M., and M. J. McCULLAGH, 1971, A simulation model of a complex recurved spit, *J. Geol.* **79**, 22–37.

KING, R. B., 1970, A parametric approach to land system classification, *Geoderma*, **4**, 37–46.

KINSMAN, B., 1965, *Wind Waves, their Generation and Propagation on the Ocean Surface* (Prentice-Hall, Englewood Cliffs).

KIRKBY, M. J., 1969, Infiltration, throughflow and overland flow; and erosion by water on hillslopes, in R. J. Chorley (ed.), *Water, Earth and Man* (Methuen, London), 215–238.

KIRKBY, M. J., and R. J. CHORLEY, 1967, Throughflow, overland flow and erosion, *Bull. Int. Assoc. Sci. Hyd.* **12**, 5–21.

KLIMASZEWSKI, M., 1956, The principles of the geomorphological survey of Poland, *Przeglad Geograficzny*, **28** (Suppl.), 32–40.

KLIMASZEWSKI, M., 1961a, Enquête sur les organismes de géomorphologie appliquée, *Rev. de Géomorph. Dynamique*, **12**, 43–5.

KLIMASZEWSKI, M., 1961b, The problems of the geomorphological and hydrographic map on the example of the Upper Silesian industrial district, *Problems of Applied Geography, Geographical Studies* (Polish Acad. Sciences Institute of Geography, Warsaw), **25**, 73–81.

KLINGEBIEL, A. A., and P. H. MONTGOMERY, 1961, Land-capability classification,

Soil Conservation Service, U.S. Dept. Agric., *Agric. Handbook*, **210**, 21 pp.

KNIGHT, C. G., and T. J. RICKARD, 1971, Perception and ethnogeography in southwestern Kansas, *Proc. Ass. Am. Geog.* **3**, 96–100.

KOLLMORGEN, W. M., 1953, Settlement control beats flood control, *Economic Geography*, **29**, 208–15.

KOMAR, P. D., 1971, The mechanism of sand transport on beaches, *J. Geophys. Res.* **76**, 713–81.

KOMAR, P. D., and D. L. INMAN, 1970, Longshore sand transport on beaches, *J. Geophys. Res.* **75**, 5,914–27.

KRUMBEIN, W. C., and W. R. JAMES, 1965, A lognormal size distribution model for estimating stability of beach fill material, U.S. Army Coastal Engineering Research Center, Washington, *Tech. Mem.* **16**, 17 pp.

KRYNINE, D. P., and W. R. JUDD, 1957, Frost and permafrost, in *Principles of Engineering Geology and Geotechnics* (McGraw-Hill, New York), 385–421.

KUIPER, E., 1965, *Water Resources Development* (Butterworth, London), 483 pp.

KUNREUTHER, H., and J. R. SHEAFFER, 1970, An economically meaningful and workable system for calculating flood insurance rates, *Water Resources Research* **6**, 659–67.

KWAAD, F. J. P. M., 1970, Experiments on the granular disintegration of granite by salt action, *Fysisch Geog. en Bodemkundig Lab.*, Publ. **16**, 67–80.

LACEY, G., 1929–30, Stable channels in alluvium, *Proc. Inst. Civil Eng.* **229**, 259–92.

LACHENBRUCH, A. H., 1966, Contraction theory of ice-wedge polygons: a qualitative discussion. *Proc. Permafrost Int. Conf.* (Nat. Academy Sciences—National Research Council), Publ. **1287**, 65–71.

LACHENBRUCH, A. H., 1970, Some estimates of the thermal effects of a heated pipeline in permafrost, *U.S. Geological Survey Circular*, **632**, 23 pp.

LAMBE, T. W., and R. V. WHITMAN, 1969, *Soil Mechanics* (Wiley, New York), 553 pp.

LANE, E. W., 1937, Stable channels in erodible materials, *Trans. Am. Soc. Civ. Engrs.* **102**, 123–94.

LANE, E. W., 1955, The importance of fluvial morphology in hydraulic engineering, *Am. Soc. Civ. Eng. Proc.* **81**, 1–17.

LANGBEIN, W. B., 1953, Flood insurance, *J. Land Economics*, 323–30.

LANGBEIN, W. B., 1962, Hydraulics of river channels as related to navigability, *U.S. Geol. Surv. Water Supply Paper*, **1539**-W, 30 pp.

LANGBEIN, W. B., and S. A. SCHUMM, 1958, Yield of sediment in relation to mean annual precipitation, *Trans. Am. Geophys. Union*, **30**, 1076–84.

LEE, T. M., 1972, Effect of transportation planning on flood plain management, *Proc. Am. Soc. Civ. Eng. J. Hyd. Div.* HY3, 475–88.

LEGGET, R. F., 1966, Permafrost in North America, *Proc. Permafrost Inter. Conf.* (Nat. Academy of Sciences—National Research Council), Publ. **1287**, 2–7.

LEGGET, R. F., and I. C. MACFARLANE, (eds.), 1972, *Canadian Northern Pipeline Research Conference, Technical Mem.* **104** (National Research Council of Canada, Ottawa), 331 pp.

LEITH, C. K., and W. J. MEAD, 1911, Origin of the iron ores of central and northeastern Cuba, *Trans. Am. Inst. Mining Engrs.* **42**, 90–102.

LENHART, W. B., 1962, Sand and gravel, *Reviews in Engineering Geology*, **1**, 187–98.

LEOPOLD, L. B., 1968, Hydrology for urban land planning—a guidebook on the hydrologic effects of urban land use, *U.S. Geol. Surv. Circular*, **554**, 18 pp.

LEOPOLD, L. B., 1969a, Quantitative comparison of some aesthetic factors among rivers, *U.S. Geological Survey Circular*, **620**, 16 pp.

LEOPOLD, L. B., 1969b, Landscape aesthetics, *Natural History* (October), 35–46.

LEOPOLD, L. B., and T. MADDOCK, Jr., 1953, The hydraulic geometry of stream channels and some physiographic implications, *U.S. Geol. Surv. Prof. Paper*, **252**, 57 pp.

LEOPOLD, L. B., and T. MADDOCK, Jr., 1954 *The Flood Control Controversy* (Ronald Press, New York), 278 pp.

LEOPOLD, L. B., and J. P. MILLER, 1956, Ephemeral streams—hydraulic factors and their relation to the drainage net, *U.S. Geol. Surv. Prof. Paper*, **282**-A, 36 pp.

LEOPOLD, L. B., and M. G. WOLMAN, 1957, River channel patterns, braided, meandering and straight, *U.S. Geol. Surv. Prof. Paper*, **282**-B.

LEOPOLD, L. B., M. G. WOLMAN, and J. P. MILLER, 1964, *Fluvial Processes in Geomorphology* (Freeman, San Francisco), 522 pp.

LEWIS, W. V., 1931, The effect of wave incidence on the configuration of a shingle beach, *Geogr. J.* **78**, 129–48.

LINELL, K. A., 1960, Frost action and permafrost, in K. B. Woods (ed.), *Highway Engineering Handbook* (McGraw-Hill, New York), section 13.

LINSLEY, R. K., M. A. KOHLER, and J. L. H. PAULHUS, 1949, *Applied Hydrology* (McGraw-Hill, New York), 689 pp.

LINTON, D. L., 1951, The delimitation of morphological regions, in L. D. Stamp and S. W. Wooldridge (eds.), *London Essays in Geography* (Longman, London), 199–217.

LINTON, D. L., 1968, The assessment of scenery as a natural resource, *Scottish Geog. Mag.* **84**, 218–38.

LITTLE, A. L., 1969, The engineering classification of residual tropical soils, *7th Proc. Int. Conf. Soil Mech. and Found. Eng.* **1**, 1–10.

LITTON, R. B., Jr., 1968, Forest landscape description and inventories—a basis for land planning and design, U.S. Department of Agriculture, *Forest Service Research Paper*, PSW–**49**, 64 pp.

LITTON, R. B., Jr., and K. H. CRAIK, 1969, Aesthetic Dimensions of the Landscape (Resources for the Future, Washington, D.C.), unpublished, 238 pp.

LOFGREN, B. E., 1961, Measurement of compaction of aquifer systems in areas of land subsidence, *U.S. Geol. Surv. Prof. Paper*, **424**-B, B49–B52.

LOFGREN, B. E., 1963, Land subsidence in the Arvin-Maricopa area, San Joaquin Valley, California, *U.S. Geol. Surv. Prof. Paper*, **475**-B, B171–B173.

LOFGREN, B. E., 1968, Analysis of stresses causing land subsidence, *U.S. Geol. Surv. Prof. Paper*, **600**-B, B219–B225.

LOFGREN, B. E., 1969, Land subsidence due to the application of water, in D. J. Varnes and G. Kiersch (eds.), *Reviews in Engineering Geology* (Geol. Soc. Am., Colorado), **2**, 271–303.

LOFGREN, B. E., and R. L. KLAUSING, 1969, Land subsidence due to ground-water withdrawal, Tulare–Wasco area, California, *U.S. Geol. Surv. Prof. Paper*, **437**-B, 103 pp.

LONGWELL, C. R., R. F. FLINT, and J. E. SANDERS, 1969, *Physical Geology* (Wiley, New York), 685 pp.

LOWENTHAL, D. (ed.), 1967, *Environmental Perception and Behaviour*, Univ. Chicago, Dept. of Geography, Research Paper, **109**, 88 pp.

LOWENTHAL, D., and H. C. PRINCE, 1965a, English landscape tastes, *Geog. Rev.* **54**, 309–46.

LOWENTHAL, D., and H. C. PRINCE, 1965b, English landscape tastes, *Geog. Rev.* **55**, 186–222; reprinted in P. W. English and R. C. Mayfield (eds.), *Man, Space and Environment* (Oxford University Press, New York, 1972), 81–114.

LUEDER, D. R., 1959, *Aerial Photographic Interpretation: Principles and application* (McGraw-Hill, New York), 462 pp.

LUNDQUIST, J., 1969, Earth and ice mounds: a terminological discussion, in T. L. Péwé (ed.), *The Periglacial Environment—Past and Present* (McGill–Queen's Univ. Press, Montreal), 203–15.

LYELL, K. A., 1970, The interpretation of water pressure factors for use in the slope theory, in P. W. J. van Rensburg (ed.), *Planning Open Pit Mines* (S. Afr. Inst. Min. and Metallurgy, Johannesburg), 73–85.

MABBUTT, J. A., and G. A. STEWART, 1963, The application of geomorphology in resource surveys in Australia and New Guinea, *Revue Géomorph. Dyn.* **14,** 97–109.

MACAR, P., P. DE BÉTHUNE, J. MAMMERICKX, and G. SERET, 1961, Travaux preparatoires a l'élaboration d'une carte géomorphologique de Belgique, *Ann. de Soc. Géol. de Belgique* (Liège), **84.**

MACDONALD, R. St. J. (ed.), 1966, *The Arctic Frontier* (Univ. Toronto Press, Toronto), 311 pp.

MCFARLANE, M. J., 1971, Lateritization and landscape development in Kyagwe, Uganda, *Q. J. Geol. Soc. London,* **126,** 501–39.

MCINTYRE, D. S., 1958a, Permeability measurements of soil crusts by raindrop impact, *Soil Science,* **85,** 185–9.

MCINTYRE, D. S., 1958b, Soil splash and the formation of surface crusts by raindrop impact, *Soil Science,* **85,** 261–6.

MCKINSTRY, H. E., 1948, *Mining Geology* (Prentice-Hall, New York), 219–32.

MCLELLAN, A. G., 1967, *The Distribution of Sand and Gravel Deposits in West Central Scotland and Some Problems Concerning Their Utilization* (University of Glasgow, Glasgow), 45 pp.

MACKAY, J. R., 1966, Pingos in Canada, *Proc. Permafrost Int. Conf.* (Nat. Acad. Sciences—National Research Council), Publ. **1287,** 71–6.

MACKAY, J. R., 1970, Disturbances to the tundra and forest environment of the western Arctic, *Canadian Geotech. J.* **7,** 420–32.

MACKAY, J. R., 1972, The world of underground ice, *Ann. Ass. Am. Geog.* **62,** 1–22.

MATTHIAS, G. F., 1967, Weathering rates of Portland Arkose tombstones, *J. Geol. Ed.* **15,** 140–4.

MAYUGA, M. N., and D. R. ALLEN, 1969, Subsidence in the Wilmington oil field, Long Beach, California, U.S.A., *Pub. Inst. Sci. Hyd.* **88,** 66–79.

MEDHURST, F., 1968, A method of regional landscape analysis, *Planning Outlook,* **4,** 61–9.

MEIGS, P., 1953, World distribution of arid and semi-arid homoclimates, in *Reviews of Research on Arid Zone Hydrology* (UNESCO, Paris), 203–9.

MELTON, M. A., 1958, Correlation structure of morphometric properties of drainage systems and their controlling agents, *J. Geol.* **66,** 442–60.

MERRILL, C. L., J. A. PIHLAINEN, and R. F. LEGGET, 1960, The new Aklavik—search for the site, *Engineering J.* **43,** 52–7.

MILLER, R. E., 1961, Compaction of an aquifer system computed from consolidation tests and decline in artesian head, *U.S. Geol. Surv. Prof. Paper,* **424**–B, B54–B58.

MILLER, R. A., J. TROXELL, and L. B. LEOPOLD, 1971, Hydrology of two small river basins in Pennsylvania before urbanization, *U.S. Geol. Surv. Prof. Paper,* **701**–A, A1–A57.

MILLER, V. C., 1961, *Photogeology* (McGraw-Hill, New York), 248 pp.

MILNE, G., 1935, Composite units for the mapping of complex soil associations,

Trans. 3rd Intern. Cong. Soil Sci. **1**, 345–7.

MINISTRY OF HOUSING AND LOCAL GOVERNMENT, 1960, *The Control of Mineral Working* (H.M.S.O., London).

MINISTRY OF OVERSEAS DEVELOPMENT, 1970, *The Work of the Land Resources Division* (Tolworth, England), 64 pp.

MINISTRY OF TOWN AND COUNTRY PLANNING, 1948, *Report of the Advisory Committee on Sand and Gravel* (H.M.S.O., London), Part 1.

MINTY, E. J., 1965, Preliminary report on an investigation into the influence of several factors on the sodium sulphate soundness test for aggregate, *Austr. Road Res.* **2**, 49–52.

MINTY, E. J., and K. MONK, 1966, Predicting the durability of rock, *Proc. 3rd Conf. Austr. Road Res. Board*, **3**, 1,316–33.

MOLLARD, J. D., 1972, Airphoto terrain classification and mapping for northern feasibility studies, *Canadian Northern Pipeline Research Conf.* (National Research Council of Canada), 105–28.

MOLLARD, J. D., and J. A. PIHLAINEN, 1966, Airphoto interpretation applied to road selection in the Arctic, *Proc. Permafrost Int. Conf.* (Nat. Academy Sciences—National Research Council), Publ. **1287**, 381–7.

MOLNAR, I., 1964, Soil conservation—economic and social considerations, *J. Austr. Inst. Agr. Sci.* **30**, 247–57.

MORGENSTERN, N. R., and V. E. PRICE, 1965, The analysis of the stability of general slip surfaces, *Géotechnique*, **15**, 79–93.

MORISAWA, M., 1968, *Streams, Their Dynamics and Morphology* (McGraw-Hill, New York), 175 pp.

MORISAWA, M., 1971, Evaluating riverscapes, in D. R. Coates (ed.), *Environmental Geomorphology* (Publ. in Geom., State Univ. New York, Binghamton), 91–106.

MOSS, R. P., 1965, Slope development and soil morphology in a part of south-west Nigeria, *J. of Soil Sci.* **16**, 192–209.

MOSS, R. P., 1968, Land use, vegetation and soil factors in south-west Nigeria: A new approach, *Pacific Viewpoint*, **9**, 107–27.

MOSS, R. P., 1969, The appraisal of land resources in tropical Africa, *Pacific Viewpoint*, **10**, 18–27.

MULCAHY, M. J., 1960, Laterites and lateritic soils in south-western Australia, *J. of Soil Sci.* **11**, 206–24.

MULLER, F., 1963, Observations on pingos, *Nat. Res. Counc. Canada, Tech. Trans.* **1073**, 117 pp.

MULLER, S. W., 1947, *Permafrost or Permanently Frozen Ground and Related Engineering Problems* (Edwards, Ann Arbor), 230 pp.

MURPHY, F. C., 1958, *Regulating Flood-Plain Development*, Univ. Chicago, Dept. of Geography, Research Paper, **56**, 204 pp.

MURRAY, H. A., 1943, *Thematic Apperception Test: Pictures and Manual* (Harvard U.P., Cambridge).

MURRAY, W. H., 1962, *Highland Landscape—a Survey* (National Trust of Scotland), 80 pp.

MUSGRAVE, G. W., 1947, Quantitative evaluation of factors in water erosion—first approximation, *J. Soil Wat. Cons.* **2**, 133–8.

NAKANO, T., 1963, Landform type analysis on aerial photographs: its principles and its technique, *Trans. Symp. Photo Interpretation (Delft)*, *Archives Internationales de Photogrammetrie*, **14**, 149–52.

NATIONAL ACADEMY OF SCIENCES—NATIONAL RESEARCH COUNCIL, 1966, *Proceedings of the Permafrost International Conference*, Publication **1287**, 563 pp.

NATIONAL COAL BOARD, 1966, Subsidence Engineers' Handbook (London), 128 pp.

NAYSMITH, J. K., 1971, *Canada North—Man and the Land* (Dept. Indian Affairs and Northern Development, Ottawa), 44 pp.

NELSON, L. M., 1970, A method of estimating annual suspended sediment discharge, *U.S. Geol. Surv. Prof. Paper*, **700**–C, 233–6.

NIEDERODA, A. W., and W. F. TANNER, 1970, Preliminary study of transverse bars, *Marine Geology*, **6**, 41–62.

NIIR, 1971, The production of soil engineering maps for roads and the storage of materials data, Nat. Inst. for Road Research, Pretoria, South Africa, *Tech. Recommendations for Highways*, **2**.

NIXON, M., 1959, A study of the bankfull discharges of rivers in England and Wales, *Proc. Inst. Civ. Eng.*, paper **6322**, 157–74.

NIXON, M., 1963, Flood regulation and river training in England and Wales, *Institution of Civ. Engineers, Symposium on Conservation of Water Resources*, Sessions III and IV, 35–48.

OCEAN WAVE SPECTRA, 1963, *Ocean Wave Spectra*, Proc. International Conf. (Prentice-Hall, Englewood Cliffs).

OLIVIER, H., 1972, *Irrigation and Water Resources Engineering* (Arnold, London), 190 pp.

OLLIER, C. D., 1965, Some features of granite weathering in Australia, *Zeit. für Geom.* **9**, 285–304.

OLLIER, C. D., 1969, *Weathering* (Oliver and Boyd, Edinburgh), 304 pp.

OLLIER, C. D., C. J. LAWRANCE, R. WEBSTER, and P. H. T. BECKETT, 1969, *The land systems of Uganda* (M.E.X.E., Christchurch, England), Rep. No. **959**, 234 pp.

OLLIER, C. D., R. WEBSTER, C. J. LAWRANCE, and P. H. T. BECKETT, 1967, The preparation of a land classification map at 1/1,000,000 of Uganda, *Actes du II*[e] *Symposium International de Photo-Interpretation* (Paris), Sect. IV. 1, 115–22.

PATTERSON, J. L., 1970, Evaluation of the streamflow data program for Arkansas, *U.S. Geol. Surv. Prof. Paper*, **700**–D, D244–D256.

PATTON, F. D., 1970, Significant geologic factors in rock slope stability, in P. W. J. van Rensburg (ed.), *Planning Open Pit Mines* (S. Afr. Inst. Min. and Metallurgy, Johannesburg), 143–51.

PECK, R. B., 1967, Stability of natural slopes, *Proc. Am. Soc. Civ. Eng. 93*, paper 5323, *Jour. Soil Mech. and Foundation Div.* **SM4**, 403–17.

PELTIER, L., 1950, The geographic cycle in periglacial regions as it is related to climatic geomorphology, *Ann. Ass. Am. Geog.* **40**, 214–36.

PERRY, R. A. (compiler), 1962, General report on lands of the Alice Springs Area 1956–7, CSIRO, Australia, *Land Research Series*, **6**, 280 pp.

PÉWÉ, T. L., 1954, Effect of permafrost on cultivated fields, Alaska, *U.S. Geological Survey Bulletin*, **989**–F, 315–51.

PÉWÉ, T. L., 1966, Ice-wedges in Alaska—classification, distribution and climatic significance, *Proc. Permafrost Int. Conf.* (Nat. Academy Sciences—National Research Council), Publ. **1287**, 76–81.

PÉWÉ, T. L. (ed.), 1969, *The Periglacial Environment, Past and Present* (McGill–Queen's U. Press, Montreal), 487 pp.

PIHLAINEN, J. A., 1962, *Inuvik, N.W.T., Engineering Site Information*, (Nat. Res. Counc. Canada, Div. Building Res.), Technical Paper, **135**, 18 pp.

PIOTROVSKI, M. V., Y. G. SIMONOV, and L. B. ARISTARKHOVA, 1972, Detailed geomorphological mapping in mineral prospecting, in J. Demek (ed.), *Manual of Detailed Geomorphological Mapping* (Academia, Prague), 267–77.

PITEAU, D. R., 1970, Geological factors significant to the stability of slopes cut in rock, P. W. J. van Rensburg (ed.), *Planning Open Pit Mines* (S. Afr. Inst. of Mining and Metallurgy), 33–53.

POCHON, J., and C. JATON, 1967, Causes of the deterioration of building materials. 2. The role of microbiological agencies in the deterioration of stone, *Chem. and Ind.* **38**, 1,587–89.

POLAND, J. F., 1969, Land subsidence in western United States, in R. A. Olson and M. M. Wallace (eds.), *Geologic Hazards and Public Problems* (Office of Emergency Preparedness, Region 7, Santa Rosa, Calif.), 77–96.

POLAND, J. F., and G. H. DAVIS, 1969, Land subsidence due to withdrawal of fluids, in D. J. Varnes and G. Kiersch (eds.), *Reviews in Engineering Geology* (Geol. Soc. Am., Colorado), **2**, 187–269.

POLLARD, E., and A. MILLAR, 1968, Wind erosion in the East Anglian Fens, *Weather*, **23**, 415–17.

PORT OF LONG BEACH, 1971, *Vertical Movement in Long Beach Harbor District* (Port of Long Beach, Long Beach), 9 pp.

POTTS, A. S., 1970, Frost action in rocks: some experimental data, *Trans. Inst. Brit. Geog.* **49**, 109–24.

PRICE, L. W., 1972, *The Periglacial Environment*, Ass. Am. Geog. Comm. College Geog., Resource Paper, **14**, 88 pp.

PRICE, R. J., 1973, *Glacial and Fluvioglacial Landforms* (Oliver and Boyd, Edinburgh), 242 pp.

PRICE, R. J., and D. E. SUGDEN (eds.), 1972, *Polar Geomorphology*, Inst. Brit. Geog., London, Sp. Pub. **4**, 211 pp.

PRICE, W. A., 1955, Correlation of shoreline types with offshore bottom conditions, *A and M College of Texas, Dept. of Oceanog. Proj.* **63**.

PRITCHARD, A. L., 1936, Factors influencing the upstream spawning migration of the pink salmon (Oncorhynchus gorbuscha Walbaum), *J. Biol. Bd. Canada*, **2**, 383–9.

PRITCHARD, G. B., 1962, Inuvik, Canada's new Arctic town, *Polar Record*, **11**, 145–54.

PROKOPOVICH, N. P., 1972, Land subsidence and population growth, *Int. Geol. Cong. 1972* (Montreal), Section **13**, 44–54.

RADLEY, J., and C. SIMS, 1967, Wind erosion in East Yorkshire, *Nature*, **216**, 20–2.

RAHN, P. H., 1971, The weathering of tombstones and its relationship to the topography of New England, *J. Geol. Ed.* **19**, 112–18.

RAPP, A., 1960, Recent development of mountain slopes in Karkevagge and surroundings, *Geog. Ann.* **42**, 71–206.

RAWIEL, R. F., 1968, Determination of volumes in opencast mining by aerial surveys, in Inst. Min. and Met., *Opencast Mining, Quarrying and Alluvial Mining* (London), 483–92.

RENSBURG, P. W. J. VAN (ed.), 1970, *Planning Open Pit Mines* (S. Afr. Inst. of Min. and Metallurgy, Johannesburg), 388 pp.

REYNOLDS, H. R., 1961, *Rock Mechanics* (Lockwood, London), 136 pp.

RITCHIE, A. M., 1958, Recognition and identification of landslides, in E. B. Eckel (ed.), *Landslides and Engineering Practice* (Highway Research Board, Special Report, **29**), 48–68.

ROBERTS, D. V., and G. E. MELICKIAN, 1970, Geologic and other natural hazards in desert areas, *Dames and Moore Engineering Bull.* **37**, 1–12.

ROBERTSON, A. M., 1970, The interpretation of geological factors for use in slope theory, in P. W. J. van Rensburg (ed.), *Planning Open Pit Mines* (S. Afr. Inst. of

Min. and Metallurgy, Johannesburg), 55–71.

ROBERTSON, R. G., 1955, Aklavik—a problem and its solution, *Canadian Geog. J.*, **50**, 196–205.

ROBINSON, A. H. W., 1968, The submerged glacial landscape off the Lincolnshire coast, *Trans. Inst. Brit. Geogr.* **44**, 199–32.

ROBINSON, D. N., 1969, Soil erosion by wind in Lincolnshire, March, 1968, *East Midland Geographer*, **4**, 351–62.

RODDA, J. C., 1967, The significance of characteristics of basin rainfall and morphometry in a study of floods in the United Kingdom, *Inst. Ass. Sci. Hydr., Leningrad Symposium*, 835–45.

RODDA, J. C., 1969, The flood hydrograph, in R. J. Chorley (ed.), *Water, Earth and Man* (Methuen, London), 405–18.

RODDA, J. C., 1970, Rainfall excesses in the United Kingdom, *Trans. Inst. Brit. Geogr.* **49**, 49–60.

ROTH, E. S., 1965, Temperature and water content as factors in desert weathering, *J. Geol.* **73**, 454–68.

RUBEY, W. W., 1952, Geology and mineral resources of the Hardin and Drussels quadrangles (in Illinois), *U.S. Geol. Surv. Prof. Paper*, **218**, 175.

RUHE, R. V., 1956, Geomorphic surfaces and the nature of soils, *Soil Science*, **82**, 441–56.

RUHE, R. V., 1971, Stream regimen and man's manipulation, in D. R. Coates (ed.), *Environmental Geomorphology* (Publ. in Geom., State Univ. of New York, Binghamton), 9–23.

RUSSELL, R. C. H., and D. H. MACMILLAN, 1952, *Waves and Tides*, 2nd edition (Hutchinson, London), 348 pp.

RUSSELL, R. J., 1943, Freeze–thaw frequencies in the United States, *Trans. Am. Geoph. Un.* **24**, 125–33.

RUSSELL, R. J., 1967, *River Plains and Sea Coasts* (Univ. California Press, Berkeley and Los Angeles), 173 pp.

RUXTON, B. P., 1968, Measure of the degree of chemical weathering of rocks, *J. Geol.* **76**, 518–27.

RUXTON, B. P., and B. L. BERRY, 1957, The weathering of granite and associated erosional features in Hong Kong, *Geol. Soc. Am. Bull.* **68**, 1,263–92.

SAARINEN, T. F., 1966, *Perception of the Drought Hazard on the Great Plains*, Univ. Chicago, Dept. of Geography, Research Paper, **106**, 183 pp.

SAUNDERS, M. K., and P. G. FOOKES, 1970, A review of the relationship of rock weathering and climate and its significance to foundation engineering, *Eng. Geol.* **4**, 289–325.

SAVIGEAR, R. A. G., 1965, A technique of morphological mapping, *Annals Assoc. Am. Geogr.* **53**, 514–38.

SAVINI, J., and J. C. KAMMERER, 1961, Urban growth and the water regimen, *U.S. Geol. Surv. Water-Supply Paper*, **1591**–A.

SCHAFFER, R. J., 1932, *The Weathering of Natural Building Stones* (H.M.S.O., London), 149 pp.

SCHAFFER, R. J., 1959, Testing building stone, *Architectural Engineering Monumental Stone*, **16** and **17**.

SCHAFFER, R. J., 1967, Causes of the deterioration of building materials. 1. Chemical and physical causes, *Chem. and Ind.* **38**, 1,584–6.

SCHICK, A. P., 1971, A desert flood: physical characteristics; effects of Man, geomorphic significances, human adaptation—a case study of the southern Arava watershed, *Jerusalem Studies in Geography*, **2**, 91–155.

SCHNACKENBERG, E. C., 1951, Slope discharge formulae for alluvial streams, *Proc. N.Z. Inst. Eng.* **37**, 340–449.

SCHOU, A., 1945, Det Marine Forland, *Folia Geog. Danica*, **4**, 1–236.

SCHUMM, S. A., 1961, Effect of sediment characteristics on erosion and deposition in ephemeral stream channels, *U.S. Geol. Surv. Prof. Paper*, 352–C.

SCHUMM, S. A., 1963, Sinuosity of alluvial rivers on the Great Plains, *Bull. Geol. Soc. Am.* **74**, 1,089–1,100.

SCHUMM, S. A., 1967, Meander wave length of alluvial rivers, *Science*, **157**, 1549–50.

SCHUMM, S. A., 1969, River metamorphosis, *J. Hydraulics, Div., Proc. Am. Soc. Civ. Eng.* **95**, 255–73.

SCHUMM, S. A., 1971, Fluvial geomorphology: The historical perspective, in H. W. Shen (ed.), *Fluvial Geomorphology in River Mechanics*, Water Resources Pub. Fort Collins, Colorado, 4.1–4.30.

SCHUMM, S. A., 1972, *River Morphology* (Dowden, Hutchinson, Ross, Inc., Stroudsburg, Pennsylvania), 429 pp.

SCHUMM, S. A., and R. W. LICHTY, 1965, Time, space and causality in geomorphology, *Am. J. Sci.* **263**, 110–19.

SCOTT, R. F., and J. J. SCHOUSTRA, 1968, *Soil Mechanics and Engineering* (McGraw-Hill, New York), 314 pp.

SEWELL, W. R. D., 1965, *Water Management and Floods in the Fraser River Basin*, Univ. Chicago, Dept. of Geography, Research Paper, **100**, 163 pp.

SHAFER, E. L., Jr., J. F. HAMILTON, Jr., and E. A. SCHMIDT, 1969, Natural landscape preferences: a predictive model, *J. Leisure Res.* **1**, 1–19.

SHARP, R. P., 1942, Soil structures in the St. Elias Range, Yukon Territory, *J. Geomorph.* **5**, 274–301.

SHARPE, C. F. S., 1938, *Landslides and Related Phenomena* (Columbia Univ. Press, New York, Reprinted 1960 by Pageant Books Inc., New Jersey), 137 pp.

SHEAFFER, J. R., 1960, *Flood Proofing: an Element in a Flood Damage Reduction Program*, Univ. Chicago, Dept. of Geography, Research Paper, **65**, 190 pp.

SHERLOCK, R. L., 1922, *Man as a Geological Agent* (Witherby, London), 372 pp.

SHREVE, R. L., 1967, Infinite topologically random channel networks, *J. Geol.* **75**, 178–86.

SHUMSKY, P. A., and B. I. VTYURIN, 1966, Underground ice, *Proc. Permafrost Inter. Conf.* (Nat. Academy of Sciences—Nat. Research Council), Publ. **1287**, 108–13.

SHVETSOV, P. F. (ed.), 1959, *Principles of Geocryology* (Academy of Science, U.S.S.R., Moscow), vol. 1, 459 pp.

SILVESTER, R., 1960, Stabilization of sedimentary coastlines, *Nature*, **188**, 467–9.

SIMONETT, D. S., 1967, Landslide distribution and earthquakes in the Bewani and Torricelli Mountains, New Guinea: A statistical analysis, in J. N. Jennings and J. Mabbutt (eds.), *Landform Studies from Australia and New Guinea* (Australian Nat. Univ. Press and C.U.P.), 64–84.

SIMONS, D. B., and E. V. RICHARDSON, 1961, Forms of bed roughness in alluvial channels, *Proc. Am. Soc. Civil Engrs. J. Hydraulics Div.* **87** (HY3 pt. 1), 87–105.

SIMPSON, J. W., and P. J. HORROBIN (eds.), 1970, *The Weathering and Performance of Building Materials* (M.T.P., Aylesbury), 286 pp.

SIMS, J., and T. F. SAARINEN, 1969, Coping with environmental threat: Great Plains farmers and the sudden storm, *Ann. Ass. Am. Geog.* **59**, 667–86.

SKEMPTON, A. W., 1948, The rate of softening of stiff, fissured clays, *Proc. 2nd Inter. Conf. Soil Mech. Found. Engr.* (Rotterdam), vol. **2**, 50–3.

SKEMPTON, A. W., 1953, Soil mechanics in relation to geology, *Proc. Yorkshire Geol. Soc.* **29**, 33–62.

SKEMPTON, A. W., 1964, Long-term stability of clay slopes (Fourth Rankin Lecture), *Géotechnique*, **14**, 77–102.

SKEMPTON, A. W. and J. N. HUTCHINSON, 1969, Stability of natural slopes and embankment foundations, *State of the Art Volume, 7th Intern. Conf. Soil Mech. and Foundation Engineering* (Mexico), 291–340.

SLAYMAKER, H. O., 1972, Patterns of present sub-aerial erosion and landforms in mid-Wales, *Trans. Inst. Brit. Geogr.* **55**, 47–68.

SMALLEY, I. J., 1967, The subsidence of the North Sea Basin and the Geomorphology of Britain, *The Mercian Geologist*, **2**, 267–78.

SMALLEY, I. J., 1970, Cohesion of soil particles and the intrinsic resistance of simple soil systems to wind erosion, *J. Soil Sci.* **21**, 154–61.

SMITH, C. T., 1969, The drainage basin as an historical basis for human activity in R. J. Chorley (ed.), *Water, Earth and Man* (Methuen, London), 101–10.

SMITH, D. D., and W. H. WISCHMEIER, 1962, Rainfall erosion, *Advances in Agronomy*, **14**, 109–48.

SMITH, K., 1972, *Water in Britain* (Macmillan, London), 232 pp.

So, C. L., 1971, Mass movements associated with the rainstorms of June 1966 in Hong Kong, *Trans. Inst. Brit. Geogr.* **53**, 55–65.

SOIL SURVEY STAFF, 1951, *Soil Survey Manual* (U.S. Dept. Agric. Handbook, **18**), 503 pp.

SOLENTSEV, N. A., 1962, Basic problems in Soviet landscape science, *Soviet Geography*, **3**, 3–15.

SPARKS, B. W., 1971, *Rocks and Relief* (Longman, London), 404 pp.

STALL, J. B., L. C. GOTTSCHALK, A. A. KLINGSBIEL, E. L. SAUER, and E. E. DE TURK, 1949, The silt problem at Spring Lake, Macomb, Illinois, *Illinois State Water Surv. Div. Dept. Invest.* **4**.

STEERS, J. A. (ed.), 1971a, *Introduction to Coastline Development* (Macmillan, London), 229 pp.

STEERS, J. A. (ed.), 1971b, *Applied Coastal Geomorphology* (Macmillan, London), 227 pp.

STEINITZ, C., 1970, Landscape resource analysis: the state of the art, *Landscape Architecture*, **69**, 101–5.

STEPHENS, J. C., 1958, Subsidence of organic soils in the Florida Everglades, *Proc. Soil Sci. Soc. Am.* **20**, 77–80.

STEPHENS, J. C., and W. H. SPEIR, 1969, Subsidence of organic soils in the U.S.A., *Pub. Int. Ass. Sci. Hyd.* **89**, 523–34.

STEWART, G. A. (ed.), 1968, *Land Evaluation* (Macmillan, Melbourne), 392 pp.

STEWART, G. A., and R. A. PERRY, 1953, Survey of Townsville-Bowen Region (1950), CSIRO, Australia, *Land Research Series*, **2**, 87 pp.

STODDART, D. R., 1969, Climatic geomorphology: review and re-assessment, *Progress in Geography*, **1**, 161–222.

STRAATEN, VAN L. M. J. U., 1965, Coastal barrier deposits in south and north Holland in particular in the area around Scheveningen and Ijmuiden, *Medel. Geol. Sticht.* N.S. **17**, 41–75.

STRAHLER, A. N., 1952, Dynamic basis of geomorphology, *Bull. Geol. Soc. Am.* **63**, 923–38.

STRAHLER, A. N., 1965, The nature of induced erosion and aggradation, in W. L. Thomas (ed.), *Man's Role in Changing the Face of the Earth* (Univ. Chicago Press, Chicago), 621–38.

SWINZOW, G. K., 1969, Certain aspects of engineering geology in permafrost, *Engineering Geology*, **3**, 177–215.

TABER, S., 1929, Frost heaving, *J. Geol.* **37**, 428–61.

TABER, S., 1930, The mechanics of frost heaving, *J. Geol.* **38**, 303–17.

TABER, S., 1952, Geology, soil mechanics and botany, *Science*, **115**, 713–14.

TANNER, W. F., 1960, Bases of coastal classification, *SE Geology*, **2**, 13–22.

TASK COMMITTEE, 1965, Sediment transportation mechanics: wind erosion and transportation, progress report, *Proc. Am. Soc. Civ. Eng., J. Hyd. Div.* **91**, HY2, 267–87.

TERZAGHI, K., 1950, Mechanism of landslides, in *Application of Geology to Engineering Practice* (Berkey Volume, Geol. Soc. Am.), 83–125.

TERZAGHI, K., 1962, Stability of steep slopes on hard unweathered rock, *Géotechnique*, **12**, 251–70.

TERZAGHI, K., and R. B. PECK, 1948, *Soil Mechanics in Engineering Practice* (Wiley, New York), 566 pp.

TERZAGHI, K., and R. B. PECK, 1967, *Soil Mechanics in Engineering Practice*, 2nd edition (Wiley, New York), 729 pp.

TERZAGHI, RUTH D., 1965, Sources of error in joint surveys, *Géotechnique*, **15**, 287–304.

THOMAS, D. M. and M. A. BENSON, 1970, Generalization of streamflow characteristics from drainage basin characteristics, *U.S. Geol. Surv. Water-Supply Paper*, **1975**, 55 pp.

THOMAS, M. F., 1966, Some geomorphological implications of deep weathering patterns in crystalline rocks in Nigeria, *Trans. Inst. Brit. Geog.* **40**, 173–93.

THOMAS, M. F., 1969, Geomorphology and land classification in tropical Africa, in M. F. Thomas and G. T. Whittington (eds.), *Environment and Land Use in Africa* (Methuen, London), 103–45.

THOMAS, W. N., 1938, *Experiments on the freezing of certain building materials*, D.S.I.R., Building Res. Tech. Paper, **17**, 146 pp.

THORARINSSON, S., 1939, Observations on the drainage and rates of denudation in the Hoffellsjokull District, *Geografiska Annaler*, **21**, 189–215.

THORNBURY, W. D., 1954, *Principles of Geomorphology* (Wiley, New York), 618 pp.

THORNTHWAITE, C. W., 1931, Climates of North America according to a new classification, *Geog. Rev.* **21**, 633–55.

THROWER, N. J. W., and R. U. COOKE, 1968, Scales for determining slope from topographic maps, *The Professional Geographer*, **20**, 181–6.

TRICART, J., 1956, Étude expérimentale du problème de la gélivation, *Biuletyn Peryglacjalny* (Łódź), **4**, 285–318.

TRICART, J., 1959, Présentation d'une feuille de la carte géomorphologique du delta du Sénégal au 1:50,000, *Rev. de Géomorph. Dynamique*, **11**, 106–16.

TRICART, J., 1961, Notice explicative de la carte géomorphologique du delta du Sénégal, *Bureau de Recherches Géologiques et Minières*, **8**, 1–137.

TRICART, J., 1965, *Principes et Méthodes de la Géomorphologie* (Masson, Paris), 496 pp.

TRICART, J., 1966, Géomorphologie et aménagement rural (example du Venezuela), *Cooperation Technique*, **44–5**, 69–81.

UGOLINI, F. C., 1966, Soil investigations in the Lower Wright Valley, Antarctica, *Proc. Permafrost Inter. Conf.* (Nat. Academy of Sciences—National Research Council), Publ. **1287**, 55–61.

UNESCO, 1969, Land Subsidence, *Publications of the Institute of Scientific Hydrology*, **88**, (2 vols).

UNESCO, 1970, *International Legend for Hydrogeological Maps*, (Paris), 101 pp.

UNITED STATES, 1957, Summary on reservoir sedimentation surveys made in the

United States through 1953, *U.S. Inter-Agency Comm. on Water Resources, Subcomm. on Sedimentation, Sedimentation Bull.* **6.**

UNSTEAD, J. F., 1933, A system of regional geography, *Geography,* **18,** 175–87.

U.S. CONGRESS, 1968, National Flood Insurance Act of 1968, P.L. 90–448, 90th Congress, 2nd Session.

U.S. DEPARTMENT OF AGRICULTURE, 1957, *Yearbook of Agriculture—Soil* (U.S.D.A., Washington), 784 pp.

U.S. DEPT. AGRIC., 1962, Agricultural land resources—capabilities, uses, conservation needs, *Agr. Inf. Bull.* **263.**

U.S. DEPT. AGRIC., 1965, *Federal Inter-Agency Sedimentation Conference, Jackson, Mississippi, 1963,* U.S. Dept. Agr. Misc. Publ. **970,** 933 pp.

U.S. DEPT. AGRIC., 1970a, First aid for flooded homes and farms, *Agriculture Handbook,* **38,** 31 pp.

U.S. DEPT. AGRIC., 1970b, WEROS: a FORTRAN IV program to solve the wind-erosion equation, *Agricultural Research Service, ARS-174,* 13 pp.

U.S. DEPT. AGRIC., 1972, How to control wind erosion, *Agri. Inf. Bull.* **354,** 22 pp.

U.S. GEOLOGICAL SURVEY, *The Alaska Earthquake, March 27, 1964, U.S. Geol. Surv. Prof. Papers,* **541–46.**

VALENTIN, H., 1954, Der Landverlust in Holderness, Ostengland, von 1852 bis 1952, *Die Erde,* **6,** 296–315.

VANN, G., 1965, Location and evaluation of sand and gravel deposits by geophysical methods and drilling, in Inst. Min. and Met. *Opencast Mining Quarrying and Alluvial Mining* (London), 3–19.

VARNES, D. J., 1958, Landslide types and processes, in E. B. Eckel, (ed.), *Landslides and Engineering Practice* (Highway Research Board, Washington), Special Report, **29**; NAS–NRC Publ. **544,** 232 pp.

VEATCH, J. O., 1933, *Agricultural land classification and land types of Michigan,* Michigan Agric. Exp. Stat. Sp. Bull. **231.**

VERSTAPPEN, H. Th., 1966, The role of landform classification in integrated surveys, *Actes du IIe Symposium International de Photo-Interpretation* (Paris), VI.35–VI.39.

VERSTAPPEN, H. Th., 1968, *Geomorphology and Environment* (Inaugural Address, Chair of Geomorphology, I.T.C. Information Publication, Delft, now at Enschede, Holland), 23 pp.

VERSTAPPEN, H. Th., 1970, Introduction to the I.T.C. system of geomorphological survey, *Koninklijk Nederlands Aardrijkskundig Genootschap Geografisch Tijdschrift Nieuwe Reeks,* **4.**1, 85–91.

VERSTAPPEN, H. TH., and R. A. VAN ZUIDAM, 1968, I.T.C. system of geomorphological survey, *I.T.C. Textbook of Photo-Interpretation* (Delft), Ch. VII. 2, 49 pp.

VISHER, G. S., 1969, Grain size distribution and depositional processes, *J. Sed. Petrol.* **39,** 1,074–1,106.

WALLWORK, K., 1960a, Some problems of subsidence and land use in the mid-Cheshire industrial area, *Geog. J.* **126,** 191–9.

WALLWORK, K., 1960b, Land-use problems and the evolution of industrial landscapes, *Geography,* **45,** 263–75.

WARD, R. C., 1971, *Small Watershed Experiments: An Appraisal of Concepts and Research Development,* Univ. of Hull, Occ. Papers in Geography, **18,** 254 pp.

WASHBURN, A. L., 1956, Classification of patterned ground and review of suggested origins, *Bull. Geol. Soc. Am.* **67,** 823–66.

WASHBURN, A. L., 1969, Weathering, frost action and patterned ground in the Mesters Vig District, northeast Greenland, *Meddelelser om Grønland,* **176,** 1–303.

WASHBURN, A. L., 1972, *Periglacial Processes and Environments* (Arnold, London), 320 pp.

WEBSTER, R., 1963, The use of basic physiographic units in air photo interpretation, *Archives Internationales de Photogrammetria*, **14**, 143–8.

WEBSTER, R., and P. H. T. BECKETT, 1970, Terrain classification and evaluation using air photography: A review of recent work at Oxford, *Photogrammetria*, **26**, 51–7.

WEDDLE, A. E., 1969, Techniques in landscape planning, *J. Town Planning Inst.* **55**, 387–98.

WEEKS, A. G., 1969, The stability of slopes in south-east England as affected by periglacial activity, *Q. J. Eng. Geol.* **2**, 49–62.

WEINERT, H. H., 1961, Climate and weathered Karroo dolerites, *Nature*, **191**, 325–9.

WEINERT, H. H., 1965, Climatic factors affecting the weathering of igneous rocks, *Agric. Met.* **2**, 27–42.

WHITE, G. F., 1945, *Human Adjustment to Floods*, Univ. Chicago, Dept. of Geography, Research Paper, **29**, 236 pp.

WHITE, G. F. (ed.), 1961, *Papers on Flood Problems*, Univ. Chicago, Dept. of Geography, Research Paper, **70**, 234 pp.

WHITE, G. F., et al., 1958, *Change in Urban Occupance of Flood Plains in the United States*, Univ. Chicago, Dept. of Geography, Research Paper, **87**, 235 pp.

WHITE, G. F., 1964, *Choice of Adjustment to Floods*, Univ. Chicago, Dept. of Geography, Research Paper, **93**, 164 pp.

WILLIAMS, L., 1964, Regionalization of freeze-thaw activity, *Ann. Ass. Am. Geog.* **54**, 597–611.

WILSON, G., and H. GRACE, 1942–3, The settlement of London due to under-drainage of the London Clay. *J. Inst. Civ. Eng.*, **19–20**, 100–27.

WILUN, Z., and K. STARZEWSKI, 1972, *Soil Mechanics in Foundation Engineering* (Intertext Books, London), 2 vols. 252 pp. and 222 pp.

WIMAN, S., 1963, A preliminary study of experimental frost weathering, *Geog. Ann.* **45**, 113–21.

WINKLER, E. M., 1966, Corrosion rates of carbonate rocks for construction, *Eng. Geol.* (Sacramento), **4**, 52–8.

WINKLER, E. M., 1970, The importance of air pollution in the corrosion of stone and metals, *Eng. Geol.* **4**, 327–34.

WINKLER, E. M., and E. J. WILHELM, 1970, Salt burst by hydration pressures in architectural stone in urban atmosphere, *Geol. Soc. Am. Bull.* **81**, 567–72.

WINSLOW, A. G., and W. W. DOYEL, 1954, Land-surface subsidence and its relation to the withdrawal of ground-water in the Houston–Galveston region, Texas, *Econ. Geol.* **49**, 413–22.

WOHLRAB, B., 1969, Effects of mining subsidences on the ground water and remedial measures, *Pub. Int. Ass. Sci. Hydr.* **89**, 502–12.

WOLMAN, M. G., and L. B. LEOPOLD, 1957, River Flood Plains: some observations on their formation, *U.S. Geol. Surv. Prof. Paper*, **282**–C, reprinted in G. H. Dury (ed.), 1970, *Rivers and River Terraces* (Macmillan, London), 166–96.

WOLMAN, M. G., and A. P. SCHICK, 1967, Effects of construction on fluvial sediment, urban and suburban areas of Maryland, *Water Resources Research*, **3**, 451–64.

WOODHOUSE, W. W., and R. E. HANES, 1967, Dune stabilization with vegetation on the Outer Banks of North Carolina, *CERC Tech. Memo.* **22**. 45 pp.

WOODRUFF, N. P., and F. H. SIDDOWAY, 1965, A wind erosion equation, *Proc. Soil*

Sci. Soc. Am. **29**, 602–8.

WOODS, K. B. (ed.), 1960, *Highway Engineering Handbook* (McGraw-Hill, New York), 1,665 pp.

WOOLDRIDGE, S. W., 1932, The cycle of erosion and the representation of relief, *Scot. Geogr. Mag.* **48**, 30–6.

WOOLDRIGE, S. W., and S. H. BEAVER, 1950, The working of sand and gravel in Britain: a problem in land use, *Geog. J.* **115**, 42–57.

WOOLDRIDGE, S. W., and D. L. LINTON, 1955, *Structure, Surface and Drainage in South-east England* (Philip, London), 176 pp.

WRIGHT, R. L., 1972a, Some perspectives on environmental research for agricultural land-use planning in developing countries, *Geoforum*, **10**, 15–33.

WRIGHT, R. L., 1972b, Principles in a geomorphological approach to land classification, *Zeit. für Geom.* **16**, 351–73.

YAALON, D. H., and E. GANOR, 1966, The climatic factor of wind erodibility and dust blowing in Israel, *Israel Journ. Earth Sci.* **15**, 27–32.

YALIN, M. S., 1972, *Mechanics of Sediment Transport* (Pergamon, Oxford), 290 pp.

YASSO, W. E., 1964, Geometry and development of spit-bar shoreline at Horseshoe Cove, Sandy Hook, New Jersey, *Tech. Rep. Office of Naval Research Gr. Br.* *388–057* Tech. Rep., **5**, 166 pp.

YASSO, W. E., 1965, Plan geometry of headland bay beaches, *J. Geol.* **73**, 702–14.

YATES, A. B., and D. R. STANLEY, 1966, Domestic water supply and sewage disposal in the Canadian north, *Proc. Permafrost Int. Conf.* (Nat. Academy Sciences —National Research Council), Publ. **1287**, 413–19.

YORKE, T. H., and W. J. DAVIS, 1971, Effects of urbanization on sediment transport in Bel Pre Creek Basin, Maryland, *U.S. Geol. Surv. Prof. Paper*, **750**-B, B218–B223.

YOUNG, A., 1969, Present rate of land erosion, *Nature*, **224**, 851–2.

YOUNG, A., 1972, *Slopes* (Oliver and Boyd, Edinburgh), 288 pp.

YOUNG, A., 1973, Rural Land Evaluation, in J. A. Dawson and J. C. Doornkamp, (eds.), *Evaluating the Human Environment, Essays in Applied Geography* (Arnold, London), 5–33.

YOUNG, E., 1968, Urban planning for sand and gravel needs, *Mineral Information Service* (California Division of Mines and Geology), **21**, 147–50.

ZARUBA, Q., and V. MENCL, 1969, *Landslides and Their Control* (Elsevier, Amsterdam), 205 pp.

ZENKOVICH, V. P., 1967, *Processes of Coastal Development*, ed. J. A. Steers (Oliver and Boyd, Edinburgh), 738 pp.

ZHIRKOV, K. F., 1964, Dust storms in the Steppes of western Siberia and Khazakhstan, *Soviet Geography*, **5**, 33–41.

ZINGG, A. W., 1940, Degree and length of land slope as it affects soil loss in runoff, *Agricultural Engineering*, **21**, 59–64.

ZINGG, A. W., 1954, The wind erosion problem in the Great Plains, *Trans. Am. Geoph. Un.* **35**, 252–8.

INDEX

(italicized items refer to illustrations)